DIMACS

Series in Discrete Mathematics
and Theoretical Computer Science

Volume 60

Algorithmic and Quantitative Real Algebraic Geometry

DIMACS Workshop
Algorithmic and Quantitative Aspects of
Real Algebraic Geometry in Mathematics
and Computer Science
March 12–16, 2001
DIMACS Center

Saugata Basu
Laureano Gonzalez-Vega
Editors

Center for Discrete Mathematics
and Theoretical Computer Science
A consortium of Rutgers University, Princeton University,
AT&T Labs–Research, Bell Labs (Lucent Technologies),
NEC Laboratories America, and Telcordia Technologies
(with partners at Avaya Labs, IBM Research, and Microsoft Research)

American Mathematical Society

This DIMACS volume contains papers from a DIMACS workshop on Algorithmic and Quantitative Aspects of Real Algebraic Geometry in Mathematics and Computer Science, held March 12–16, 2001.

2000 *Mathematics Subject Classification.* Primary 14P05, 14P10, 14P15, 14P20, 14P25, 14P99, 14Q20, 68W30.

Library of Congress Cataloging-in-Publication Data

Algorithmic and quantitative real algebraic geometry : DIMACS workshop, Algorithmic and quantitative aspects of real algebraic geometry in mathematics and computer science, March 12–16, 2001, DIMACS Center / Saugata Basu, Laureano Gonzalez-Vega, editors.

p. cm. — (DIMACS series in discrete mathematics and theoretical computer science, ISSN 1052-1798 ; v. 60)

Includes bibliographical references.

ISBN 0-8218-2863-0 (alk. paper)

1. Geometry, Algebraic—Congresses. 2. Computer science—Mathematics—Congresses. I. Basu, Saugata, 1968– II. Gonzalez-Vega, Laureano. III. Series.

QA564.A455 2003
516.3′5—dc21 2002038522

∞ The paper used in this book is acid-free and falls within the guidelines
established to ensure permanence and durability.
Visit the AMS home page at http://www.ams.org/

10 9 8 7 6 5 4 3 2 1 08 07 06 05 04 03

Contents

Foreword

This volume consists of papers resulting from the DIMACS Workshop on Algorithmic and Quantitative Aspects of Real Algebraic Geometry, held at DIMACS on March 12–16, 2001. We would like to express our appreciation to Saugata Basu and Laureano Gonzalez-Vega for putting together the volume and Ricky Pollack, Marie-Francoise Roy and Micha Sharir for organizing the workshop.

The workshop was part of the Special Focus on Mathematics and the Foundations of Computer and Information Science. We extend our thanks to Michael Fredman, Janos Komlos, and Fred Roberts for their work as special focus organizers.

The workshop brought together researchers in algorithmic and quantitative aspects of real algebraic geometry and other mathematicians and computer scientists who are concerned with related problems, aiming at promoting more productive interactions between them. The conference and this volume have succeeded in emphasizing such areas of commonality as interactions between real algebraic geometry and computational geometry, efficient recent algorithms in real algebraic geometry and their implementation, quantitative aspects of basic semi-algebraic sets, lower bounds and real algebraic geometry, combinatorial characterization of real algebraic varieties, combinatorial patchworking for constructing real algebraic curves and hypersurfaces, and real aspects of enumerative geometry.

DIMACS gratefully acknowledges the generous support that makes these programs possible. The National Science Foundation, through its Science and Technology Centers program, the New Jersey Commission on Science and Technology, DIMACS's partners at Rutgers, Princeton, AT&T Labs-Research, Bell Labs, NEC Laboratories America, and Telcordia Technologies, and DIMACS affiliates at Avaya Labs and Microsoft Research generously supported the special focus.

Fred S. Roberts
Director

Robert Tarjan
Co-Director for Princeton

Preface

Algorithmic and quantitative aspects are becoming increasingly important areas of research in Real Algebraic Geometry because of their roles in other areas of mathematics and computer science. The papers in this volume collectively span several different areas of Real Algebraic Geometry which are of current research interest. These include deciding basic algebraic properties of real semi-algebraic sets, application of quantitative results in real algebraic geometry towards investigating the computational complexity of various problems, algorithmic and quantitative questions in real enumerative geometry, new approaches towards solving decision problems in semi-algebraic geometry as well as computing algebraic certificates, and applications of real algebraic geometry to concrete problems arising in robotics and computer graphics.

Subsets of finite dimensional real spaces, defined by Boolean formulas whose atoms are polynomial inequalities and equalities, are called semi–algebraic sets, and they are one of the most important geometric objects in Real Algebraic Geometry.

Semi–algebraic sets which are defined by a conjunction of strict inequalities are called basic open semi–algebraic sets. The paper by Carlos Andradas aims at characterizing the basic open semi–algebraic sets, by presenting procedures for recognizing whether a given open semi algebraic set is basic or not and, if it is the case, finding a minimal (in the sense of shortest) description.

In the same vein, the paper by Clint McCrory gives a technique for proving that a given semi–algebraic set is not a real algebraic set in dimensions up to four, by constructing a list of local topological obstructions using the link operator on the ring of constructible functions (introduced by C. McCrory and A. Parusinski). These functions are used in the paper by Isabelle Bonnard to show how to reformulate a condition due to S. Akbulut and H. King, for a compact semi-algebraic set X to be homeomorphic to a real algebraic set when $\dim X \leq 3$, and use it to produce new necessary conditions in dimension 4.

The paper by Benoit Chevallier uses Viro's method with quadratic transforms, to solve several instances of the problem of deciding if there exists a real algebraic hyper-surface in $\mathbb{R}^n$ defined by a polynomial in $\mathbb{R}[x_1, \ldots, x_n]$ of degree d which is isotopic to a given smooth hyper-surface in $\mathbb{R}^n$.

Semi–Pfaffian sets are useful generalizations of semi–algebraic sets obtained by enlarging the family of polynomial functions by including Pfaffian functions which are analytic functions satisfying certain triangular systems of differential equations. Semi–Pfaffian sets share many of the finiteness properties of semi–algebraic sets, and this leads to the problem of determining uniform bounds on their topological invariants, such as the number of connected components, Betti numbers etc. The family of semi–Pfaffian sets are distinguished from semi–algebraic sets by the fact that it is not closed under projection. Recently, A. Gabrielov introduced the notion of relative closure in order to enlarge the family of semi-Pfaffian sets and obtain a family which is stable under projection. A different definition of the same family was given by A. J. Wilkie earlier, but the new definition makes it simpler to estimate the topological complexity of sets in the family. The paper by Andrei Gabrielov and Thierry Zell provides the first uniform bounds on the number of connected components of the relative closure of a pair of semi–Pfaffian sets, which is an important first step in this direction.

The problem of computing algebraic certificates is becoming an exciting new research area in Real Algebraic Geometry: if S is the basic closed semi-algebraic set in $\mathbb{R}^n$ defined by $f_1 \geq 0, \ldots, f_k \geq 0$ and a polynomial p can be written as a finite sum of products of the f_i's and sums of squares in $\mathbb{R}[x_1, \ldots, x_n]$ then clearly p is non–negative on S. A "representation theorem" or an algebraic certificate is a converse to this fact. Namely, such a theorem would state that if a polynomial p is positive on S, then there is a representation of p in terms of the f_i's and (possibly) sums of squares making evident the positivity of p on S. The paper by Dionne Bailey and Victoria Powers presents a constructive approach to the Kadison–Dubois Theorem (giving an algebraic certificate for the non–negativity of p over S under one additional hypothesis implying the compactness of S), with degree bounds and an algorithm for finding such a representation. This is then used to give a constructive representation theorem in certain non-compact cases.

The paper by Bruce Reznick is devoted to another quantitative aspect of Real Algebraic Geometry: if $F = \{f_1, \ldots, f_r\}$ is a family of polynomials in $\mathbb{R}[x_1, \ldots, x_n]$ and $T(F)$ denote the set of integers m such that $\{f_1^m, \ldots, f_r^m\}$ is linearly dependent, then it is proven that $|T(F)| \leq \binom{r-1}{2}$. Many concrete examples are also discussed in the paper. The elements in $T(F)$ are called the tickets of F and they generalize several classical constructions such as Fermat's Last Theorem for univariate polynomials ($r = 3$) and have connections with Nevanlinna Theory and with Waring's Problem for polynomials.

The paper by Pablo Parrilo and Bernd Sturmfels describes a new approach towards global optimization of polynomial functions in many variables by using sums of squares and a convex relaxation to semi–definite programming. The resulting algorithm seems to (heuristically) work very well relative to other exact methods and it is the first practical algorithm for computing algebraic certificates for the Positivstellensatz.

The critical point method provided the first single exponential algorithm for solving the problem of deciding emptiness of a given semi–algebraic set. The paper by Fabrice Rouillier describes recent advances in the efficient implementations of critical point method in solving real polynomial systems. The theoretically best algorithms are designed from the point of view of asymptotic complexity. Efficient practical implementation requires important modifications to the theoretically optimal algorithms, and several approaches in this direction are described in Rouillier's paper.

The problem of proving lower and upper bounds on the number of *real* (as opposed to complex) solutions of algebraic problems is often considerably more difficult than the problem of bounding (or even counting) complex solutions. The classical tool in this area is that of Schubert calculus, which provides the answer to complex enumerative problems. The paper by Frank Sottile is an exhaustive survey of results from enumerative Real Algebraic Geometry showing that the upper bounds regarding real solutions and arising from Schubert calculus are usually sharp. It also describes how a new notion of degree for real Grassmann varieties, introduced by A. Eremenko and A. Gabrielov, helps to prove new improved lower bounds for some enumerative problems.

The paper by Thorsten Theobald presents one application of Real Algebraic Geometry to computer graphics and computational geometry: the investigation of visibility computations with moving viewpoints. The considered problems are of discrete and algorithmic nature, but even for simple classes of objects (such as balls and polytopes), they lead to interesting and difficult problems from Enumerative Real Algebraic Geometry. In particular, a new sweep algorithm in dimension 2 is introduced, together with the extension of several algebraic–geometric results regarding the tangency problems involving lines and spheres in $\mathbb{R}^3$.

An important application of Real Algebraic Geometry in Theoretical Computer Science has been in proving lower bounds in certain models of computations such as algebraic computation trees or decision trees. Upper bounds on the geometric and topological complexity of semi–algebraic sets gets translated into lower bounds in these models of computation. Often, these lower bounds are tight and thus the exact computational complexity of certain problems gets determined. The paper by Peter Bürgisser provides a comprehensive survey of these results.

The paper by Ileana Streinu deals with the algorithmic problem of motion planning of a planar robot arm with many links (open or closed polygonal chains). Translating this problem into the language of configurations spaces, one obtains algorithmic problems in Real Algebraic Geometry mainly related with the computation of real solutions of a polynomial system of equations. Connections with pseudo–triangulations are also discussed.

The papers appearing in this volume are based on the talks given at the *DIMACS Workshop on Algorithmic and Quantitative Aspects of Real Algebraic Geometry* held at DIMACS Center (Rutgers University), March 12–16, 2001. The editors of this volume would like to thank Ricky Pollack, Marie–Françoise Roy and Micha Sharir for providing us with an optimal set of initial conditions to start with.

The first editor acknowledges support from NSF grants CCR-0049070 and CCR-0133597. The second editor acknowledges the support of the Spanish grant PB 98–0713–C02–02.

Saugata Basu and Laureano Gonzalez–Vega

DIMACS Series in Discrete Mathematics
and Theoretical Computer Science
Volume **60**, 2003

Characterization and description of basic semialgebraic sets

C. Andradas

ABSTRACT. We give a short survey of the characterization of basic semialgebraic sets and their description by minimal systems of polynomial inequalities. We pay attention to the lack of efficient algorithmic approachs to find such descriptions, in spite of that the characterization of basicness is known to be decidible over the real numbers.

1. Introduction

Semialgebraic sets are rather ubiquitous in mathematics and nature. The reason is that since they are described by polynomials they are at the same time "easy" to manipulate and adequate to model most geometric objects due to the density of polynomials in compact ambients.

Let $V \subset R^n$ be an algebraic set, that is, the set of solutions of a finite system of polynomial equations, where R is a real closed field. Recall that a subset $S \subset V$ is called semialgebraic if it is of the form

$$S = \bigcup_{i=1}^{t} \{x \in V \mid f_i(x) = 0, g_{i1} > 0, \dots, g_{in_i} > 0\}$$

for some polynomial functions f_i, g_{ij} on V. The pieces $\{x \in V \mid g_{i1} > 0, \dots, g_{in_i} > 0\}$, i.e., the solution set of a system of strict inequalities are called *open basic semialgebraic*. Open basic semialgebraic are, somehow, the building stones of semialgebraic sets, so that they have deserved some attention in the last years. In particular we have the following two main questions:

(P1): Recognize whether a given open semialgebraic set is basic.

(P2): Given a basic open semialgebraic set find a minimal (in the sense of shorter) description.

Both questions have precise answers, mainly due to the work of L. Bröcker in the 80's, and also C. Scheiderer, by means of the theory of spaces of orderings, cf. [**AnBrRz**]. To be more precise, we denote by A the ring of polynomial functions on V. We will assume that V is irreducible so that A is a domain and we will

1991 *Mathematics Subject Classification.* Primary 14P10.
Key words and phrases. Semialgebraic sets, basicness.
Partially supported by DGES PB98-0756-C02-01.

denote by K its field of fractions. In particular K is a finitely generated extension of R of transcendence degree $d = \dim V$. We have:

THEOREM 1.1. ([**Br1**]) *An open semialgebraic set S is basic if and only if for any 4-element fan $F \subset \mathrm{Spec}_r A$ we have $\#(F \cap S) \neq 3$.*

THEOREM 1.2. ([**Br1**]) *There is a constant $s(d)$ depending only on $d = \dim(V)$ such that any basic open set can be written with $s(d)$ inequalities.*

In fact the bound given in the quoted paper was $s(d) = d(d-2)(d-4)\cdots 1$, but later C. Scheiderer, [**Sch**], and L. Bröcker, [unpublished], proved independently that $s(d) = d$ and that this bound is optimal.

All the above results share several characteristics: first they are beautiful; second, they are highly non constructive. In the late 90's, Acquistapace et al, [**AcBgVe**], showed that question P1 is indeed algorithmically decidable over the real numbers and produce a theoretical algorithm to decide it. However it is highly unpractical. Their work is inspired in previous work in [**AcAnBg**] and [**AnRz1**],[**AnRz2**],[**AnRz3**]. Concerning question P2, not many attempts have been taken towards giving algorithms for finding explicit inequations of minimal length. Only some work of A. Bernig in the 2 dimensional case, [**Be**] gave explicit construction of the 2 inequalities to describe open, convex, basic semialgebraic sets, and recently [**AnVe**] have given an algorithm for basic semilinear sets (i.e. described by linear inequalities), showing that in this case basicness and convexity are equivalent. The main attempt toward finding a constructive minimal description was given by Buresi and Mahé, [**BuMa**] but they only could find explicit upper bounds for the degrees of the polynomials appearing in it.

In the following pages we will present a short survey of Buresi and Mahé's proof, as well as of Bröcker's fan criterion and Acquistapace et al algorithm. As pointed above, a lot of work has still to be done towards an efficient answer to problems P1 and P2. The purpose of this paper is mainly enlightening and intuitive, concentrating in the concepts, trying to serve as bridge between the Real Algebraic Geometry and the Computer Science communities. Complete proofs can be found in the papers quoted in the references. I want to thank the organizers of the DIMACS Workshop, for the opportunity to lecture there and enjoy the nice atmosphere of the institute and the conference.

2. Changing inequalities

Suppose that we want to manipulate a system of strict inequalities in a similar way to what we made in Linear Algebra to transform a linear system in a triangular one, that is, defining some admissible operations to transform the initial system into an equivalent one and hoping that the latter is simpler (in some way) than the older.

Here is the first "rule of thumb", that can be checked by direct inspection: *the systems*

$$\left.\begin{array}{rcl} h_1 & > & 0 \\ h_2 & > & 0 \end{array}\right\} \qquad\qquad \left.\begin{array}{rcl} h_1 + h_2 & > & 0 \\ h_1 h_2 & > & 0 \end{array}\right\}$$

are equivalent.

The second rule states that the system does not change when equations are multiplied by polynomials of precise positivity point sets. Again a direct inspection shows that *the systems*

$$\left.\begin{array}{rcl} h_1 & > & 0 \\ h_2 & > & 0 \\ \cdots & & \\ h_{n-1} & > & 0 \\ h_n & > & 0 \end{array}\right\} \qquad \left.\begin{array}{rcl} h_1 u_0 & > & 0 \\ h_2 u_1 & > & 0 \\ \cdots & & \\ h_{n-1} u_{n-2} & > & 0 \\ h_n u_{n-1} & > & 0 \end{array}\right\}$$

are equivalent, where $u_0 > 0$ everywhere and $\{u_j > 0\} \supset \{h_1 > 0, \ldots, h_j > 0\}$ for $j \geq 1$.

Combining these two operations we get the following one which collects both at once (the u_j's are as above): *the systems*

$$\left.\begin{array}{rcl} h_1 & > & 0 \\ h_2 & > & 0 \\ \cdots & & \\ h_{n-2} & > & 0 \\ h_{n-1} & > & 0 \\ h_n & > & 0 \end{array}\right\} \qquad \left.\begin{array}{rcl} h_1 & > & 0 \\ h_2 & > & 0 \\ \cdots & & \\ h_{n-2} & > & 0 \\ h_{n-1} u_{n-2} + h_n u_{n-1} & > & 0 \\ h_{n-1} h_n u_{n-1} & > & 0 \end{array}\right\}$$

are equivalent.

Repeating the rule with the rows $n-1$ and $n-2$ and so on, we get that the initial system of inequalities is equivalent to

$$\left.\begin{array}{rcl} h_1 u_0 + h_2 u_1 + \cdots + h_n u_{n-1} & > & 0 \\ h_1(h_2 u_1 + \cdots + h_n u_{n-1}) & > & 0 \\ \cdots & & \\ h_{n-2}(h_{n-1} u_{n-2} + h_n u_{n-1}) & > & 0 \\ h_{n-1} h_n u_{n-1} & > & 0 \end{array}\right\} \qquad (*)$$

Now, if $u_0, u_1, \ldots, u_{n-1}$ can be chosen such that $h_1 u_0 + h_2 u_1 + \cdots + h_n u_{n-1} > 0$ is trivial, for instance $1 = h_1 u_0 + h_2 u_1 + \cdots + h_n u_{n-1}$, then we get that our system is equivalent to one with one less equation, namely $\{g_1 > 0, \ldots, g_{n-1} > 0\}$ with $g_k = h_k(h_{k+1} u_k + \cdots + h_n u_{n-1})$.

This will be the strategy to shorten the system of inequalities: try to find $u_0, u_1, \ldots, u_{n-1}$ such that $h_1 u_0 + h_2 u_1 + \cdots + h_n u_{n-1}$ becomes a strict positive function (for instance 1). Of course the hard point is to choose the u_i's and for that we need to know how the polynomials which are positive on a certain basic semialgebraic set look like. The description of such polynomials is given by the Positivstellensatz, cf. [**St**], [**BCR**]:

THEOREM 2.1. *A polynomial u is strictly positive on $\{f_1 > 0, \ldots, f_n > 0\}$ if and only if $u(\prod f_i)^{2\ell} + t_1 = (\prod f_i)^{2r} + t_2$, where t_1, t_2 are in the preordering generated by $f_1, \ldots, f_n$, that is, they are linear combinations of products of the f_i's with coefficients sums of squares of polynomials.*

Solving for u, we find that the natural candidates for the u_i's are linear combinations of $f_1, \ldots, f_{i-1}$ with coefficients *quotients of sums of squares*, or equivalently elements of the preordering generated by $f_1, \ldots, f_{i-1}$ in the field K of rational functions. However in doing so, we loose some information since we are not paying attention to the precise form of numerator and denominator of such coefficients. The price we are paying off is that the equivalence between the original and new systems of inequalities holds only up to a subset of codimension 1. We will refer to this situation by saying that the two systems are generically equal or equivalent.

On the other hand we do gain something: as the reader will have noticed the u_i are now elements represented by the Pfister form $\ll f_1, \ldots, f_i \gg$ over K. Therefore the rich theory of Pfister forms, cf. [**Lm**], is at our disposition.

3. Pfister's form attached to a semialgebraic set

Let $S = \{f_1 > 0, \ldots, f_n > 0\} \subset R^d$ be an open semialgebraic set, with $f_i \in A$. Consider the Pfister form

$$\varphi = \ll f_1, \ldots, f_n \gg := < f_1^2, f_1 > \otimes \cdots \otimes < f_n^2, f_n >$$

For any $x \in V$ set

$$\hat{\varphi}(x) = \text{signature}(\varphi(x)) = \prod_{i=1}^{d+1} (\text{sign}(f_i^2(x)) + \text{sign}(f_i(x)))$$

where signature$(\varphi(x))$ stands for the signature of the quadratic form $\varphi(x)$ over R and sign(x) stands for the sign of the element x in the real closed field R. Then $\hat{\varphi}$ is a constructible function and

$$\hat{\varphi}(x) = \begin{cases} 2^n & \text{if } x \in S \\ 0 & \text{otherwise} \end{cases}$$

Therefore the form φ acts as a characteristic function for the set S and the number of inequalities coincides with the dimension of φ and the exponent of the value of $\hat{\varphi}$. Now, if $S = \{g_1 > 0, \ldots, g_{n-1} > 0\}$ for some $g_i \in A$ and we set $\psi = \ll g_1, \ldots, g_{n-1} \gg$, we have $\hat{\varphi} = 2\hat{\psi}$ so that by the definition of equality of forms in the reduced theory of forms we have $\varphi = 2\psi := \psi \perp \psi$. In other words, the question of deciding whether a system of inequalities can be shortened, that is, $\{f_1 > 0, \ldots, f_n > 0\} = \{g_1 > 0 \ldots, g_{n-1} > 0\}$ is equivalent to check whether $\ll f_1, \ldots, f_n \gg\hat{} = \ll g_1, \ldots, g_{n-1} \gg\hat{}$, that is, a question of arithmetic of Pfister forms. Unfortunately this arithmetic works well only over fields (or more generally over rings with many units, but we will not enter into this here). So, from now on, we work over he field of fractions K of A. In this case we have

$$\varphi = \ll f_1, \ldots, f_n \gg := < 1, f_1 > \otimes \cdots \otimes < 1, f_n > = < 1 > \perp \varphi',$$

where the $2^n - 1$ dimension form

$$\begin{aligned} \varphi' = < f_1 > \perp f_2 \ll f_1 \gg \perp f_3 \ll f_1, f_2 \gg &\perp \cdots \\ \cdots \perp &\; f_n \ll f_1, f_2, \ldots, f_{n-1} \gg \end{aligned}$$

is called the pure part of φ.

The rules for manipulating systems of inequalities have a well known translation in terms of Pfister forms. For that we remember that an element $b \in K^*$ is said to be represented by the form $\rho = < a_1, \ldots, a_n >$ if there is a vector $v \in (K^*)^n$ such that $b = \rho(v) = \sum a_i v_i^2$. This implies that there are $b_2, \ldots, b_n \in K^*$ such that $\rho = < b, b_2, \ldots, b_n >$ in the sense defined above (i.e., there signatures coincide), but we recall that the converse is not true: the later equality implies only that a multiple of ρ represents b, that is, $b = \rho(v_1) + \cdots + \rho(v_k)$ for some vectors $v_1, \ldots, v_k$, cf. [**Lm**]. In particular, if u is represented by the Pfister form $\ll f_1, \ldots, f_i \gg$ then u is positive on the semialgebraic set $\{f_1 > 0, \ldots, f_i > 0\}$. One of the most useful properties of Pfister forms is that the sets of elements represented by them are multiplicative closed, and therefore a subgroup of K^* since $\varphi(v)^{-1} = \varphi(\frac{v}{\varphi(v)})$.

Coming back to the translation of the $(*)$ above to Pfister forms we have, cf. [**BCR**, Lemma 6.4.14]:

LEMMA 3.1. *Assume that u_0 is a square in K^* and that for $i = 1, \ldots, n-1$, $u_i \in K$ is represented by the form $\varphi_i = \ll f_1, \ldots, f_i \gg$. For $i = 0, \ldots, n-1$, set $g_i = f_{i+1}u_i + f_{i+2}u_{i+1} + \cdots + f_n u_{n-1}$ and assume $g_i \in K^*$. Then*

$$\varphi = \ll f_1, \ldots, f_n \gg = \ll g_0, f_1 g_1 \ldots, f_{n-1} g_{n-1} \gg$$

Notice that an element of K^* has the shape of g_0 above if and only if it is represented by the pure form φ' of φ. Thinking in terms of systems of inequalities we are saying that we can always replace our original system $\{f_1 > 0, \ldots, f_n > 0\}$ by a new one (of rational functions) which is generically equivalent to it and that contains the inequality $g_0 > 0$ where g_0 is any polynomial represented by φ'. In particular, if 1 is represented by φ' we get an spurious inequality and therefore, after clearing denominators, a new system of inequalities, generically equivalent to the former, and with one less inequality. This is exactly the content of the following result which is the fundamental step in Buresi and Mahé proof.

4. Buresi and Mahé proof

Recall that $d = \dim(V)$. In particular the function field K has transcendence degree d over the real closed field R.

THEOREM 4.1. *Given $\varphi = \ll f_1, \ldots, f_{d+1} \gg$ then $1 = \varphi(x)$ for some vector $x \in K^{2^{d+1}}$.*

PROOF. We only sketch the proof. Consider the quadratic equation over $K(i)$

$$\sum_{\varepsilon=0,1} f_1^{\varepsilon_1} \cdots f_{d+1}^{\varepsilon_{d+1}} X^2_{(\varepsilon_1, \ldots, \varepsilon_{d+1})} = 0$$

in the 2^{d+1} variables $X_{(\varepsilon_1, \ldots, \varepsilon_{d+1})}$. Since $R(i)$ is algebraically closed, by the Tsen-Lang theorem it has a non trivial solution $u = \mathrm{Re}(u) + i\mathrm{Im}(u)$ in $K(i)^{2^{d+1}}$. Thus,

$$0 = \varphi(u) = \varphi(\mathrm{Re}(u)) - \varphi(\mathrm{Im}(u)) + 2iB_\varphi(\mathrm{Re}(u), \mathrm{Im}(u)),$$

where B_φ stands for the symmetric bilinear form associated to φ. Thus

$$\left.\begin{array}{l} \varphi(\mathrm{Re}(u)) = \varphi(\mathrm{Im}(u)) \\ B_\varphi(\mathrm{Re}(u), \mathrm{Im}(u)) = 0 \end{array}\right\}$$

Since the set of elements represented by φ is multiplicative closed there is $z \in K^{2^{d+1}}$ such that

$$\varphi(z) = \varphi(\mathrm{Re}(u))\varphi(\mathrm{Im}(u)) = \varphi(\mathrm{Re}(u))^2$$

Moreover, z can be constructed explicitely and so that its first component is $B_\varphi(\mathrm{Re}(u), \mathrm{Im}(u)) = 0$, i.e. $z = (0, z')$, with $z' \in K^{2^d}$. Thus,

$$1 = \varphi(\frac{1}{\varphi(\mathrm{Re}(u))} z) = \varphi(0, \frac{1}{\varphi(\mathrm{Re}(u))} z') = \varphi'(\frac{1}{\varphi(\mathrm{Re}(u))} z')$$

□

THEOREM 4.2. ([**Sch**]. Cf. also [**AnBrRz**]) *Any basic open semialgebraic subset $S = \{f_1 > 0, \ldots, f_n > 0\} \subset V$ can be described generically by d inequalities.*

What can be said about the degrees of the shorter description? It is well known that they may increase: for instance, take the regular n-polygon described by the n linear inequalities of its sides. Then the product of all these linear factors must divide the product of the two equations of the shorter generic description. Thus one of them must have degree al least $n/2$. On the other hand, Buresi and Mahé proved that there exists an upper bound for the degrees of a generic minimal description in terms of the dimension d and degrees and number of the initial description.

Let's sketch his approach: the general idea is to keep track as close as possible of the construction of the vector in K^{2^n} that produces the representation of 1 by φ'. First they observe, by a simple consideration on the degrees, that the solution $u \in K(i)^{2^{d+1}}$ to the equation

$$\sum_{\varepsilon=0,1} f_1^{\varepsilon_1} \cdots f_{d+1}^{\varepsilon_{d+1}} X^2_{(\varepsilon_1,\dots,\varepsilon_{d+1}} = 0,$$

guaranteed by Tsen-Lang theorem can be taken in $A[i] = A \otimes_R R[i]$ of degree (i.e. in all its components) $\leq d\alpha$, where α is the sum of the degrees of the f_i's. The next difficult point is constructing a vector that represents the product of two represented elements, namely, given $x, y \in K^{2^{d+1}}$ find z such that $\varphi(x)\varphi(y) = \varphi(z)$. For that they use Pfister's construction that gives

$$\varphi(x)\varphi(y) = \varphi(T_x(y)),$$

where T_x is the linear transformation defined recursively as follows: set $\varphi_i = \ll f_1, \dots, f_i \gg$, $i = 1, \dots, d+1$, $\varphi_0 =< 1 >$ and denote by A_i the (diagonal) matrix of φ_i. In particular $A_i = \begin{pmatrix} A_{i-1} & \\ & f_i A_{i-1} \end{pmatrix}$. Now divide $x \in K^{2^m}$ into to halves, $x = (u, \overline{u})$ with $u, \overline{u} \in K^{2^{m-1}}$. Then T_x is defined starting by $T_x = x$ if $x \in R$, and

$$T_x = \begin{cases} \begin{pmatrix} T_u & \\ & T_u \end{pmatrix} & \text{if } \varphi_{m-1}(\overline{u}) = 0 \\ \\ \begin{pmatrix} T_u & f_m T_{\overline{u}} \\ -T_{\overline{u}} & U_{m-1} \end{pmatrix} & \text{if } \varphi_{m-1}(\overline{u}) \neq 0 \end{cases}$$

where

$$U_{m-1} = \frac{1}{\varphi_{m-1}(\overline{u})} T_{\overline{u}} A_{m-1}^{-1} T_u^t A_{m-1} T_{\overline{u}}.$$

These precise formulas allow following up the denominators appearing in the process, so that in the end we get

THEOREM 4.3. *Given $f_1, \dots, f_{d+1}$ as above, we can construct $g_1, \dots, g_d$ such that $\{f_1 > 0, \dots, f_{d+1} > 0\}$ and $\{g_1 > 0, \dots, g_d > 0\}$ are generically equal and* $\deg(g_i) \leq (3^d + 1)d\alpha$

5. Bröcker's fan characterization of basicness

Let's try to understand what fans are. We start with an example. Consider a line H in $\mathbb{R}^2$ and two points $a_1, a_2 \in H$. Now, at each point take one of the two half branches of H at it, say ξ_1 at the point a_1 and ξ_2 at a_2. Again at each a_i consider the two ultrafilters α_k of open semialgebraic sets of $\mathbb{R}^2$ "adherent" to ξ_i, i.e., those whose elements contain ξ_i in its closure. This way we obtain four ultrafilters $\alpha_1, \alpha_2, \alpha_3, \alpha_4$ with the property that we cannot separate one of them from the other three, that is, if a polynomial is positive in three of them, it is also positive in the fourth. Here, of course, we say that a $f \in \mathbb{R}[X,Y]$ is positive in the ultrafilter α if is positive in some $S \in \alpha$. In this situation we say that $\alpha_1, \alpha_2, \alpha_3, \alpha_4$ are a fan of $\mathbb{R}(X,Y)$. The following picture summarizes the situation:

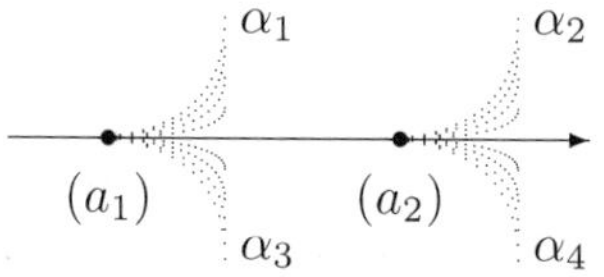

Of course the same construction can be made in $\mathbb{R}^d$ starting with any algebraic hypersurface $H \subset \mathbb{R}^d$, two points $a_1, a_2 \in H$ and two ultrafilters ξ_1, ξ_2 of open semialgebraic subsets of H with a_1 and a_2 in its closure. Again the resulting $\alpha_1, \alpha_2, \alpha_3, \alpha_4$ are "inseparable" in the sense above and build a fan of $\mathbb{R}(X_1, \ldots, X_d)$.

Let's make this definition more precise. By Artin-Lang Theorem, ultrafilters of open semialgebraic subsets of an irreducible algebraic set V correspond to orderings of the function field $R(V)$ of this set, and it is in this context of orderings of a field, where the original definition of fans takes place.

DEFINITION 5.1. *We say that four orderings $\alpha_1, \alpha_2, \alpha_3, \alpha_4$ are a* 4-element fan *of K if any element of K which is positive in three of them is also positive in the fourth. Thinking in the orderings of K as mappings assigning a sign $+1$ or -1 to the elements of K, this property is stated by saying that the product of three of the α's is the fourth:*

$$\alpha_1\alpha_2\alpha_3 = \alpha_4.$$

More generally, a finite subset F of orderings of K are a *fan* if for any three $\alpha_1, \alpha_2, \alpha_3 \in F$ their product is also an ordering of F. Thus, 4-element fans are the simplest (non-trivial) examples of fans. Since we are interested in the geometric situation, the field K will always be the field $R(W)$ of rational functions of an algebraic subvariety $W \subset V$. In this situation we will say that the orderings (resp. fans) are supported in the subvariety W or simply, when no confusion is possible that they are orderings in W.

The set of all orderings in all possible subvarieties of V is the real spectrum $\mathrm{Spec}_r R[V]$ of the ring of polynomial functions on V, or simply the real spectrum of V, and will be also denoted by $\widetilde{V}$. In particular, $\widetilde{V}$ contains the orderings supported at the points $x \in V$ which are identified with the points, so that it contains V. Also it contains the orderings (ultrafilters) centered at infinity, so that $\widetilde{V}$ results a compactification of V. The set of all fans in all subvarieties of V constitutes the family of fans of $\mathrm{Spec}_r R[V]$. Notice in particular, that a fan F is always a set of orderings supported in the same subvariety $W \subset V$. Given any ordering $\alpha \in \widetilde{V}$ and a polynomial $f \in R[V]$ we say that $\alpha \in \{f > 0\}$ if $f|_W$ is positive in α, and analogously for $f < 0$. In case f vanishes on W we say that $\alpha \in \{f = 0\}$. Translating this back to the geometric language, if we identify α with its corresponding ultrafilter, we are just saying that $\alpha \in \{f > 0\}$ if and only if $\{f > 0\}$ is in the ultrafilter α.

In general fans have remarkable combinatorial properties:

PROPOSITION 5.2. *Let F be a fan in $\widetilde{V}$.*

a) *for any $f \in R[V]$ then $\{f > 0\} \cap F$ is either trivial (i.e. $\emptyset$, or F) or contains exactly half of the elements of F.*

b) *$\#F = 2^k$ for some k.*
c) *If $S = \{f_1 > 0, \ldots, f_n > 0\}$, then either $\#(F \cap S) \geq 2^{k-n}$ or $(F \cap S) = \emptyset$. In particular if there is a fan f with 2^n elements such that $\#(F \cap S) = 1$, then S cannot be written with less than n inequalities.*

Fans are a very important class of families of sets of orderings since they are like building stones for more general subspaces and also serve as a testing family for some properties like for instance the property of a semialgebraic set to be basic as we will see in the next section. Although combinatorially they are very simple, from the point of view of its geometric translation, arbitrary fans can be very weird. However, there is an easy way to produce a special type of fans in the spirit of the example quoted at the beginning of the section: take a flag of subvarieties $W_k \subset W_{k-1} \subset \cdots \subset W_1 \subset W_0 = V$, where $\dim W_j = d - j$, $1 \leq k \leq d-1$, and take two orderings (i.e. two ultrafilters) in W_k. Then pulling back these ultrafilters along the flag of subvarieties as in the example above, i.e., considering in W_j all the ultrafilters adherent to the ultrafilters in W_{j+1}, we get a fan with 2^{k+1} elements. In the example at the beginning of the section we have done this for the case $k = 1$ to produce 4-element fans.

From the algebraic point of view we are considering a discrete valuation ring B of $R(V)$ of rank k and lifting through it two fixed orderings of its residue field by giving all possible signs to the regular parameters of B. With this valuation point of view it is immediate to check that the set of orderings obtained is indeed a fan. We call such fans *geometric* because of its obvious geometric meaning. The nice fact is that *when the ground field is the field of real numbers $\mathbb{R}$, geometric fans are enough to check most properties of semialgebraic sets since they are dense in the set of all fans*, cf. [**AnRz1**].

6. Characterization of Basicness

Let's go back to the characterization of basic semialgebraic sets. We denote by $\partial_Z S$ the Zariski closure of the boundary, $\overline{S} \setminus S$, of S. A necessary condition for an open semialgebraic set to be basic is that $S \cap \partial_Z S = \emptyset$. Indeed, if $S = \{f_1 > 0, \ldots, f_n > 0\}$ then $\partial_Z S \subset \{\prod f_i = 0\}$. Therefore we assume from now on that S verifies this topological condition, which in particular implies that it is open. We have the following celebrated theorem of Bröcker:

THEOREM 6.1. [**Br1**]) *Assume that $S \cap \partial_Z S = \emptyset$. Then S is basic if and only if for any 4-elements fan F of $\widetilde{V}$, $\#(S \cap F) \neq 3$.*

One of the directions of the if and only if condition in the theorem is immediate from the properties of fans enumerated above: if S is basic its intersection with F must have cardinality $2^0 = 1$, $2^1 = 2$ or $2^2 = 4$. However the converse is not at all obvious although intuitively we are saying that if S fails to be basic this failure can be detected in $\widetilde{V}$ in the minimal possible spot where basicness is not trivial: 4-elements fans. Remark that in the Theorem we are considering all fans in all possible subvarieties W of V. If instead the statement holds only for fans in V, i.e., supported in the whole ambient variety V that is consisting of orderings in $R(V)$, we get that S is basic as a subset of the space of orderings of $R(V)$, or, again by applying Artin-Lang theorem, up to a subset of codimension 1. In this case we say that S is *generically basic.* Therefore Bröcker's theorem asserts that $S \subset V$ is basic if and only $S \cap W$ is generically basic for any subvariety $W \subset V$ (including V

itself). Hence, to check basicness we only need to check generic basicness, although apparently we need to do it over an infinite family of subvarieties. However, only a finite family is required:

PROPOSITION 6.2. *If S fails to be basic either it fails to be generically basic (in V) or $S \cap W$ is not basic for one of the irreducible components W of the set* $\partial_Z S \cup \mathrm{Sing} X$.

PROOF. (Sketch) Assume that S is not basic but $S \cap W$ is basic for any irreducible component of $\partial_Z S \cup \mathrm{Sing} X$. In particular $S \cap Y$ is basic for any subvariety Y of this set. Thus, our assumption on S implies that there is a subvariety Y not contained in $\partial_Z S \cup \mathrm{Sing} X$ such that $S \cap Y$ is not generically basic, i.e. there is a fan $\overline{F}$ in Y such that $\#(S \cap \overline{F}) = 3$. But, since $Y \not\subset \mathrm{Sing}(X)$, the localization of $R[V]$ at the ideal of Y is a discrete valuation ring B, and we can pull back $\overline{F}$ to a 4-elements fan in V by giving suitable signs to the regular parameters of B in such a way that $\#(S \cap F) = 3$. Therefore S is not generically basic. □

This proposition shows that the first (and main) step in proving basicness is to show generic basicness for V. After that we have to prove basicness for $S \cap W$ in the irreducible components W of $\partial_Z S \cup \mathrm{Sing} X$ which again yields to check generic basicness on $S \cap W$ and basicness on some smaller subvarieties. This way we produce a finite cascade of subvarieties so that generic basicness of S on them implies basicness of S.

So, let us concentrate on showing generic basicness. From now on we suppose that the ground field is $\mathbb{R}$, the field of real numbers. Then, using the density of geometric fans, we get

THEOREM 6.3. *S is generically basic if and only if $\#(S \cap F) \neq 3$ for any 4-elements geometric fan F in V.*

Remember that geometric 4-elements fans appear attached to a hypersurface W and that, by the proposition above, after resolving singularities, W must be a component of $\partial_Z S$. Then, the condition that S contains only three of the elements of the fan means that W enters into S. We make this idea more precise: set $S^* = \mathrm{Int}(\overline{S})$. The set $\partial_Z S^* = \overline{(\overline{S^*} \setminus S^*)}^{Z}$ is called the generic Zariski boundary of S, and any irreducible component of codimension 1 of it is called a *wall* of S. We will say that $\partial_Z S^*$ *crosses* S if they intersect in a piece of dimension $d-1$, or equivalently if S contains some regular point of some wall. Then,

THEOREM 6.4. (Universal obstruction to basicness, [**AnRz2**]) *S is generically basic if and only if in any birational model of V the generic Zariski boundary $\partial_Z S^*$ does not cross S.*

Here is the paradigmatic example of a non basic semialgebraic set. Consider the set S of the picture below. Notice that the 4-element fan constructed over the horizontal axis at the beginning of section 5 has three elements in S and therefore it follows from Bröcker's theorem that S is not basic. On the other hand, the axis itself (that is, the subvariety to which the fan is attached) produces a wall of S that enters into S.

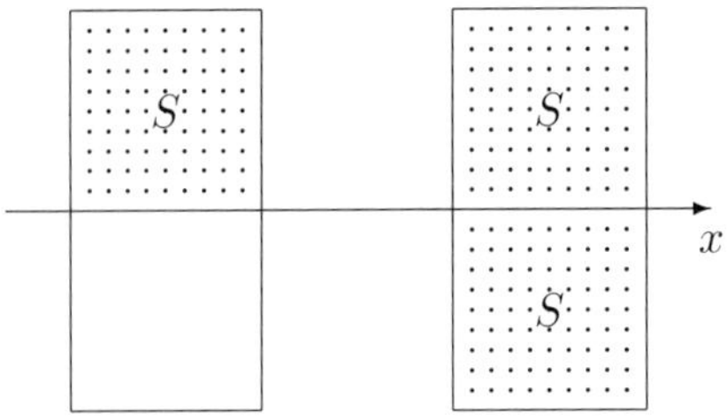

The above result shows that this is always the situation: an irreducible component of $\partial_Z S^*$ must enter into S, although this component may appear only after blowing-up V. However this result can be improved yielding to a decision method to check basicness if we impose some extra condition on the walls of S.

7. Geometric decision method

Given a subvariety W of V. We define the *shadow* of S in W as $S_W := \overline{S} \cap W$. We assume that V is compact (which is not a restriction since any real algebraic variety can be easily compactified).

THEOREM 7.1. ([**AcBgVe**]) *Assume that V is non-singular and that the Zariski boundary of S is at normal crossings. Then S is not generically basic if and only if there is a wall W of S such that either W crosses S or the shadow of S into W is not generically basic.*

PROOF. (Sketch) If S is not generically basic there is a 4-elements geometric fan F in V such that $\#(F \cap S) \neq 3$. Let B be the rank 1 discrete valuation ring attached to F. If B is centered at a hypersurface in V then it must be a wall W of S and we get that W crosses S. If the center Z of B in V has codimension greater than 1 then it is contained in the intersection of at least two walls, say W_1, W_2, whose local equations $x_1 = 0$, $x_2 = 0$ are part of a system of parameters of Z. Then, we can lift the orderings induced by F in Z first to W_2 giving different signs to x_2 and then to W_1 keeping fix the sign of x_1 so that we get a 4-elements fan which can be read also in W_1 and produces an obstruction for the shadow of S in W_1 to be basic. □

The above result can be generalized to check s–basicness. A generically basic semialgebraic set S is called *generically s–basic* if it coincides, up to a proper algebraic set, with a set of the form $\{f_1 > 0, \ldots, f_s > 0\}$. Then we have:

THEOREM 7.2. ([**AcBgVe**]) *Assume that M is non-singular and that the Zariski boundary of S is at normal crossings. Then S is generically s–basic if and only if for any wall N of S, the shadow S_N is $(s-1)$–basic.*

EXAMPLE 7.3. *In the following picture the union of any two cubes is a basic set, while the union of the three is not generically basic, since its shadow on the wall W is not generically basic, as one can easily see.*

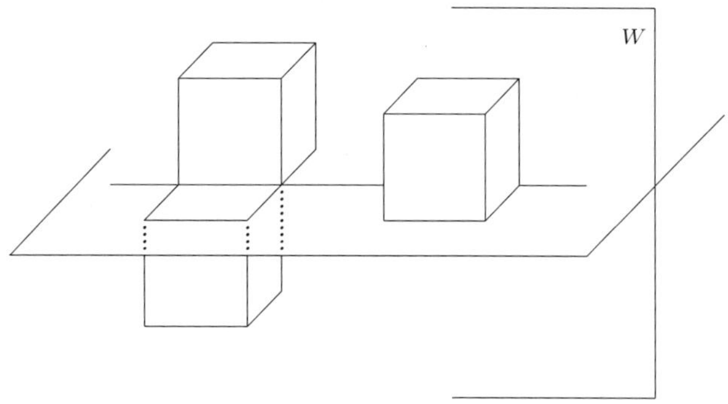

The simplest example for Theorem 3.5 is the n-ant $= \{x \in \mathbb{R}^n \mid x_1 > 0, \dots, x_n > 0\} \subset \mathbb{R}^n$. The theorem shows that it is n-basic and for each wall $x_i = 0$ its shadow on it is a $n-1$-ant which is $n-1$ basic.

8. A concrete example

We finish this note with a concrete example: showing that the set S of quartic polynomials with no real roots is not basic, cf. [**Dz**]. Set,

$$S = \{(a_1, \dots, a_4) \mid t^4 + a_1 t^3 + \cdots + a_4 \text{ has no real root } \} \subset \mathbb{R}^4.$$

We will check that the intersection of S with the plane $H = \{a_1 = a_3 = 0\}$ of biquadratic polynomials is not basic. Indeed, the polynomial $t^4 + a_2 t^2 + a_4$ has no real root if and only if $s^2 + a_2 s + a_4$ has either no real roots or its two roots negative. This yields $S \cap H = \{a_2^2 - 4a_4 < 0\} \cup \{a_2 > 0, a_4 > 0\}$, which is not basic: consider the two half-branches ξ_1, ξ_2 of the parabola $a_2^2 - 4a_4 = 0$ at the origin, and let F be the 4-elements fan $\{\alpha_1, \alpha_2, \alpha_3, \alpha_4\}$ of ultrafilters of $\mathbb{R}^2$ adherent to these half branches. We have that $\#(F \cap (S \cap H)) = 3$.

References

[AcAnBg] F. Acquistapace, C. Andradas, F. Broglia: *Separation of Semialgebraic Sets.* Journal of the AMS **12**, (1999), 703-728.

[AcBgVe] F. Acquistapace, F. Broglia, P. Vélez: *Basicness of Semialgebraic Sets.* Geometriae Dedicata **78**, (1998), 229-240.

[AnBrRz] C. Andradas, L. Bröcker, J.M. Ruiz: *Constructible sets in real geometry*, Ergeb. Math. **33**, Springer-Verlag Berlin-Heidelberg-New York, (1996).

[AnRz1] C. Andradas, J.M. Ruiz: *More on basic semialgebraic sets*, in Real Algebraic and analytic geometry, Lect. Notes in Math. **1524**, Springer-Verlag, New York, (1992), 128–139.

[AnRz2] C. Andradas, J.M. Ruiz: *Low dimensional sections of basic semialgebraic sets*, Illinois J. of Math. **38** (1994), 303–326.

[AnRz3] C. Andradas, J.M. Ruiz: *Ubiquity of Lojasiewicz example of a non-basic semialgebraic set*, Michigan Math. J. **41** (1994), 465–472.

[AnVe] C. Andradas, P. Vélez: *Convex polyhedra over ordered fields.* Preprint (2000).

[Be] A. Bernig: *Constructions for the Theorem of Bröcker and Scheiderer.* Master's Thesis, Universität Dortmund, 1998.

[BCR] J. Bochnak, M. Coste, M.F. Roy: *Real Algebraic Geometry*, Ergeb. Math. **36**, Springer-Verlag Berlin-Heidelberg-New York, (1998).

[Br1] L. Bröcker: *Minimale Erzeugung von Positivbereichen* Geom. Dedicata **16** (1984) 335–350.

[Br2] L. Bröcker: *On basic semialgebraic sets.* Expo. Math. **9**, (1988) 289–334.

[BuMa] J. Buresi, L. Mahé: *Reducing inequalities with bounds* Math. Z. **227**, (1998) 231–243
[Dz] C. Delzell: *Non-existence of analytically varying solutions to Hilbert's 17th problem.* Contemporary Math. **155**, (1994) 107–117.
[Lm] T. Y. Lam: *The algebraic theory of quadratic forms* New York, Benjamin (1973).
[Sch] C. Scheiderer: *Stability index of real varieties*, Invent. Math. **97** (1989) 467–483.
[St] G. Stengle: *A Nullstellensatz and a Positivstellensatz in semialgebraic geometry.* Math. Ann. **207** (1974) 87–97.

Departamento de Algebra, Univ. Complutense, Madrid 28040, Spain
E-mail address: carlos_andradas@mat.ucm.es

DIMACS Series in Discrete Mathematics
and Theoretical Computer Science
Volume **60**, 2003

Constructive Approaches to Representation Theorems in Finitely Generated Real Algebras

Dionne Bailey and Victoria Powers

ABSTRACT. A constructive approach to the Kadison-Dubois Representation Theorem is given, with degree bounds and an algorithm for finding a representation. This is applied to give a constructive approach to M. Marshall's representation theorem for certain non-compact semialgebraic sets. Our approach is based on work of M. Schweighofer.

1. Introduction

Given $\{f_1, \ldots, f_k\}$, a finite set of polynomials in $\mathbb{R}[X] := \mathbb{R}[x_1, \ldots, x_n]$, let S be the basic closed semialgebraic set $\{\alpha \in \mathbb{R}^n \mid f_1(\alpha) \geq 0, \ldots, f_k(\alpha) \geq 0\}$. Suppose that a polynomial p can be written as a finite sum of products of the f_i's and sums of squares in $\mathbb{R}[X]$, then clearly p is non-negative on S. A "representation theorem" is a converse to this fact, namely, a theorem that says that if a polynomial p is positive on S, then there is a representation of p in terms of the f_i's and (possibly) sums of squares. Such a representation can be thought of as a "certificate of positivity" from which the positivity condition is immediately apparent.

In [**13**], K. Schmüdgen proved a remarkable and far-reaching representation theorem, namely, he showed that if S as above is compact and $p > 0$ on S, then p is in the **preorder** generated by the f_i's, i.e., the set of finite sums of elements of the form $s_\epsilon f_1^{\epsilon_1} \cdot \cdots \cdot f_k^{\epsilon_k}$, where s_ϵ is in $\sum \mathbb{R}[X]^2$ and $\epsilon = (\epsilon_1, \ldots, \epsilon_k) \in \{0,1\}^k$. The proof, which uses functional analysis techniques, is not constructive. Thus no information is obtained on the degree of the sums of squares involved, nor on how to find a representation for a specific p. T. Wörmann [**15**] gave an algebraic proof of this result but this proof is also non-constructive. The main tool used in Wörmann's proof is the Kadison-Dubois representation theorem for archimedean partially ordered rings. Recently, M. Schweighofer [**14**] found a constructive approach to Schmüdgen's Theorem, using instead of Kadison-Dubois a representation theorem of Pólya for polynomials positive on the standard simplex. Unlike the Kadison-Dubois Theorem, degree bounds for the "output data" are known for Pólya's Theorem [**12**].

In this paper we use the ideas of Schweighofer to give a constructive approach to the Kadison-Dubois Theorem, with degree bounds and an algorithm for finding a representation. We then apply this to give a constructive approach to a result of M. Marshall [**9**] which gives a representation theorem for certain non-compact semialgebraic sets. As in Schweighofer's work, our algorithm depends on having representations of polynomials of the form $N \pm x_i$ for all i, and thus is not completely constructive. We discuss one possible approach to finding such representations in specific cases.

Schmüdgen's theorem was proven for a real polynomial ring, however Wörmann's proof and Schweighofer's algorithm both generalize easily to the finitely generated real algebra case. Our results will also hold in this more general case. The full generality will be needed for the proof of Marshall's theorem.

Some of the results in this paper are part of the first author's Ph.D thesis at Emory University, written under the direction of the second author.

2. Preliminaries on Archimedean Preprimes and Preorders

We work in the coordinate ring $R = \mathbb{R}[V]$ of an algebraic set V in $\mathbb{R}^n$. Then R is an $\mathbb{R}$-algebra generated by the coordinate functions $x_i : V \to \mathbb{R}$. Given $f_1, \dots, f_k \in R$, let $S(f_1, \dots, f_k)$ be the **basic closed semialgebraic set** generated by the f_i's, i.e., $S(f_1, \dots, f_k) := \{\alpha \in V \mid f_i(\alpha) \geq 0 \text{ for } 1 \leq i \leq k\}$.

A subset P of R is a **preprime** if

$$P + P \subseteq P, \quad P \cdot P \subseteq P, \quad \mathbb{R}^+ \subseteq P, \quad -1 \notin P.$$

Let $\sum$ denote the set of sums of squares in R. A preprime P is a **preorder** if $\sum \subseteq P$. Given $f_1, \dots, f_k \in R$, $PP(f_1, \dots, f_k)$ denotes the preprime generated by the f_i's, i.e., the set of finite sums of elements in R of the form $a_\epsilon f_1^{\epsilon_1} \dots f_k^{\epsilon_k}$, where $\epsilon = (\epsilon_1, \dots, \epsilon_k) \in \mathbb{N}^k$ and $a_\epsilon \in \mathbb{R}^+$.

DEFINITION 2.1. Given a finitely-generated preprime $P := PP(f_1, \dots, f_k)$ and suppose $g \in P$. A P**-representation** of g is a an equation of the form

$$g = \sum_{\epsilon \in \mathbb{N}^k} a_\epsilon f_1^{\epsilon_1} \cdot \dots \cdot f_k^{\epsilon_k},$$

where $a_\epsilon \in \mathbb{R}^+$ for all ϵ. The **degree** of the representation is the maximum of $\{|\epsilon|\}$.

REMARK 2.2. Another way to view elements of $PP(f_1, \dots, f_k)$ is as the set of all elements of R which can be written $F(f_1, \dots, f_k)$, where F is a polynomial in k variables with non-negative coefficients. If we have a representation $g = F(f_1, \dots, f_k)$ of $g \in PP(f_1, \dots, f_k)$, then the degree of the representation is the degree of F as a polynomial.

The preorder generated by the f_i's, denoted $PO(f_1, \dots, f_k)$, is the set of all elements of the form

$$\sum_{\epsilon \in \{0,1\}^k} s_\epsilon f_1^{\epsilon_1} \cdot \dots \cdot f_k^{\epsilon_k},$$

where $s_\epsilon \in \sum$ for all ϵ.

REMARK 2.3. Preorders and preprimes are the fundamental algebraic objects associated to semialgebraic sets. In some sense, preorders play a role in semialgebraic geometry that is comparable to the role played by ideals in algebraic geometry. Computationally, a finitely generated preprime is much simpler to work with than a finitely generated preorder, since sums of squares are not involved. However, for the most general cases, the preprime is not sufficient, as we shall see.

We say that a subset P of R is **archimedean** if for each $g \in R$ there exists a non-negative integer m such that $m - g \in P$. Suppose $P := PP(f_1, \dots, f_k)$ and $S := S(f_1, \dots, f_k)$ and P is archimedean. Then it is easy to see that S is compact: There must be some $N \in \mathbb{N}$ such that $N \pm x_i \in P$ for all i, from which it follows that S is contained in the compact set $\{\alpha \in V \mid N \pm x_i(\alpha) \geq 0 \text{ for all } i\}$ and hence is also compact. For preorders, the converse also holds:

THEOREM 2.4. *Let S be as above and $P := PO(f_1, \dots, f_k)$. Then S is compact if and only if P is archimedean.*

PROOF. The proof that P archimedean implies S compact was given above. The converse follows from [**15**]. □

REMARK 2.5. In general, the theorem is not true if we replace the preorder P by the preprime generated by the f_i's. This can be shown by the following example, which is due to D. Handelman [**5**]: In $\mathbb{R}[x, y]$, consider $PP(x, y, 1 - x^2 - y^2)$ then the corresponding semialgebraic set is the semi-circle $\{x \geq 0, y \geq 0, x^2 + y^2 \leq 1\}$, which is obviously compact. On the other hand, an elementary argument shows that for any positive integer N, it is not possible to write $N - x$ in the form $F(x, y, 1 - x^2 - y^2)$, where F is a polynomial in three variables with non-negative coefficients. Thus the preprime is not archimedean.

It turns out that in the linear case, the preprime is enough:

PROPOSITION 2.6. *Given linear functions $l_1, \dots, l_k \in R$, suppose $S = S(l_1, \dots, l_k)$ is compact. Then $PP(l_1, \dots, l_k)$ is archimedean.*

PROOF. This follows from [**5**, 1.3], see also [**12**]. □

Now suppose that P is an archimedean preprime. Then there is some non-negative integer N such that for $N \pm x_i \in P$ for $i = 1, \dots, n$. We say that a minimal such N is the **archimedean bound** of P. We fix representations of $N \pm x_i$ in P.

REMARK 2.7. For a preorder, we can construct from representations of $N \pm x_i$, a representation of $nN^2 - \sum x_i^2$, and, conversely, we can construct representations of $\frac{N+1}{2} \pm x_i$ in P, using a representation of $N - \sum x_i^2$. This follows from the identity

$$\frac{N+1}{2} \pm x_i = \frac{1}{2}\Big((x_i \pm 1)^2 + (N - \sum_i x_i^2) + \sum_{j \neq i} x_j^2\Big) \tag{2.1}$$

Thus a preorder P is archimedean iff $M - \sum x_i^2 \in P$ for some $M \in \mathbb{N}$. However, if P is a preprime it is not true that $M - \sum x_i^2$ in P for some M implies that P is archimedean. This follows from the example $PP(x, y, 1 - x^2 - y^2)$ above.

THEOREM 2.8. *Suppose P is an archimedean preprime, with archimedean bound N and we have fixed representations of $\{N \pm x_i\}$ in P. Let D be the maximum degree of the fixed representations.*

(1) *For any $g \in R$, $M - g \in P$ for some $M \in \mathbb{N}$ and a P-representation of $M - g$ in P can be constructed explicitly, in terms of the representations of $\{N \pm x_i\}$.*
(2) *Suppose we have $g = F(x_1, \dots, x_n)$, where F is a real polynomial with $\deg F = d$ and L is the sum of the absolute values of the coefficients of F. Then we can take $M = LN^d$ and the degree of the P-representation will be at most dD.*

PROOF. We prove this for a real polynomial ring and then, using an epimorphism $\mathbb{R}[X_1, \dots, X_n] \to R$, obtain the general result. So suppose g is a real polynomial with degree d. We will proceed by induction on d. The proposition is obviously true for $d = 0$. Since preprimes preserve addition, it is enough to prove the theorem for g a monomial. Suppose $g = aX^\epsilon$, where $\epsilon = (\epsilon_1, \dots, \epsilon_n) \in \mathbb{N}^n$ and assume w.l.o.g. that $\epsilon_1 \neq 0$. Then $g = X_1 \cdot aX^\beta$, where $|\beta| = |\epsilon| - 1$. Then

$$(2.2) \quad |a|N^{|\epsilon|} - g = \frac{1}{2}\Big((N + X_1)(|a|N^{|\beta|} - aX^\beta) + (N - X_1)(|a|N^{|\beta|} + aX^\beta)\Big).$$

Thus we are done by our induction hypothesis. Further, this gives us a recursive algorithm for constructing a P-representation for a specific g. □

COROLLARY 2.9. *Suppose P is a preprime in R, then P is archimedean iff there exists $N \in \mathbb{N}$ such that $N \pm x_i \in P$ for all i.*

Generally speaking, we will give algorithms for constructing a representation of an element in an archimedean preprime P in terms of the polynomials $\{N \pm x_i\}$. For ease of exposition, we want to enlarge the generating set of the preprime by adding these elements. Note that adding to the generators of P a polynomial p which is in P does not change P or S. Also, by Theorem 2.8, if $\{f_1, \dots, f_k\}$ is the set of generators of P, then we have $M - \sum f_i \in P$ for some M. Replacing each f_i by f_i/M and adding $1 - \sum f_i/M$, we have a set of generators for which $\sum f_i = 1$. For our constructions, it will be necessary to add to our set of generators elements of the form $N/M - x_i/M$, where $N, M \in \mathbb{N}$ and to have $\sum f_i = 1$. It will be convenient to replace the generators of P with this (possibly) enlarged set. As a final step in any construction, we can get a representation in terms of the original generators by using our fixed representations of the set $\{N \pm x_i\}$ and the representation of $1 - \sum f_i/M$ from Theorem 2.8.

DEFINITION 2.10. Suppose $P = PP(f_1, \dots, f_k)$ is an archimedean preprime. We say that the set of generators $\{f_1, \dots, f_k\}$ is **full** if

(1) The first n generators $\{f_1, \dots, f_n\}$ are $\{N/M - x_1/M, \dots, N/M - x_n/M\}$ for some $N, M \in \mathbb{N}$.
(2) $\sum f_i = 1$

3. Pólya's Theorem and the Kadison-Dubois Theorem

Pólya's Theorem is a representation theorem for homogeneous polynomials positive on the standard n-simplex. In this section, we show how other representation theorems can be reduced to Pólya's Theorem, in particular, the Kadison-Dubois Theorem. Since explicit bounds for the degree of a representation have been given for Pólya's Theorem, this yields degree bounds for the theorems reduced to Pólya. This technique was first used by M. Schweighofer [**14**] to study Schmüdgen's Theorem. In [**12**], it was used to study representations of polynomials positive on compact polyhedra.

We write Δ_n for the n-simplex $\{(x_1, \dots, x_n) \mid x_i \geq 0, \sum_i x_i = 1\}$. Pólya's Theorem ([**11**], [**6**, pp.57-59]) says that if $f \in \mathbb{R}[X]$ is homogeneous and positive on Δ_n, then for sufficiently large N all the coefficients of $(x_1 + \cdots + x_n)^N f(x_1, \dots, x_n)$ are positive. In [**12**], an explicit bound for N is given:

THEOREM 3.1. *Given homogeneous $f \in \mathbb{R}[X]$ of degree d, say*

$$f(X) = \sum_{|\alpha|=d} a_\alpha X^\alpha = \sum_{|\alpha|=d} c(\alpha) b_\alpha X^\alpha,$$

where $c(\alpha) := \frac{d!}{\alpha_1! \cdots \alpha_n!}$. Let $L = L(f) := \max_{|\alpha|=d} |b_\alpha|$ and $\lambda = \lambda(f) := \min_{X \in \Delta_n} f(X)$. If

$$N > \frac{d(d-1)}{2} \frac{L}{\lambda} - d,$$

then $(x_1 + \cdots + x_n)^N f(x_1, \ldots, x_n)$ has positive coefficients.

We now fix the following set-up: Given an archimedean preprime P in R with $P := PP(f_1, \ldots, f_k)$ and assume that the set of generators is full. Let $S := S(f_1, \ldots, f_k)$. Let $\mathbb{R}[Y]$ denote the polynomial ring $\mathbb{R}[y_1, \ldots, y_k]$ in k variables and define $\phi : \mathbb{R}[Y] \to R$ by $\phi(y_i) = f_i$. Since the generators contain $\{N/M - x_i/M\}$, the map ϕ is onto. Let $\{r_1, \ldots, r_t\}$ be a basis for $\ker \phi$. To construct such a basis, we can compute a Gröbner Basis B for the ideal generated by $\{y_1 - f_1, \ldots, y_k - f_k\}$ in the polynomial ring $\mathbb{R}[x_1, \ldots, x_n, y_1, \ldots, y_k]$ using lex order with $x_1 > \cdots > x_n > y_1 > \cdots > y_k$. Then $B \cap \mathbb{R}[y_1, \ldots, y_k]$ is the desired basis.

Given $g \in R$, we can construct explicitly $\tilde{g} \in \mathbb{R}[Y]$ such that $\phi(\tilde{g}) = g$ and $\deg \tilde{g} = \deg g$. For example, if the first n generators of P are $\{N/M - x_1/M, \ldots, N/M - x_n/M\}$, then we can set $\tilde{g} = g(N - My_1, \ldots, N - My_n)$.

The following lemma is the key to reducing to Pólya's Theorem. It was first proven in [**14**] for preorders in polynomial rings and in [**12**] it was proven for preprimes associated to polyhedra. The argument generalizes immediately to our setting.

LEMMA 3.2. *Suppose S, P, and ϕ are above. Given $g \in R$ such that $g > 0$ on S. Then there exists a homogeneous $G \in \mathbb{R}[Y]$ such that $G > 0$ on Δ_k and $\phi(G) = g$ Furthermore, there is an algorithm for constructing G and there is a constant D which depends only on the generators of P such that* $\deg G \leq \max\{\deg g, D\}$.

PROOF. Let $\tilde{g}$ be as above, then $\phi(\tilde{g}) = g$. Also, let $\{r_1, \ldots, r_t\}$ be a basis for $\ker \phi$, which can be constructed as above. Set $r := \sum r_i^2$ and let $D = \deg r$. The argument of [**14**, 3.1] shows that $r > 0$ on the compact set $U := \Delta_k \cap \{\tilde{g} \leq 0\}$. Now choose a real number $c > \max\{0, \frac{-m_1}{m_2}\}$, where m_1 is the minimum of $\tilde{g}$ on Δ_k and m_2 is the minimum of r on U. Then $\tilde{g} + cr > 0$ on Δ_k. Note that $\deg(\tilde{g} + cr) = \max\{\deg g, D\}$. Finally, we let G be the homogeneous polynomial obtained from $\tilde{g} + cr$ by multiplying monomials by appropriate powers of $(y_1 + \cdots + y_k)$. □

The above theorem shows that instead of working in the archimedean preprime in R, we can use the map ϕ to transform to a polynomial ring and then apply Pólya's Theorem. In particular, we want to apply this to the Kadison-Dubois Theorem.

The Kadison-Dubois Theorem [**7**], [**4**] is a representation theorem for archimedean preprimes in commutative rings; it can be viewed as a far-reaching generalization of the fact that a field with an archimedean order has an order-preserving embedding into $\mathbb{R}$. For the most general statement of the theorem, see [**1**].

In our setting, the theorem is as follows:

THEOREM 3.3 (Kadison-Dubois). *Given $f_1, \ldots, f_k \in R$ and suppose that $P := PP(f_1, \ldots, f_k)$ is archimedean. Then for any $h \in R$, the following are equivalent:*

(1) $h(\alpha) \geq 0$ *for each* $\alpha \in S := S(f_1, \ldots, f_k)$
(2) $h + \gamma \in P$ *for all positive* $\gamma \in \mathbb{R}$.

PROOF. It is clear that (ii) implies (i); we prove (i) implies (ii). Given $h \geq 0$ on S and real $\gamma > 0$, let $g = h + \gamma$. We can assume that the given set of generators for P is full, by the discussion in the previous section. Let $\mathbb{R}[Y]$ and ϕ be as above. Given $g > 0$ on S, by Lemma 3.2, there is homogeneous G in $\mathbb{R}[Y]$ such that $G > 0$ on Δ_k and $\phi(G) = g$. By Pólya, there is some non-negative integer M such that $(\sum y_i)^M G$ has only positive coefficients in $\mathbb{R}[Y]$ and hence, with $m = M + \deg G$, we have

$$(\sum y_i)^M G = \sum_{\epsilon \in \mathbb{N}^k, |\epsilon| = m} a_\epsilon y_1^{\epsilon_1} \cdot \cdots \cdot y_k^{\epsilon_k},$$

where $a_\epsilon \in \mathbb{R}^+$. Applying ϕ to both sides, we obtain a representation of g in P, of degree m. □

EXAMPLE 3.4. Let $S = \{1 - x^2 \geq 0\} = [-1, 1]$, and

$$P = PP(\frac{1-x}{3}, \frac{1-x^2}{3}, \frac{1+x+x^2}{3}).$$

Note $2 + x = (1 - x^2) + (1 + x + x^2) \in P$ and so P is archimedean. The given set of generators $\{f_1, f_2, f_3\} = \{\frac{1-x}{3}, \frac{1-x^2}{3}, \frac{1+x+x^2}{3}\}$ is full.

Let $f = 7x^4 + 7x^3 + 7x^2 + 6$, then by Kadison-Dubois, $f \in P$. We want to find a representation of f in P. Set $g(y_1, y_2, y_3) = 27(y_1^2 + y_2^2 + y_3^2) - 9(y_1y_2 + y_1y_3 + y_2y_3)$, then $g(f_1, f_2, f_3) = f$ and $g > 0$ on Δ_3.

It turns out that g has Pólya exponent 2:

$$(y_1 + y_2 + y_3)^2 g =$$
$$27y_1^4 + 45y_1^3y_2 + 36y_1^2y_2^2 + 45y_1y_2^3 + 27y_2^4 + 45y_1^3y_3 + 9y_1^2y_2y_3 + 9y_1y_2^2y_3 +$$
$$45y_2^3y_3 + 36y_1^2y_3^2 + 9y_1y_2y_3^2 + 36y_2^2y_3^2 + 45y_1y_3^3 + 45y_2y_3^3 + 27y_3^4.$$

This yields the following representation of f in P:

$$7x^4 + 7x^3 + 7x^2 + 6 =$$
$$27(\frac{1-x}{3})^4 + 45(\frac{1-x}{3})^3(\frac{1-x^2}{3}) + 36(\frac{1-x}{3})^2(\frac{1-x^2}{3})^2 + 45(\frac{1-x}{3})(\frac{1-x^2}{3})^3$$
$$+ 27(\frac{1-x^2}{3})^4 + 45(\frac{1-x}{3})^3(\frac{1+x+x^2}{3}) + 9(\frac{1-x}{3})^2(\frac{1-x^2}{3})(\frac{1+x+x^2}{3})$$
$$+ 9(\frac{1-x}{3})(\frac{1-x^2}{3})^2(\frac{1+x+x^2}{3}) + 45(\frac{1-x^2}{3})^3(\frac{1+x+x^2}{3}) + 36(\frac{1-x}{3})^2(\frac{1+x+x^2}{3})^2$$
$$+ 9(\frac{1-x}{3})(\frac{1-x^2}{3})(\frac{1+x+x^2}{3})^2 + 36(\frac{1-x^2}{3})^2(\frac{1+x+x^2}{3})^2 + 45(\frac{1-x}{3})(\frac{1+x+x^2}{3})^3$$
$$+ 45(\frac{1-x^2}{3})(\frac{1+x+x^2}{3})^3 + 27(\frac{1+x+x^2}{3})^4,$$

as is easily checked!

REMARKS 3.5. (1) Note that the preprime $PP(1-x^2)$ is not archimedean. We could start with the preorder $PO(1 - x^2)$, so that Schmüdgen's Theorem applies. This must be archimedean, in fact $1-x = \frac{1}{2}[(1-x^2)+(1-x)^2]$ and $1+x = \frac{1}{2}[(1-x^2) + (1+x)^2]$. Let $P = PP(1-x^2, (1-x)^2, (1+x)^2)$. Then P contains $1 \pm x$ and thus is archimedean by Corollary 2.9. We now proceed as above. This shows that we do not need the "full" preorder to obtain the conclusion of Schmüdgen's Theorem.

In general, if we start with compact $S = S(f_1, \dots, f_k)$ and the preorder $P = PO(f_1, \dots, f_k)$, we know that P is archimedean. Suppose we have representations of $N \pm x_i$ in P, and say that $\{s_1, \dots, s_m\}$ are all the sums of squares that occur in the representations. Then the *preprime* $PP(f_1, \dots, f_k, s_1, \dots s_m)$ is archimedean and hence Kadison-Dubois applies.

(2) The proof of the Kadison-Dubois Theorem, along with Lemma 3.2, yields an algorithm for constructing a representation of $h+\gamma$ in P, assuming the set of generators for P is full. Thus we have an algorithm for the Kadison-Dubois Theorem modulo representations of $\{N \pm x_i\}$ in P. However, it should be noted that to calculate the constant c needed in Lemma 3.2 is in general computationally difficult, probably as difficult as finding a P-representation. In [**14**], an alternate method is suggested: Instead of trying to calculate c, make it a parameter. For some c and some N, $(\sum y_i)^N(\tilde{g} + cr)$ has only positive coefficients, and the coefficients of this are linear in c. We can then write this as a linear programming problem for a fixed N and increase N until a solution exists. We hope to explore this idea in further work.

(3) We do not know an algorithm for finding the archimedean bound N and representations of $N \pm x_i$ in archimedean P. There is a procedure that could be used in specific cases, which comes from the fact that the problem of finding a P-representation of fixed degree for an element of P can be rewritten as a semidefinite programming (SDP) problem, which can then be solved using SDP software. Such problems can involve parameters, thus the archimedean bound N could be taken as a parameter. Then we could look for the existence of P-representations of $N \pm x_i$ of fixed degree for non-negative N. If no such P-representations exist, then one could raise the degree bound and try again. This would yield a sequence of SDP problems which must eventually have a solution. For details on the relationship between representations in preorders and preprimes and semidefinite programming, the reader should consult the work of J. B. Lasserre [**8**] and P. Parillo [**10**].

4. A Representation Theorem in the Non-compact Case

Recently, M. Marshall proved a generalization of Schmüdgen's Theorem to the non-compact case. The proof uses a generalization of Wörmann's proof, i.e., an algebraic argument is used to reduce to the archimedean case so that the Kadison-Dubois Theorem applies. In this section, we apply our proof of Kadison-Dubois to give a constructive version of this result.

Given $f_1, \dots, f_k \in R$, set $S = S(f_1, \dots, f_k)$ and $P := PO(f_1, \dots, f_k)$. We do not assume that S is compact, hence P need not be archimedean. We need an element p in $1 + P$ such that there exists non-negative integers M and l with $Mp^l \pm x_1, \dots, Mp^l \pm x_k \in P$. Note that such a p always exists, e.g., we can always let $p = 1 + \sum x_i^2$ and $M = l = 1$. Also note that if S is compact, then we can take $p = 1$.

The main theorem of this section is:

THEOREM 4.1 (Marshall). *Given P, S, p, M, and l as above. Then, for any $f \in R$ of degree D, the following are equivalent.*

(1) *$f(\alpha) \geq 0$ for each $\alpha \in S(p_1, \dots, p_k)$*
(2) *For each $\epsilon > 0$ there is some sufficiently large integer $t \geq 0$ such that $p^t(f + \epsilon p^{lD}) \in P$*

PROOF. The proof of [**9**] shows that we can reduce to the archimedean case by localizing. We include the proof here for completeness.

Consider the following $\mathbb{R}$-algebra

$$C := \mathbb{R}[\frac{x_1}{p^l}, \dots, \frac{x_n}{p^l}, \frac{1}{p}].$$

Let $d_i = \deg p_i$. Then we can decompose each p_i as $\sum_{j=0}^{d_i} p_{i_j}(x_1, \dots, x_n)$ where each p_{i_j} is a homogeneous polynomial of degree j. It follows that

$$\frac{p_i}{p^{ld_i}} = \sum_{j=0}^{d_i} \frac{1}{p^{l(d_i - j)}} p_{i_j}(\frac{x_1}{p^l}, \dots, \frac{x_n}{p^l}) \in C \tag{4.1}$$

Define the following preprime in C:

$$\tilde{P} := PP(\frac{p_1}{p^{ld_1}}, \dots, \frac{p_k}{p^{ld_k}}, \frac{1}{p}, \frac{Mp^l \pm x_1}{p^l}, \dots, \frac{Mp^l \pm x_n}{p^l}, \frac{p \pm 1}{p}) \tag{4.2}$$

By Corollary 2.9, $\tilde{P}$ is archimedean in C since we have $M \pm \frac{x_i}{p^l} = \frac{Mp^l \pm x_i}{p^l} \in \tilde{P}$ $(1 \leq i \leq n)$ and $1 \pm \frac{1}{p} = \frac{p \pm 1}{p} \in \tilde{P}$.

Decompose f given in our hypothesis as $f = \sum_{j=0}^{D} f_j(x_1, \dots, x_n)$ where each f_j is homogeneous of degree j. Then

$$\tilde{f} := \frac{f}{p^{lD}} = \sum_{j=0}^{D} \frac{1}{p^{l(D-j)}} f_j(\frac{x_1}{p^l}, \dots, \frac{x_n}{p^l}) \in C. \tag{4.3}$$

Define the semialgebraic set associated to $\tilde{P}$ as follows:

$$\tilde{S} := S(\frac{p_1}{p^{ld_1}}, \dots, \frac{p_k}{p^{ld_k}}, \frac{1}{p}, \frac{Mp^l \pm x_1}{p^l}, \dots, \frac{Mp^l \pm x_n}{p^l}, \frac{p \pm 1}{p}) \subseteq \mathbb{R}^n$$

then we will show that $\tilde{f} \geq 0$ on $\tilde{S}$. Let $\alpha \in \tilde{S}$ then $\frac{p_i(\alpha)}{p(\alpha)^{ld_i}} \geq 0$ for each i, and $\frac{1}{p(\alpha)} > 0$ and $p(\alpha) \neq 0$ since p is positive definite. Hence, $p_i(\alpha) \geq 0$ for every i which yields that $\alpha \in S$. Thus by our hypothesis $f(\alpha) \geq 0$. From equation (4.3), we obtain $\tilde{f}(\alpha) \geq 0$. We then complete the proof by applying Kadison-Dubois to $\tilde{f} + \epsilon$ with $\tilde{P}$ and $\tilde{S}$. □

References

[1] E. Becker, *Valuations and real places in the theory of formally real fields*, in: Géométrie Algébrique Réelle et Formes Quadratiques, Proceedings Rennes 1981, Springer Lecture Notes in Mathematics **959**, Springer Verlag, 1982, 1–40.

[2] J. A. de Loera and F. Santos, *An effective version of Pólya's theorem on positive definite forms*, J. Pure Appl. Alg. **108** (1996), 231–240.

[3] J. A. de Loera and F. Santos, *Corrections to An effective version of Pólya's theorem on positive definite forms*, J. Pure Appl. Alg., to appear.

[4] D. W. Dubois, *A note on David Harrison's theory of preprimes*, Pac. J. Math. **21** (1967), 15–19.

[5] D. Handelman, *Representing polynomials by positive linear functions on compact convex polyhedra*, Pac. J. Math. **132** (1988), 35–62.
[6] G. H. Hardy, J. E. Littlewood, and G. Pólya, *Inequalities*, 2nd ed., Camb. Univ. Press, 1952.
[7] R. V. Kadison, *A representation theorem for commutative topological algebra*, Mem. Amer. Math. Soc. **7** (1951).
[8] J. B. Lasserre, *Global optimization with polynomials and the problem of moments*, SIAM J. Optimization **11** (2001), 796-817.
[9] M. Marshall, *Extending the Archimedean Positivstellensatz to the non-compact case*, Canad. Math. Bull., to appear.
[10] P. Parillo, Structured Semidefinite Programs and Semialgebraic Geometry Methods in Robustness and Optimization, Ph.D thesis, California Institute of Technology, 2000.
[11] G. Pólya, *Über positive Darstellung von Polynomen Vierteljschr*, Naturforsch. Ges. Zürich **73** (1928), 141–145. In *Collected Papers* **2** (1974), MIT Press, 309-313.
[12] V. Powers and B. Reznick, *A new bound for Pólya's Theorem with applications to polynomials positive on polyhedra*, Proceedings of the MEGA 2000 conference, J. Pure and Applied Algebra, to appear.
[13] K. Schmüdgen, *The K-moment problem for compact semialgebraic sets*, Math. Ann.**289** (1991), 203-206.
[14] M. Schweighofer, *An algorithmic approach to Schmüdgen's Positivstellensatz*, J. Pure and Applied Algebra, to appear.
[15] T. Wörmann, *Short algebraic proofs of theorems of Pólya and Schmüdgen*, preprint.

Department of Mathematics, Angelo State University, P.O. Box 10900, San Angelo, TX 76909

Department of Mathematics and Computer Science, Emory University, Atlanta, GA 30322
E-mail address: vicki@mathcs.emory.edu

DIMACS Series in Discrete Mathematics
and Theoretical Computer Science
Volume **60**, 2003

Combinatorial characterizations of algebraic sets

Isabelle Bonnard

ABSTRACT. We first present conditions for a compact semialgebraic set X to be homeomorphic to a real algebraic set. A first necessary condition is that the Euler characteristic of the link of X at any point is even (Sullivan). This condition is sufficient if the dimension of X is less than or equal to two (Akbulut-King, Benedetti-Dedò), but not if it is greater. In dimension 3, Akbulut and King establish a necessary and sufficient condition, using invariants modulo 2. A generalization of Sullivan's condition, and new invariants modulo 2^k, are given by Coste and Kurdyka (in any dimension).

In the second part we introduce McCrory and Parusiński's algebraically constructible functions. We show how to use these functions to reformulate Akbulut-King's conditions if $\dim X \leq 3$, and to get new necessary conditions in dimension 4. We end by generalizing these new conditions to the characterization of arc-symmetric sets.

We consider the following question. Let X be a topological space; is X homeomorphic to a real algebraic set? By real algebraic set we mean the zero set in $\mathbb{R}^n$ of some polynomials in $\mathbb{R}[x_1, \ldots, x_n]$. We will take here for X a closed semialgebraic subset of $\mathbb{R}^N$. If X is compact then there exists a triangulation of X ([**BCR**, Th.9.2.1]) and we can as well work on the corresponding polyhedron. This will lead to combinatorial conditions.

The first necessary condition for X to be homeomorphic to a real algebraic set has been given by Sullivan in 1970 ([**S**]). It states that the Euler characteristic of the link of X at any point must be even. Ten years later, Akbulut and King on one hand, and Benedetti and Dedò on the other, proved independently that if X is compact and $\dim X \leq 2$ then this condition is also sufficient ([**AK1**],[**BD**]). In dimension 3 this is no more true. In [**AK2**], for X of dimension 3, Akbulut and King introduce invariants modulo 2 which vanish if and only if X is homeomorphic to a real algebraic set. In [**CK1**] Coste and Kurdyka give a generalization of Sullivan's condition. Namely, they prove that if W is an irreducible algebraic subset of a real algebraic set V, then for a generic point x in W, the link of W in V at x has constant Euler characteristic modulo 4. They also introduce invariants modulo 2^k, which in dimension 3 are equivalent to the Akbulut-King condition.

2000 *Mathematics Subject Classification.* Primary 14P25.

Key words and phrases. real algebraic set, semialgebraic set, Euler characteristic, constructible function.

New formulations of these conditions in dimension ≤ 3 have been given in the nineties by McCrory and Parusiński, using algebraically constructible functions and their stability by the "half-link operator" (see [**MP1**]). In dimension 4, using again these functions and their stability by the operator $\varphi \mapsto \frac{1}{2}(\varphi^4 - \varphi^2)$, these authors give new necessary conditions ([**MP3**]). They get $2^{43} - 43$ independent invariants!

In the first part we will present Sullivan's condition, and the results of Akbulut and King, and Coste and Kurdyka. The second part is devoted to the approach using algebraically constructible functions, its last paragraph dealing with the characterization of arc-symmetric sets.

1. First approach

1.1. Sullivan's condition. Let $X \subset \mathbb{R}^N$ be a closed semialgebraic set.

DEFINITION 1.1. If x is a point of X, then the *link* of X at x is defined as

$$\mathrm{lk}(x, X) = S(x, \varepsilon) \cap X$$

where $S(x, \varepsilon)$ is the sphere centered at x of radius ε, and $\varepsilon > 0$ is small enough.

The link of X at x is well-defined up to semialgebraic homeomorphism ([**CK1**, Prop.1]).

THEOREM 1.2 (Sullivan's condition, [**S**, Ex.ii]). *If X is homeomorphic to a real analytic set, then for every $x \in X$ we have*

$$\chi\big(\mathrm{lk}(x, X)\big) \equiv 0 \bmod 2.$$

The idea of the proof is the following. Assume X is a real analytic set. Denote by $X_{\mathbb{C}}$ a complexification of X. If $x \in X$, then it can be seen that $\chi\big(\mathrm{lk}(x, X_{\mathbb{C}})\big) = 0$ since $\mathrm{lk}(x, X_{\mathbb{C}})$ is a compact stratified set with only odd-dimensional strata. Consider a triangulation of $\mathrm{lk}(x, X_{\mathbb{C}})$ compatible with $\mathrm{lk}(x, X)$ and with the conjugation. The numbers of simplices in $\mathrm{lk}(x, X)$ and $\mathrm{lk}(x, X_{\mathbb{C}})$ are then congruent modulo 2, so we get that $\chi\big(\mathrm{lk}(x, X)\big) \equiv \chi\big(\mathrm{lk}(x, X_{\mathbb{C}})\big) \bmod 2 \equiv 0 \bmod 2$.

EXAMPLE 1.3. a) If $\dim X = 1$, this condition means that there must be an even number of branches at any singular point.

b) We recall the well-known construction of *suspension*. If V is an algebraic set, the suspension of V is the set $\Sigma V = (V \times [-1, 1])/\sim$ where the relation $\sim$ is defined by $(x, -1) \sim (x', -1)$ and $(x, 1) \sim (x', 1)$ for all $x, x' \in V$. The *suspension points* are the classes of $(x, -1)$ and $(x, 1)$.

Take $V = \mathbb{RP}^2$, and consider the link of $X = \Sigma\mathbb{RP}^2$ at a suspension point p. This link is homeomorphic to $\mathbb{RP}^2$, and $\chi\big(\mathrm{lk}(p, X)\big) = \chi(\mathbb{RP}^2) = 1$. So X is not homeomorphic to a real algebraic set.

DEFINITION 1.4. The closed semialgebraic set X is said to be *Euler* if Sullivan's condition holds at any point of X.

A necessary condition for X to be homeomorphic to a real algebraic set is that X is Euler. A natural question is to ask if it is also a sufficient condition.

REMARK 1.5. We may assume that X is compact. Indeed, X is homeomorphic to a real algebraic set if and only if its one-point compactification is so.

1.2. Akbulut and King's results. The following result answers to the previous question if $\dim X \leq 2$. It was proved at the beginning of the eighties by Akbulut and King, and independently by Benedetti and Dedò.

THEOREM 1.6 ([**AK1**, Th.6.1], [**BD**, Th.3.2]). *Let $X \subset \mathbb{R}^N$ be a compact semialgebraic set such that* $\dim X \leq 2$. *Then X is homeomorphic to a real algebraic set if and only if X is Euler.*

As X is triangulable, there is only a finite number of possible semialgebraic homeomorphism types for $\text{lk}(x, X)$. So this theorem leads to a finite number of conditions.

EXAMPLE 1.7. a) In dimension 1, the condition on the parity of the number of branches guarantees that the set is homeomorphic to a real algebraic set.

b) Every compact real analytic set of dimension ≤ 2 is homeomorphic to a real algebraic set.

REMARK 1.8. The result of theorem 1.6 is no more true if $\dim X \geq 3$. The first known counter-example in dimension 3 has been given by Akbulut and King (see [**Ki**]). It is the suspension X of the set Y drawn below. The set Y is the one-point compactification of the union in $\mathbb{R}^3$ of the Cartan umbrella of equation $z^2 = xy^2$, and of the circle in the plane $y = 0$ centered at $(-1, 0, 0)$ of radius 1. As Y is an algebraic set whose Euler characteristic is 0, the set X is Euler. We will see later that X is not homeomorphic to a real algebraic set (see also the article of Clint McCrory in this volume).

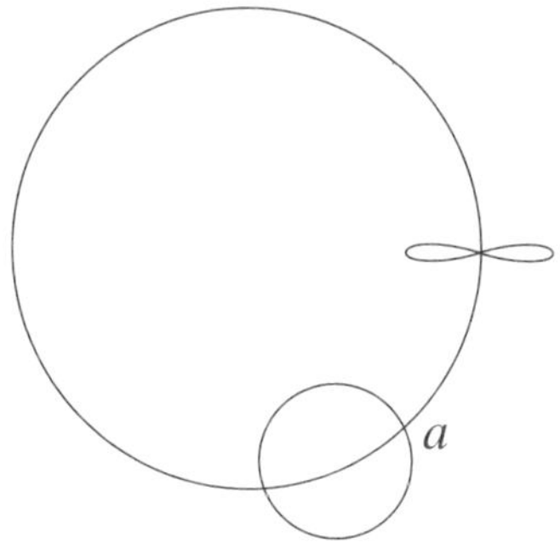

Consider now a compact semialgebraic subset X of dimension 3 of $\mathbb{R}^N$, and a triangulation $\mathcal{T}$ of X. Assume X is Euler. Denote by X_0 the set of vertices of $\mathcal{T}$. Akbulut and King define new invariants $\theta_i(x) \in \mathbb{Z}/2\mathbb{Z}$, for every $x \in X_0$ and every $i \in \{0, 1, 2, 3\}$, such that

THEOREM 1.9 ([**AK2**, Th.7.1.1]). *The set X is homeomorphic to a real algebraic set if and only if for every $x \in X_0$ and every $i \in \{0, 1, 2, 3\}$, we have $\theta_i(x) = 0$.*

Let's present the definition of the θ_i's. Consider an element W of $\mathcal{T}$. If $x \in W$, then there exists an open neighbourhood of x in X which is isomorphic to $U \times Z$ where U is an open neighbourhood of x in W, and Z is an open cone of basis L. This set L is independent of the choice of $x \in W$, it is called *link of W in X*. We denote it by L_W.

If W is an element of dimension 1 of $\mathcal{T}$, and if $i \in \{0, \ldots, 7\}$, define

$$\alpha_i(W) = \text{Card}\,\{x \text{ vertex in } L_W \mid \chi\big(\text{lk}(x, L_W)\big) \equiv i \bmod 8\}.$$

Then, for $a, b, c, d \in \{0, 1\}$, let $\mathcal{E}_{abcd}$ be the closure of the union of the elements W of dimension 1 of $\mathcal{T}$ such that

$$\alpha_0(W) \equiv a \bmod 2, \quad \alpha_6(W) \equiv b \bmod 2,$$
$$\alpha_0(W) + \alpha_4(W) \equiv 2c \bmod 4, \quad \alpha_2(W) + \alpha_6(W) \equiv 2d \bmod 4.$$

Now, set $\mathcal{Z}_0 = \mathcal{E}_{1110} \cup \mathcal{E}_{1111}$, and if $i = 1, 2, 3$, set $\mathcal{Z}_i = \mathcal{E}_{ab00} \cup \mathcal{E}_{ab01}$ where ab is the binary representation of i. The sets $\mathcal{Z}_i$ are independent of $\mathcal{T}$.

Finally define $\theta_i(x) \equiv \chi\big(\mathrm{lk}(x, \mathcal{Z}_i)\big) \bmod 2$ for $x \in X_0$ and $i \in \{0, 1, 2, 3\}$. The theorem means that X is homeomorphic to a real algebraic set if and only if the $\mathcal{Z}_i$'s are Euler.

EXAMPLE 1.10. For the set X of remark 1.8, it can be proved that the spaces $\mathcal{Z}_1$, $\mathcal{Z}_3$ and $\mathcal{Z}_4$ are empty, and $\mathcal{Z}_2$ is the suspension of the point a represented on the drawing. If p is a suspension point, the link of $\mathcal{Z}_2$ at p is formed by one point and $\mathcal{Z}_2$ is not Euler.

There is also a stratified version of Theorem 1.9 (see [**AK2**, Th.7.1.2]), to check whether a stratified space is isomorphic to the singular stratification of a real algebraic set. It uses seven new invariants.

1.3. Coste and Kurdyka's results.

DEFINITION 1.11. Let $V \subset \mathbb{R}^N$ be a real algebraic set, and let $W \subset V$ be an algebraic subset. Consider a regular point x in W, and denote by $N_x(W)$ the normal space to W at x in $\mathbb{R}^N$. Then $N_x(W) \cap V$ is an algebraic set containing x, and we define the *link of W in V at x* by

$$\mathrm{lk}_x(W, V) = \mathrm{lk}\big(x, V \cap N_x(W)\big).$$

Sullivan's condition states that the Euler characteristic of $\mathrm{lk}_x(W, V)$ is even. The following theorem is an improvement of this condition.

THEOREM 1.12 ([**CK1**, Th.1′]). *Let W be an irreducible algebraic subset of a real algebraic set $V \subset \mathbb{R}^N$. Then the function $x \mapsto \chi\big(\mathrm{lk}_x(W, V)\big)$ is generically constant modulo 4 on W.*

"Generically" means "outside an algebraic subset of W of strictly smaller dimension".

EXAMPLE 1.13. ([**CK1**]). We will now use this theorem to show that the space X we define in remark 1.8 is not homeomorphic to a real algebraic set. Assume X as an algebraic model; we still denote by X this model. Let Z be the following subset of Y:

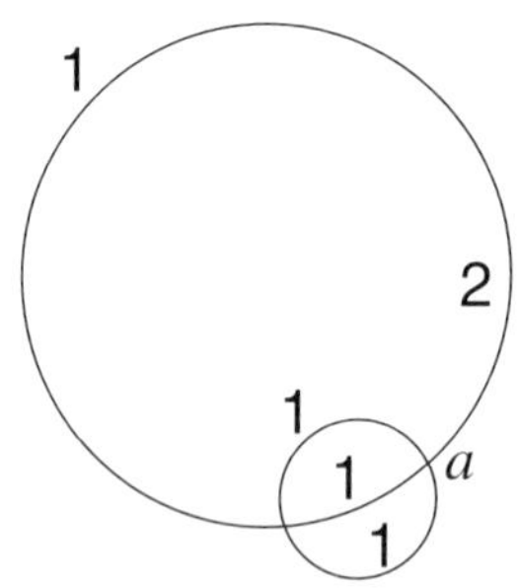

The suspension ΣZ is a semialgebraic subset of dimension 2 of ΣX. Denote by $\overline{\Sigma Z}^Z$ its Zariski closure: $\overline{\Sigma Z}^Z$ is an algebraic set of dimension 2. Let x be a regular point of $\overline{\Sigma Z}^Z$.

If x is in the suspension of the part 1 (resp. part 2) of the drawing, then $\mathrm{lk}_x(\overline{\Sigma Z}^Z, X)$ is empty (resp. formed by four points). In both cases we get

$$\chi\big(\mathrm{lk}_x(\overline{\Sigma Z}^Z, X)\big) \equiv 0 \bmod 4.$$

If now $x \in \overline{\Sigma Z}^Z \setminus \Sigma Z$, then $\mathrm{lk}_x(\overline{\Sigma Z}^Z, X)$ is formed by two points and its Euler characteristic is 2. By Theorem 1.12, we get that $\dim\big(\overline{\Sigma Z}^Z \setminus \Sigma Z\big) < 2$.

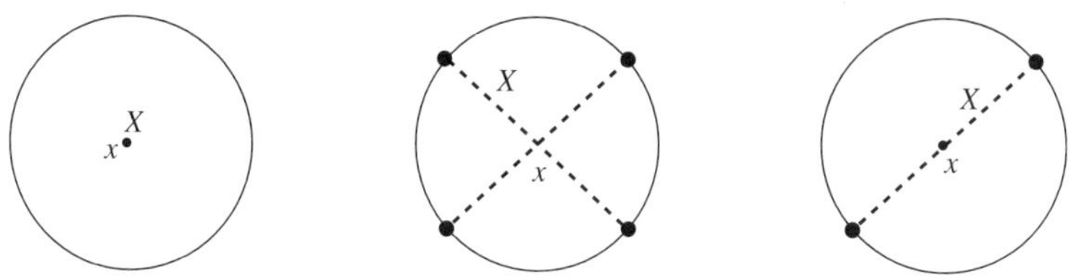

x in suspension of part 1 $\quad$ x in suspension of part 2 $\quad$ x not in ΣZ

Consider now the suspension Σa of the point a. It is a semialgebraic arc contained in ΣZ. Denote by $\overline{\Sigma a}^Z$ its Zariski closure. Consider a regular point x of $\overline{\Sigma a}^Z$ which belongs to Σa. The link $\mathrm{lk}_x(\overline{\Sigma a}^Z, \overline{\Sigma Z}^Z)$ is formed by four points, since $\dim\big(\overline{\Sigma Z}^Z \setminus \Sigma Z\big) < 2$. The link $\mathrm{lk}_x(\overline{\Sigma a}^Z, X)$ is formed by the union of three points and the figure 8.

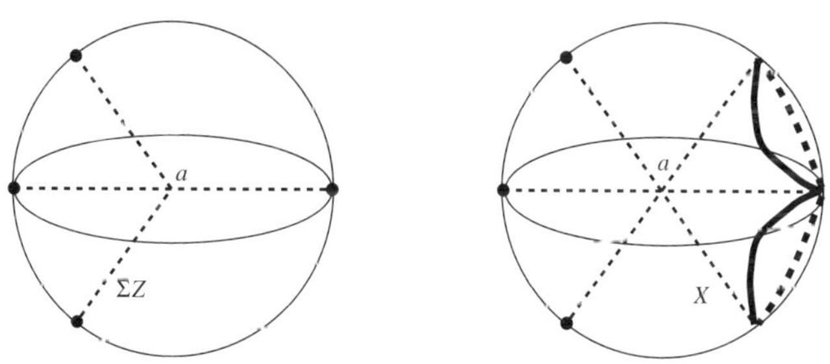

So $\chi\big(\mathrm{lk}_x(\overline{\Sigma a}^Z, \overline{\Sigma Z}^Z)\big) \equiv 0 \bmod 4$ and $\chi\big(\mathrm{lk}_x(\overline{\Sigma a}^Z, X)\big) \equiv 2 \bmod 4$. By Theorem 1.12, these values are the same for every point $x \in \overline{\Sigma a}^Z$ except a finite number of points. But it can be checked by systematic verification that such a point cannot exist outside Σa. We deduce from this that $\overline{\Sigma a}^Z \setminus \Sigma a$ is formed by a finite number of points.

As in Example 1.10, we conclude that $\overline{\Sigma a}^Z$ is not Euler, and we get a contradiction. So X is not homeomorphic to a real algebraic set.

For the proof of Theorem 1.12, Coste and Kurdyka use stratifications given by *stratifying families of polynomials*. These are families $(f_{i,j})_{1\le i\le n, 1\le j\le l_i}$ such that

(1) For a fixed i, the family $(f_{i,j})_{1\le j\le l_i}$ is a family of non constant elements of $\mathbb{R}[X_1, \dots, X_i]$. Each $f_{i,j}$ has a constant leading coefficient if considered

as a polynomial in X_i, and its derivative with respect to X_i is constant or belongs to the family.

(2) For a fixed $k > 1$, if S is a connected subset of $\mathbb{R}^{k-1}$ on which the sign of any $f_{i,j}$, $i < k$, is constant, then the set $\{f_{k,j} = 0 \ \forall j\} \subset S \times \mathbb{R}$ is the union of graphs of a finite number of continuous functions $S \to \mathbb{R}$.

The associated stratification is formed by all non empty sets of the form

$$\bigcap_{i=1}^{k} \bigcap_{j=1}^{l_i} \{f_{i,j} \varepsilon_{i,j} 0\}$$

where $\varepsilon_{i,j}$ is one of the signs $>$, $<$ or $=$.

If S, T are strata of a stratification given by a stratifying family of polynomials, we denote by $T \xrightarrow{1} S$ the condition $S \subset \overline{T}$ and $\dim T = \dim S + 1$.

Theorem 1.12 is not as strong as the condition of Akbulut and King in dimension 3. But Coste and Kurdyka also define the following invariants modulo 2^k.

THEOREM 1.14 ([**CK1**, Th.4]). *Let $V_1 \subset \ldots \subset V_k$ be real algebraic sets such that $\dim V_i = d + i$ for $i = 1, \ldots, k$. Consider a stratification of V_k given by a stratifying family of polynomials, compatible with all the V_i's. If S is a stratum of dimension d contained in V_1, we denote*

$$\Phi(S) = \mathrm{Card}\,\{T_k \xrightarrow{1} T_{k-1} \xrightarrow{1} \ldots \xrightarrow{1} T_1 \xrightarrow{1} S \mid T_i \textit{ stratum } \subset V_i \textit{ for } i = 1, \ldots, k\}.$$

Then

- $\Phi(S) \equiv 0 \bmod 2^k$,
- *if $S' \subset V_1$ is a stratum of dimension d of same Zariski closure as S then* $\Phi(S) \equiv \Phi(S') \bmod 2^{k+1}$.

In dimension 3, Theorem 1.12 and the invariant mod 8 give an equivalent condition to the one of Akbulut and King.

REMARK 1.15. a) These invariants have been formulated in terms of monodromy and generalized by McCrory and Parusiński in [**MP2**].

b) This theorem is not very useful practically, since it is necessary to know the stratification to calculate the value of Φ.

2. Algebraically constructible functions

2.1. Definitions. These functions have been introduced by McCrory and Parusiński in [**MP1**] in 1997. Let $V \subset \mathbb{R}^N$ be a real algebraic set, and let $\varphi : V \to \mathbb{Z}$ be a function.

DEFINITION 2.1. The function φ is *constructible* if it can be written as a finite sum

$$\varphi = \sum_i m_i \mathbf{1}_{X_i}$$

where the m_i's are integers, the X_i's semialgebraic subsets of V and $\mathbf{1}_{X_i}$ denotes the characteristic function of X_i.

In other words, φ is constructible if and only if φ is constant on each element of a finite semialgebraic partition of V.

DEFINITION 2.2. The function φ is said to be *algebraically constructible* if $\varphi(x)$ is a finite sum

$$\varphi(x) = \sum m_i \chi(f_i^{-1}(x))$$

for every x in X, with $m_i \in \mathbb{Z}$ and $f_i : Z_i \to V$ a proper regular morphism defined on an algebraic set Z_i for each i. (Here χ denotes the Euler characteristic.)

Equivalently, algebraically constructible functions may be defined as sums of signs of polynomials on V, where the sign of a function f on V is the function sgn f defined by

$$(\operatorname{sgn} f)(x) = \begin{cases} 1 & \text{if } f(x) > 0, \\ 0 & \text{if } f(x) = 0, \\ -1 & \text{if } f(x) < 0 \end{cases}$$

([**PS1**], [**PS2**], [**CK2**]).

Constructible functions on V form a ring for the usual operations of addition and multiplication. Algebraically constructible functions form a subring of this ring.

EXAMPLE 2.3. a) If $X \subset V$ is algebraic then $\mathbf{1}_X$ is algebraically constructible. This is the way we will use algebraically constructible functions to characterize algebraic sets.

b) If $\dim V = 1$ and V is irreducible, then the algebraically constructible functions on V are exactly the constructible functions which are generically constant modulo 2 on V. In greater dimension, algebraically constructible functions are generically constant modulo 2 on each irreducible component, but there exist non algebraically constructible functions with the same property.

c) If φ is constructible on V and constant modulo $2^{\dim V}$ then φ is algebraically constructible. This is a consequence of the fact that if an open semialgebraic subset of V can be described with a conjunction of strict polynomial inequalities, then it can be described with a conjunction of at most $\dim V$ inequalities.

Let φ be a constructible function on V. In the representation $\varphi = \sum m_i \mathbf{1}_{X_i}$, we can always choose the X_i's closed. If C is a compact semialgebraic subset of V, the *Euler integral of* φ *on* C is defined by

$$\int_C \varphi d\chi = \sum m_i \chi(X_i \cap C)$$

(where $X_i \cap C$ is compact for every i). This value does not depend on the chosen representation of φ with closed X_i's.

We also define, for any constructible function φ on V, the function $\widetilde{\Lambda}\varphi$ on V by

$$\forall x \in V, \quad \widetilde{\Lambda}\varphi(x) = \frac{1}{2} \int_{S(x,\varepsilon)} \varphi d\chi,$$

where $S(x, \varepsilon)$ is the sphere centered at x of radius ε, and $\varepsilon > 0$ is small enough. The value of $\widetilde{\Lambda}\varphi(x)$ is independent of ε for $\varepsilon > 0$ small enough.

Fix $x \in V$, and let $C \subset V$ be a compact semialgebraic set such that x is in the interior of C. Consider a triangulation of C, on each of whose elements φ is constant, and such that x is a vertex. Denote by $T_1, \ldots, T_s$ the elements T of the triangulation such that $x \in \overline{T}$ and $T \neq \{x\}$. Then it can be easily checked that

$$\widetilde{\Lambda}\varphi(x) = \frac{1}{2}\sum_{i=1}^{s}(-1)^{\dim T_i - 1}\varphi(T_i).$$

EXAMPLE 2.4. If $\varphi = \mathbf{1}_X$ where X is a closed semialgebraic set, then

$$\widetilde{\Lambda}\varphi(x) = \frac{1}{2}\chi\big(S(x,\varepsilon) \cap X)\big) = \frac{1}{2}\chi\big(\mathrm{lk}(x, X)\big).$$

Remark that there exists a finite semialgebraic partition of V such that $\widetilde{\Lambda}\varphi$ is constant on each element of the partition, but that $\widetilde{\Lambda}\varphi$ is not a priori integer-valued.

DEFINITION 2.5. The constructible function φ is said to be *Euler* if the function $\widetilde{\Lambda}\varphi$ is integer-valued. (Then $\widetilde{\Lambda}\varphi$ is constructible.)

The function φ is *completely Euler* if all functions built from φ using the usual addition, subtraction, multiplication and the operator $\widetilde{\Lambda}$, are integer-valued.

EXAMPLE 2.6. If X is a closed semialgebraic subset of V, then the function $\mathbf{1}_X$ is Euler if and only if the set X is Euler.

THEOREM 2.7 ([**MP1**, Th.2.5]). *If φ is algebraically constructible on V then φ is Euler. Moreover, $\widetilde{\Lambda}\varphi$ is algebraically constructible.*

In particular, φ is completely Euler.

The first point of this theorem is only a reformulation of Sullivan's condition, but the second one is much stronger.

2.2. Dimension ≤ 3. Let X be a compact semialgebraic set of dimension ≤ 3. McCrory and Parusiński reformulate Akbulut and King's conditions of algebraicity for X in terms of Euler functions.

THEOREM 2.8 ([**MP1**, Th.4.2]). *The set X is homeomorphic to a real algebraic set if and only if $\mathbf{1}_X$ is completely Euler.*

This result is no longer true in dimension 4, see next section.

REMARK 2.9. There is also a stratified version of this theorem, corresponding to the Akbulut and King's condition for stratified sets.

In fact, to know if X is homeomorphic to a real algebraic set, it is sufficient to check that a finite number of functions are Euler. Namely:

PROPOSITION 2.10. *If $\dim X \leq 2$ then X is homeomorphic to a real algebraic set if and only if $\mathbf{1}_X$ is Euler.*

PROPOSITION 2.11. *Assume $\dim X = 3$ and denote $\varphi = \mathbf{1}_X - \widetilde{\Lambda}\mathbf{1}_X$. Then X is homeomorphic to a real algebraic set if and only if the five functions $\mathbf{1}_X$, $\varphi\widetilde{\Lambda}(\varphi^2)$, $\varphi\widetilde{\Lambda}(\varphi^3)$, $\widetilde{\Lambda}(\varphi^2)\widetilde{\Lambda}(\varphi^3)$, $\varphi\widetilde{\Lambda}(\varphi^2)\widetilde{\Lambda}(\varphi^3)$ are Euler.*

Remark that in proposition 2.11, the support of φ is of dimension 2, and the supports of the four last functions are of dimension 1. The verification that these five functions are Euler can easily be done practically.

EXAMPLE 2.12. If X is the set considered in remark 1.8 and example 1.13, then it can be checked that the function $\varphi\widetilde{\Lambda}(\varphi^2)$ is not Euler at the suspension points. This calculation is explained with all details in [**C**] and in the paper of Clint McCrory in this volume.

2.3. Dimension 4. McCrory and Parusiński give new necessary conditions for the algebraicity of a compact semialgebraic set of dimension 4. In order to work in lower dimension, we consider a single link.

Let $X \subset \mathbb{R}^N$ be a compact semialgebraic set of dimension 4, and let L be the link of X at some point. Then $\dim L \leq 3$. Remember that there is only a finite number of semialgebraic homeomorphism types of links in X.

If X is homeomorphic to a real algebraic set, then $\mathbf{1}_X$ is completely Euler. The authors reformulate this condition as the fact that $\mathbf{1}_L$ is completely Euler (which is given by five invariants, see proposition 2.11) and the nullity of $2^{29} - 29$ other invariants modulo 2.

These $2^{29} - 29$ invariants are independent. This means that for any subset $\mathcal{C}$ of this set of invariants, there exists a completely Euler compact semialgebraic set L whose the vanishing invariants are exactly the elements of $\mathcal{C}$.

To get more conditions, McCrory and Parusiński use an operator P on constructible functions, defined by

$$P(\varphi) = \frac{1}{2}(\varphi^4 - \varphi^2).$$

LEMMA 2.13. *If φ is algebraically constructible then $P(\varphi)$ is algebraically constructible.*

PROOF. We use that algebraically constructible functions are sums of signs of polynomials. Write $\varphi = \sum \mathrm{sgn} g_i$, with g_i polynomial. Then $\varphi^2 = \sum \mathrm{sgn}(g_i^2) + 2\sum_{i<j} \mathrm{sgn}(g_i g_j)$ and

$$\varphi^4 = \sum \mathrm{sgn}(g_i^4) + 2\sum_{i<j} \mathrm{sgn}(g_i^2 g_j^2) + 4\sum \mathrm{sgn}(g_i^2) \sum_{i<j} \mathrm{sgn}(g_i g_j) + 4\left(\sum_{i<j} \mathrm{sgn}(g_i g_j)\right)^2.$$

Since $\mathrm{sgn}(g_i^2) = \mathrm{sgn}(g_i^4)$, we get that $\frac{1}{2}(\varphi^4 - \varphi^2)$ is a sum of signs of polynomials. □

So, if X is homeomorphic to a real algebraic set, then all functions obtained from $\mathbf{1}_X$ using the usual addition, subtraction, multiplication and the operators $\widetilde{\Lambda}$ and P, are Euler.

The authors define $2^{43} - 48$ invariants modulo 2, expressed as the Euler integral on L of certains of the functions obtained from $\mathbf{1}_L$ using the usual addition, subtraction, multiplication and the operators $\widetilde{\Lambda}$ and P. They get the following:

THEOREM 2.14 (**[MP3**, Th.4.7]). *Denote $\varphi = \mathbf{1}_L \quad \widetilde{\Lambda}\mathbf{1}_L$, $\beta = \widetilde{\Lambda}(\varphi^2)$ and $\gamma = \widetilde{\Lambda}(\varphi^3)$. Assume X is homeomorphic to a real algebraic set. Then*

1. *the functions $\mathbf{1}_L$, $\varphi\beta$, $\varphi\gamma$, $\beta\gamma$, $\varphi\beta\gamma$ are Euler,*
2. *the Euler integral on L of each of these five functions is even,*
3. *the $2^{43} - 48$ additional invariants modulo 2 vanish.*

Condition 1. means that $\mathbf{1}_L$ is completely Euler. Conditions 2. and 3. give a list of $2^{43} - 43$ independent invariants modulo 2. So we cannot verify practically if all the conditions hold. Moreover, it is not known if these necessary conditions are also sufficient for X being homeomorphic to a real algebraic set.

REMARK 2.15. The authors prove that the ring $\mathcal{P}$ of polynomials in $\mathbb{Q}[t]$ which preserve the ring of all algebraically constructible functions is generated modulo 8 by $\mathbb{Z}[t]$ and the polynomial $\frac{1}{2}(t^4 - t^2)$. As any constructible function on L which is divisible by 8 is algebraically constructible (since $\dim L \leq 3$), in our case the polynomial $\frac{1}{2}(t^4 - t^2)$ give the whole information contained in $\mathcal{P}$.

2.4. Arc-symmetric sets. The results we state in parts 2.2 and 2.3 can be transposed to the case of arc-symmetric sets.

DEFINITION 2.16 ([**Ku**, Déf.1.1]). Let X be a semialgebraic subset of $\mathbb{R}^N$. The set X is *arc-symmetric* if for every analytic arc $\gamma :]-1,1[\to \mathbb{R}^N$ we have

$$\gamma(]-1,0[) \subset X \Rightarrow \gamma(]-1,1[) \subset X.$$

EXAMPLE 2.17. Algebraic sets and Nash sets are arc-symmetric. But there are also arc-symmetric sets which are not Nash. For instance consider the plane $\{z = 1\}$ in $\mathbb{R}^3$ and a quartic Q in this plane with two compact connected components Q_1 and Q_2. Build the cone C (respectively C_1) based on Q (respectively Q_1) with the origin for vertex. Then C_1 is clearly arc-symmetric, but it is not Nash. Indeed, if f is a Nash function vanishing on C_1, then f is a polynomial (since it is an analytic function vanishing on a cone). We conclude that f vanishes on C, and C_1 is not the zero set of some Nash functions.

To transpose the previous results, we need some equivalent of algebraically constructible functions for algebraic sets. This is the notion of Nash constructible function.

DEFINITION 2.18 ([**MP1**]). Let $V \subset \mathbb{R}^N$ be an algebraic set. A function $\varphi : V \to \mathbb{Z}$ is *Nash constructible* if it can be written as

$$\varphi(x) = \sum m_i \chi\big(f_i^{-1}(x) \cap T_i\big)$$

where the sum is finite, m_i is an integer, f_i a proper regular map from an algebraic set Z_i to V, and T_i a connected component of Z_i.

PROPOSITION 2.19 ([**MP1**, Prop.5.4]). *Let $S \subset \mathbb{R}^N$ be a closed semialgebraic set. Then S is arc-symmetric if and only if $\mathbf{1}_S$ is Nash constructible.*

The proof of Proposition 2.19 is based on the fact that after desingularization, an irreducible arc-symmetric set corresponds exactly to one connected component of algebraic set.

THEOREM 2.20 ([**MP1**]). *If φ is a Nash constructible function on an algebraic set, then $\widetilde{\Lambda}\varphi$ is Nash constructible.*

In particular, if a closed semialgebraic set $X \subset \mathbb{R}^N$ is homeomorphic to an arc-symmetric set, then $\mathbf{1}_S$ is completely Euler. Using Theorem 2.8 we get the following:

COROLLARY 2.21 ([**MP1**, Cor.5.6]). *Every arc-symmetric set of dimension ≤ 3 is homeomorphic to a real algebraic set.*

To transpose the condition of algebraicity in dimension 4 to the arc-symmetric case, we need to know that the operator P preserves Nash constructible functions. For this, we write Nash constructible functions as sums of signs of functions of a certain class, namely, the class of arc-analytic functions.

DEFINITION 2.22 ([**Ku**, Déf.4.1]). A semialgebraic function defined on a real algebraic set V is *arc-analytic* if for every analytic arc $\gamma :]-1,1[\to V$, the function $f \circ \gamma$ is analytic.

THEOREM 2.23 ([**B**, Th.3.6]). *Let $V \subset \mathbb{R}^N$ be a compact real algebraic set. Nash constructible functions on V are exactly sums of signs of arc-analytic functions on V.*

It is not known if this result is true in the non compact case.

COROLLARY 2.24. *If φ is Nash constructible on a compact real algebraic set, then $P(\varphi) = \frac{1}{2}(\varphi^4 - \varphi^2)$ is Nash constructible.*

COROLLARY 2.25. *If $X \subset \mathbb{R}^N$ is a compact semialgebraic set of dimension 4 which is homeomorphic to an arc-symmetric set, then conditions* 1., 2. *and* 3. *of Theorem 2.14 hold.*

References

[AK1] S.Akbulut, H.King, *The topology of real algebraic sets*, Enseign. Math. **29** (1983) 221-261

[AK2] S.Akbulut, H.King, *Topology of real algebraic sets*, MSRI Publ. **25**, Springer (1992)

[BD] R.Benedetti, M.Dedò, *The topology of two-dimensional real algebraic varieties*, Ann. Mat. Pura Appl. **127** (1981) 141-171

[BCR] J.Bochnak, M.Coste, M-F.Roy, *Real Algebraic Geometry*, Ergebnisse der Math. 3.Folge Vol.36 Springer (1998)

[B] I.Bonnard, *Nash constructible functions*, Prépublications de l'Université d'Angers **128**

[C] M.Coste, *Reconnaître Effectivement les Ensembles Algébriques Réels*, Bulletin d'information de l'IREM de Rennes **41** (1999). Also available on the web page *http://www.maths.univ-rennes1.fr/~coste/articles.html*

[CK1] M.Coste, K.Kurdyka, *On the link of a stratum in a real algebraic set*, Topology **31** (1992) 323-336

[CK2] M.Coste, K.Kurdyka, *Le discriminant d'un morphisme de variétés algébriques réelles*, Topology **37** (1998) 393-400

[Ki] H.King, *The topology of real algebraic sets*, Proc. Symp. Pure Math. **40** Part I (1983) 641-654

[Ku] K.Kurdyka, *Ensembles semi-algébriques symétriques par arcs*, Math. Ann. **282** (1988) 445-462

[MP1] C.McCrory, A.Parusiński, *Algebraically constructible functions*, Ann. Scient. Éc. Norm. Sup. **30**(4) (1997) 527-552

[MP2] C.McCrory, A.Parusiński, *Complex monodromy and the topology of real algebraic sets*, Comp. Math. **106** (1997) 211-233

[MP3] C.McCrory, A.Parusiński, *Topology of real algebraic sets of dimension 4: necessary conditions*, Topology **39** (2000) 495-523

[PS1] A.Parusiński, Z.Szafraniec, *Algebraically constructible functions and signs of polynomials*, Manuscripta Math. **93** (1997) 443-456

[PS2] A.Parusiński, Z.Szafraniec, *On the Euler characteristic of fibres of real polynomial map*, Singularities Symposium - Łojasiewicz 70, Banach Center Publications **44** (1998) 175-182

[S] D.Sullivan, *Combinatorial invariants of analytic spaces*, Proc. of Liverpool Singularities Symposium I, Lecture Notes in Math. **192** (1971) 165-168

MAX-PLANCK-INSTITUT FÜR MATHEMATIK, VIVATSGASSE 7, 53111 BONN, GERMANY
E-mail address: ibonnard@mpim-bonn.mpg.de

DIMACS Series in Discrete Mathematics
and Theoretical Computer Science
Volume **60**, 2003

Lower Bounds and Real Algebraic Geometry

Peter Bürgisser

ABSTRACT. Many relevant problems in symbolic computation and computational geometry can be formalized as semi-algebraic computation or decision problems, to be solved by means of algebraic computation trees. It is a general paradigm that a complicated geometrical, topological, or combinatorial structure of a problem should result in high computational complexity. This survey outlines some of the ideas leading to lower bounds in terms of degree, Betti numbers, and number of faces of an arrangement or a polyhedron, emphasizing the role of real algebraic geometry in this endeavour. The impact of randomization on decision complexity is briefly discussed.

1. Introduction

Computational decision problems can often be cast in the following form: Given real numbers $x_1, \ldots, x_n$, determine whether they satisfy some fixed system of polynomial equalities and inequalities. In other words, we have to decide for a given point $x \in \mathbb{R}^n$, whether it is contained in a fixed semi-algebraic subset W of $\mathbb{R}^n$. Our goal is the systematic study of the computational complexity to solve such problems. In Section 2.1 we list some basic computational problems, which can be formulated in such a way. We then indicate that typically, these problems can be algorithmically solved much faster than one would expect naively. The optimality proof of these algorithms is one of the great successes of algebraic complexity theory. In this survey, we present some of the most important ideas leading to nonlinear complexity lower bounds and optimality proofs. As the underlying model of computation, we use algebraic computation trees, which are introduced in Section 2.2.

Given the semi-algebraic formulation of the problems considered, it is not surprising, that the main tools to understand their intrinsic complexity are provided by real algebraic geometry. The concept of the degree of a (real) algebraic variety, and methods to estimate this degree play a key role throughout these investigations. By the degree of a real variety we will understand the geometric degree of its complexification.

We start our journey with geometrical lower bounds in Section 3. Strassen's Degree Bound 3.2 [**54**] is historically the first result, which provides nonlinear lower

1991 *Mathematics Subject Classification.* 68Q17; 14Pxx, 52B05, 55-xx.

Key words and phrases. algebraic complexity, lower bounds, real algebraic geometry, algebraic topology, arrangements and polyhedra.

complexity bounds. This fundamental insight bounds the nonscalar complexity of a set of rational functions from below by the logarithm of the degree of the graph of the associated rational map. The proof is a rather straightforward consequence of Bézout's Inequality 3.1. Another important idea is the Derivative Inequality 3.3 due to Baur and Strassen [**1**], which relates the complexity of a rational function with those of its gradient.

So far, peculiarities of the reals played no role. This changes, when we move on to the discussion of the decision complexity of an irreducible, real algebraic set W. An algebraic computation tree, which recognizes W in $\mathbb{R}^n$, does not necessarily recognize the complexification of W. Nevertheless, it is possible to derive a lower bound on the decision complexity of W in terms of the logarithm of the degree of W. This Real Degree Bound 3.4 is due to Lickteig and Roy [**36**].

In Section 4 we address lower bounds on the decision complexity of a semi-algebraic set in terms of its topological invariants. The major tool to obtain such results is the Milnor-Thom Bound 4.2 [**46, 42, 57**], which provides upper bounds on the sum of Betti numbers of semi-algebraic sets in terms of the degrees of the defining polynomials. Steele and Yao [**53**] recognized that the Milnor-Thom Bound 4.2 can be applied to obtain nontrivial lower bounds in terms of the number of connected components for the model of algebraic decision trees. This was extended to algebraic computation trees by Ben-Or [**2**], who proved the Connected Component Bound 4.1.

The extension of this result to higher Betti numbers took quite a while. It was first established for polyhedral sets W in the model of linear decision trees by Björner et al. [**6**], who proved a lower bound in terms of the Euler characteristic of W, and then generalized to higher Betti numbers by Björner and Lovász [**5**]. Finally, Yao [**62, 63**] succeeded in extending these results to the model of algebraic computation trees by proving the Betti Number Bound 4.4. The key idea is to replace the Betti numbers by those with compact supports, which behave subadditively with respect to the union of certain sets. This issue is dicussed rather detailed in Section 4.2. We remark that Montaña et al. [**44**] have proved an extension of the Betti Number Bound 4.4 to parallel complexity.

In Section 5 we investigate the complexity to test membership to polyhedra or to unions of hyperplane arrangements. For polyhedra, the Betti Number Bound 4.4 does not yield good lower bounds since polyhedra are contractible and thus have trivial topology. On the other hand, it seems plausible that membership testing to a polyhedron with many faces is difficult. This was confirmed by Grigoriev et al. [**27**], who proved a lower bound in terms of the number of faces. Their proof is based on concepts from differential geometry and the Milnor-Thom Bound 4.2. It exploits the idea that when deforming a polyhedron to a smooth hypersurface H_ϵ, the points of H_ϵ close to a vertex of the polyhedron must have large principal curvatures. We do not go into the details of this complicated argument, but instead sketch the proof of a stronger lower bound (Face Bound 5.4), which is a consequence of the methods developed in the work by Grigoriev [**21, 23**] about randomized computation trees. Some of the new ideas are best explained by first treating the situation of complex arrangements, which is done in Section 5.1 following Grigoriev [**22**]. The Degree Bound 3.2 and the Derivative Inequality 3.3 play a key role in this argument.

In Section 6 we briefly address randomized decision complexity. The problem to prove nonlinear lower bounds for the real knapsack problem in the randomized computation tree model was open for about a decade. It was attacked and finally solved in a sequence of papers authored by Grigoriev, Karpinski, Meyer auf der Heide, and Smolensky [**25, 24, 26, 22, 23**] culminating in the paper [**23**] by Grigoriev. The proof technique outlined in Section 5 is a by-product of this endeavour.

Acknowledgments. I thank Martin Lotz for discussions and helpful comments.

2. Problems, Fast Algorithms, and Models of Computation

2.1. Fast Algorithms. We list here a couple of fundamental computational or decisional problems over the real numbers and indicate a rapid algorithmical solution for each of these problems. In the complexity bounds mentioned, the symbol $M(n)$ will stand for an upper bound on the complexity to multiply two univariate polynomials of degree n over $\mathbb{R}$. We have $M(n) = O(n \log n)$, when counting all arithmetic operations, and $M(n) = O(n)$, when only the nonscalar multiplications and divisions are counted.

1. Element Distinctness. One has to decide for given real numbers $x_1, \ldots, x_n$, whether they are pairwise distinct. Note that this means to test membership to the complement of a certain arrangement of real hyperplanes. It can be solved by sorting the given real numbers with $O(n \log n)$ comparisons. An algebraic solution is to first compute the discriminant $\prod_{i<j}(x_i - x_j)^2$ with $O(M(n) \log n)$ arithmetic operations [**1**] and then testing the result for zero.

2. Convex Hull. We want to compute the convex hull of n given points in the real plane. The output of the algorithm should consist of those input points, which are extremal points of the convex hull. A related decision problem is to find out, whether all the input points are extremal. Basic algorithms in computational geometry show that this can be done with $O(n \log n)$ arithmetic operations and tests (cf. [**47**]).

3. Root Verification. Given n real numbers $x_1, \ldots, x_n$ and a real polynomial $p(t) = \sum_{i=0}^{n} (-1)^i y_i t^{n-i}$ by its coefficients $y_0 = 1, y_1, \ldots, y_n$. We have to verify whether the collection of the x_i forms the complete set of roots of $p(t)$ (including multiplicities). If we denote by σ_i the ith elementary symmetric polynomial in the x-variables, then this question amounts to test whether $y_i = \sigma_i(x_1, \ldots, x_n)$ for all i. There are algorithms, which compute all the elementary symmetric polynomials from the x-variables with $O(M(n) \log n)$ arithmetic operations [**54**].

4. Real Knapsack. This is a membership problem to the union of a certain arrangement of real hyperplanes. One has to decide for given real numbers $x_1, \ldots, x_n$ whether there is a set of indices $I \subseteq \{1, \ldots, n\}$ such that $\sum_{i \in I} x_i = 1$. The corresponding problem over the rationals is NP-complete [**31**]. Therefore, it is quite astonishing that the Real Knapsack Problem can be solved by linear decision trees with $O(n^4 \log n)$ tests, as shown by Meyer auf der Heide [**39**]. (Extensions to arbitrary hyperplane arrangements have been obtained by Meiser [**38**].) It seems that the computation tree model is too powerful to capture the "true" complexity of this problem.

5. Continued Fraction and GCD: Given are univariate real polynomials A_1 and A_2 of the degrees $n = \deg A_1 \geq \deg A_2$ by their coefficients. We are required to compute the (coefficients of the) quotients $Q_1, \ldots, Q_{t-1}$ occuring in the Euclidean algorithm

$$A_i = Q_i A_{i+1} + A_{i+2},\ \deg A_{i+2} < \deg A_{i+1},\ A_{i+1} \neq 0,\ 1 \leq i \leq t-1,\ A_{t+1} = 0,$$

as well as the greatest common divisior A_t of A_1 und A_2. Note that the sequence $(Q_1, \ldots, Q_{t-1})$ constitutes the continued fraction expansion of A_1/A_2. By the Euclidean representation of the pair (A_1, A_2) we understand the extension of this sequence by the gcd A_t and we call

$$d := (d_1, \ldots, d_{t-1}, d_t) := (\deg Q_1, \ldots, \deg Q_{t-1}, \deg A_t)$$

the corresponding degree pattern.

We remark that by Sturm's theorem (cf. [**7**]) one can compute the number of zeros of a squarefree polynomial A_1 in any real interval from the Euclidean representation of A_1 and its derivative $A_2 = A_1'$ by just $O(n)$ arithmetic operations. For more information about the Euclidean algorithm and its countless applications, we refer to [**19**].

The Knuth-Schönhage algorithm [**33, 32, 48, 43**] is a sophisticated efficient algorithm for computing the Euclidean representation of given polynomials A_1, A_2 with $O(M(n)\log n)$ arithmetic operations. A more detailed analysis of this algorithm was done by Strassen [**55**]. Let $H(d) := -\sum_{d_i>0} \frac{d_i}{n} \log \frac{d_i}{n}$ denote the entropy of the degree pattern d. Then the Knuth-Schönhage algorithm in fact uses $O(n(1+H(d)))$ arithmetic operations on a pair of input polynomials (A_1, A_2), whose Euclidean representation has degree pattern d. (For a comprehensive account of these results see also [**10**].)

2.2. Models of Computation. Algebraic computation trees over the reals provide a formal framework to describe computations involving arithmetic operations, as well as branchings according to sign tests. They can be defined as finite ternary trees whose simple nodes are labeled by arithmetic instructions, while their branching nodes are labeled by sign tests. On an input $x \in \mathbb{R}^n$, the arithmetic operations and sign tests are successively executed and accordingly, the computation follows a unique path in the tree from the root to a leaf (if no attempt to divide by zero is made). In general, the output consists of some of the real numbers computed along this path and of the information, in which leaf of the tree the computation has ended up. Over the complex numbers, the computation trees are binary, as only equality tests are allowed. For formal definitions we refer to [**10**]. (Note that the model there is slightly different as only binary trees are considered.)

When studying decision problems, which is our main focus, we require that the leaves carry a 'yes' or 'no'-label, which provides the corresponding information. It is obvious that the sets of inputs accepted by such computation trees are just the semi-algebraic sets. We also remark that when studying decision problems, then we may assume without loss of generality that there are no divisions, since a quotient p/q can be encoded by the pair (p, q), and one arithmetic operations on such quotients can be simulated by at most 4 arithmetic operations on the numerators and denominators.

By the *decision complexity* $C(W)$ of a semi-algebraic set W, we will understand the minimal number of nonscalar multiplications and branchings sufficient to solve this membership problem by means of algebraic computation trees in the worst case. Thus additions, subtractions, and multiplications with real scalars are not counted, which is motivated by the lower bounds we are going to prove. In other words, $C(W)$ is the minimal depth of algebraic computation trees solving the membership problem to W, where depth refers to the number of nonscalar multiplication and branching nodes.

A more restricted (and less realistic) model is that of algebraic decision trees. These are ternary trees whose inner nodes v are labeled by polynomials f_v. On input x, the computation starts at the root and branches at node v according to the sign of $f_v(x)$. Again the leaves carry a 'yes' or 'no'-label. In this model, we count only the number of branchings, but we put a restriction on the f_v by requiring that their degree is less than or equal to some a priori bound d. The model of linear decision trees is obtained by the restriction $d = 1$. Usually, lower bounds are considerably easier to prove in these restricted models.

3. Geometrical Lower Bounds

3.1. Strassen's Degree Bound. We consider the algorithmical problem to compute real polynomials $f_1, \dots, f_m$ in $\mathbb{R}[X_1, \dots, X_n]$ from variables $X_1, \dots, X_n$ (considered as the inputs) and real constants by means of straight-line programs. For simplicity of exposition, we assume that there are no divisions. A straight-line program performing this task produces a sequence $g_{-n+1} = X_1, \dots, g_0 = X_n, g_1, \dots, g_r$ of intermediate results such that for all $1 \le k \le r$ there are $i, j < k$ satisfying

$$g_k = g_i \circ g_j \text{ or } g_k = \lambda g_i, \quad \circ \in \{+, -, *\},\ \lambda \in \mathbb{R},$$

and such that all f_i occur among the intermediate results. The special treatment of scalar multiplications is motivated by the lower bound we are going to prove, which in fact holds for the minimal number of nonscalar multiplications sufficient for such a computation. This quantity is called the *(nonscalar) complexity* $L(f_1, \dots, f_m)$ of the polynomials to be computed.

Since the degree can at most double in a multiplication step, it is obvious that $\deg g_k \le 2^{\mu_k}$, where μ_k denotes the number of nonscalar multiplication steps in the initial segment of the computation up to g_k. Therefore, the degree bound $L(f_m) \ge \log_2 \deg f_m$ holds. Our goal will be an extension of this elementary observation.

We first note that any straight-line program solving the computation problem over the reals, also works over the complex numbers, hence we may assume without loss of generality that $f_i \in \mathbb{C}[X_1, \dots, X_n]$ and that the computation takes place in the polynomial ring over $\mathbb{C}$. However, we remark that for decision problems solved by algebraic computation trees, it makes a big difference whether we work over $\mathbb{R}$ or $\mathbb{C}$, see Section 3.2.

The *degree* $\deg V$ of an irreducible, affine variety $V \subseteq \mathbb{C}^n$ of dimension d can be characterized as the finite number of intersection points of V with a generic affine subspace of $\mathbb{C}^n$ having the complementary dimension $n - d$. This is a well-defined notion, as shown in algebraic geometry (cf. [**50, 45, 29**]). In algebraic complexity theory, it is common to define the degree of an affine variety as the sum of the degrees of all of its irreducible components (as opposed to the sum of the degrees of the components of maximal dimension). This is advantageous for degree estimations.

Bézout's theorem relates the degree of the intersection of two varieties with the degrees of the varieties themselves. In algebraic complexity theory, the Bézout Inequality 3.1 given below is a fundamental tool. It can be derived in a straightforward way from the version of Bézout's theorem treating the intersection of an irreducible variety with an irreducible hypersurface, see for instance [**29**, Thm. I.7.7] and [**10**]. We remark that Bézout's Inequality 3.1 does also hold for the intersection of any two locally closed subsets in projective space.

BÉZOUT INEQUALITY 3.1. *Let V be an affine variety in $\mathbb{C}^n$ and H be a hypersurface. Then we have* $\deg(V \cap H) \leq \deg V \cdot \deg H$.

We assign to a sequence $(f_1, \ldots, f_m)$ of polynomials the graph of the corresponding polynomial map $f\colon \mathbb{C}^n \to \mathbb{C}^m$ and define the degree $\deg(f_1, \ldots, f_m)$ as the degree of this graph. Note that this clearly extends the usual notion of degree for polynomials ($m = 1$). What is the growth of the degrees $d_k := \deg(g_{-n+1}, \ldots, g_k)$ during a straight-line computation with intermediate results g_i? We first note that $d_0 = 1$. If $g_k = g_i * g_j$, then we can write $G_k := \text{graph}(g_1, \ldots, g_{k-1}, g_k)$ as the intersection of $G_{k-1} \times \mathbb{C}$ with the quadric given by the equation $Y_k - Y_i Y_j = 0$, where the Y_i denote the corresponding coordinate variables. Since a quadric has degree two, we conclude with the Bézout Inequality 3.1 that $d_k \leq 2d_{k-1}$. In the case $g_k = \lambda g_i + \mu g_j$, $\lambda, \mu \in \mathbb{C}$, we obtain by intersecting with a linear subspace that $d_k \leq d_{k-1}$ (in fact, equality holds). We conclude that $d_k \leq 2^{\mu_k}$, where again μ_k denotes the number of nonscalar multiplication steps in the initial segment of the computation up to g_k. Finally, one can show that the degree does not increase under projections, which implies that $\deg(f_1, \ldots, f_m) \leq d_r$. We therefore obtain the following fundamental result:

DEGREE BOUND 3.2 (Strassen [**54**], 1973). *For polynomials f_i over $\mathbb{C}$ we have*

$$L(f_1, \ldots, f_m) \geq \log_2 \deg(f_1, \ldots, f_m).$$

We remark that this lower bounds remain true for the computation of rational functions when allowing divisions.

The Degree Bound 3.2 implies the optimality with respect to nonscalar complexity of numerous basic algorithms [**54**], see also [**10**]. For instance, for the elementary symmetric polynomials σ_i, it is not hard to see that $\deg(\sigma_1, \ldots, \sigma_n) = n!$. The Degree Bound 3.2 implies $L(\sigma_1, \ldots, \sigma_n) \geq \log_2(n!) \geq n(\log_2 n - 2)$, which shows the optimality of the corresponding algorithm mentioned in Section 2.1 (up to a constant factor).

One of the most beautiful applications of the Degree Bound 3.2 is the optimality of the Knuth-Schönhage algorithm (compare Section 2.1). Let $D(d)$ denote the set of all pairs (A_1, A_2) of complex polynomials, whose Euclidean representation has the degree pattern $d = (d_1, \ldots, d_t)$. Strassen [**55**] proved that any algebraic computation tree over $\mathbb{C}$ computing the Euclidean representation of given polynomials needs at least $n(H(d) - 2)$ nonscalar operations on all inputs (A_1, A_2) in a Zariski dense subset of $D(d)$. Schuster [**49**] proved a similar lower bound for the corresponding decision problem to test whether the Euclidean representation of a given input (A_1, A_2) has the degree pattern d. (This is based on the degree bound for decision complexity (3.3), cf. Section 3.2.)

The Degree Bound develops its full strength only in combination with the so-called Derivative Inequality, which relates the complexity of a polynomial f with the complexity of its gradient $\text{grad} f$ as follows:

DERIVATIVE INEQUALITY 3.3 (Baur and Strassen [**1**], 1983).

$$L(f, \text{grad} f) \leq 3L(f).$$

In combination with the Degree Bound 3.2 we obtain the lower bound

$$L(f) \geq \frac{1}{3} \log_2 \deg(f, \text{grad} f) \tag{3.1}$$

for the complexity of a polynomial f. This implies for instance the lower bound $L(\prod_{i<j}(X_i - X_j)^2) = \Omega(n \log_2 n)$ for the discriminant as a function of the roots [**1**], which we know to be optimal up to a constant factor from the discussion in Section 2.1.

3.2. Degree Bound over the Reals. Let $W \subset \mathbb{R}^n$ be an irreducible real algebraic set. We study the algorithmical problem to decide membership to W of input points $x \in \mathbb{R}^n$. Assume we are given an algebraic computation tree solving the membership problem to W. To each leaf v of this tree we can associate the *leaf set* D_v consisting of those inputs x in $\mathbb{R}^n$, whose computation in the tree ends up with the node v. The condition $x \in D_v$ can be expressed by sign conditions on the polynomials evaluated at x along the path to the leaf v. Each leaf set D_v may thus be written in the form

$$D_v = \{x \in \mathbb{R}^n \mid f_1(x) = 0, \ldots, f_p(x) = 0, g_1(x) > 0, \ldots, g_q(x) > 0\},$$

where the polynomials f_i and g_j are computed along the path of v and therefore have a nonscalar complexity bounded by $C(W) - p - q$. The set W is the union of all the leaf sets corresponding to 'yes'-leaves. Hence there exists at least one leaf set D_v, which is Zariski dense in W. If we denote by U the open semi-algebraic set given by the corresponding conditions $g_j > 0, 1 \leq j \leq q$, and by $Z(f_1, \ldots, f_p)$ the zero set of the f_i, we obtain that

$$W \cap U = D_v = Z(f_1, \ldots, f_p) \cap U. \tag{3.2}$$

Over the complex numbers an analogous observation is true, we just have to replace the conditions $g_j > 0$ by the conditions $g_j \neq 0$, thus U is Zariski open in this case. In this situation, it is easy to obtain a lower bound on the decision complexity $C(W)$. Indeed, Equation (3.2) implies that W is an irreducible component of $Z(f_1, \ldots, f_p)$, which relies on the general fact that a nonempty Zariski open subset of an irreducible algebraic set is Zariski dense in this algebraic set. By our definition of degree and the Bézout Inequality 3.1 we conclude $\deg W \leq \deg Z(f_1, \ldots, f_p) \leq \deg(f_1, \ldots, f_p)$. Thus from the Degree Bound 3.2, we we obtain the following degree bound for the decision complexity $C(W)$ of an irreducible algebraic subset W of $\mathbb{C}^n$:

$$C(W) \geq \log_2 \deg W. \tag{3.3}$$

(Note that this lower bound is even valid for the number of nonscalar multiplications, irrespective of the number of branchings.)

This argument cannot be immediately transferred to the reals. In fact, a nonempty Euclidean open subset of an irreducible algebraic set V is Zariski dense in V iff it contains a regular point of V (cf. [**7**, §7.6]). (For instance, think of an irreducible, real elliptic curve having an isolated point; this point forms a Euclidean open subset of the curve.) Therefore, our argument gets stuck in the case where W is contained in the singular locus of $Z(f_1, \ldots, f_p)$. As a way out, we define $f := f_1^2 + \cdots + f_p^2$ and note that $L(f) \leq C(W)$ (we have traded the branchings for squaring operations). Note that $Z(f_1, \ldots, f_p) = Z(f) = Z(f, \mathrm{grad} f)$. The key observation is the following degree estimation

$$\deg W \leq \deg(f, \mathrm{grad} f). \tag{3.4}$$

Hereby, the degree of the real irreducible algebraic set W is defined as the degree of its complexification $W_{\mathbb{C}}$, which we define as the Zariski closure of W in $\mathbb{C}^n$.

From this observation and the Derivative Inequality 3.3 we immediately obtain the desired degree bound over the reals:

REAL DEGREE BOUND 3.4 (Lickteig and Roy [**36**], 1996). *The decision complexity $C(W)$ of an irreducible real algebraic subset W of $\mathbb{R}^n$ satisfies*

$$C(W) \geq \frac{1}{3} \log_2 \deg W.$$

The proof of the estimate (3.4) in [**36**] is based on a heavy machinery using the notion of the real spectrum. We are going to briefly explain the geometric idea behind the proof, without attempting to supply the subtle technical details. The basic idea is as follows: we perturb the equation $f = 0$ in order to achieve a nondegenerate situation (transversality) which can be easily handled, and then use a lower semi-continuity property of the degree.

We need the following general observation: Assume that V is an irreducible real algebraic set of dimension d and let $h_1, \ldots, h_{n-d}$ be polynomials vanishing on V, which intersect transversally in some point $\xi \in V$. Using the implicit function theorem, we see that the complex zero set of the given polynomials is a smooth, complex manifold of dimension d around ξ, and hence its complexification $V_{\mathbb{C}}$ is an irreducible component of the complex zero set $Z_{\mathbb{C}}(h_1, \ldots, h_{n-d})$. By our definition of the real degree and the Bézout Inequality 3.1 we get

$$\deg V = \deg V_{\mathbb{C}} \leq \deg Z_{\mathbb{C}}(h_1, \ldots, h_{n-d}) \leq \deg(h_1, \ldots, h_{n-d}).$$

Let now W be a d-dimensional irreducible real algebraic set as in (3.2) and put $f := f_1^2 + \cdots + f_p^2$. One can show that after some linear coordinate transformation, the system of perturbed equations

$$f = \epsilon, \partial_1 f = 0, \ldots, \partial_{n-d-1} f = 0$$

defines for almost all $\epsilon > 0$ some d-dimensional irreducible component W_ϵ, in some of whose points these equations intersect transversally. (If the zero set of $f - \epsilon$ is compact, then this can be shown as in the proof of the Milnor-Thom Bound 4.2, see Section 4.1 and [**10**, Prop. 11.5].) Moreover, one can achieve that the family W_ϵ of real irreducible varieties "converges" to W as ϵ goes to zero. This convergence means that the limit of a sequence of polynomials vanishing on W_ϵ must vanish on W. One can show that the degree is lower semi-continuous with this respect. This can be done by replacing ϵ by an infinitesimal and by using the characterization of the degree of varieties by the leading coefficient of the corresponding Hilbert polynomials. (See [**36**] or [**10**, §8.5], where a similar problem is treated.) On the other hand, we have by the general observation from before that

$$\deg W_\epsilon \leq \deg(f - \epsilon, \partial_1 f, \ldots, \partial_{n-d-1} f) \leq \deg(f, \mathrm{grad} f),$$

which finishes our sketch of the proof.

As an application of the Real Degree Bound 3.4, we obtain the lower bound $\Omega(n \log n)$ for the complexity of the Root Verification problem, showing the optimality of the algorithm mentioned in Section 2.1.

The Real Degree Bound 3.4 allows to transfer the optimality results for the Knuth-Schönhage algorithm by Strassen [**55**] and Schuster [**49**] to the real setting. Thus it follows that any algebraic computation tree over $\mathbb{R}$ computing the Euclidean representation of given polynomials needs at least $n(H(d) - 2)$ nonscalar operations on all inputs (A_1, A_2) in $D(d)$ outside a semi-algebraic subset of smaller dimension. This answers a question posed by Strassen [**56**].

To finish this section, we mention some further lower bound results involving concepts of real algebraic geometry. Lickteig [**34, 35**] proved a degree based lower bound on the decision complexity of irreducible real hypersurfaces, which is an extension of the lower bound (3.1) to the corresponding decison problem. In the papers [**12, 13, 9**] linear lower bounds are proved, which allow the determination of decision complexities in generic situations.

4. Topological Lower Bounds

4.1. Connected Components. Our goal is now to derive lower bounds on the decision complexity $C(W)$ of a semi-algebraic set W in $\mathbb{R}^n$ in terms of topological invariants of the set W. The simplest such invariant is the the number $b_0(W)$ of connected components of W. If we focus on polyhedral sets W and restrict the model of computation to linear decision trees, then it is an easy exercise to show that at least $\log_3 b_0(W)$ sign tests are needed for deciding membership to W (each sign test decomposes the set under consideration into at most three connected components). It is a nontrivial result that this lower bound can be extended to the model of algebraic computation trees.

CONNECTED COMPONENT BOUND 4.1 (Ben-Or [**2**], 1983). *The decision complexity of a semi-algebraic set W in $\mathbb{R}^n$ satisfies*

$$C(W) \geq \frac{1}{2}(\log_3 b_0(W) - n).$$

This result implies lower bounds of order $\Omega(n \log n)$ for the Element Distinctness or the Convex Hull problem and thus proves the optimality of the algorithms mentioned in Section 2.1. For the Knapsack problem, one obtains the lower bound $\Omega(n^2)$. For the determination of the number of connected components in these cases we refer to [**15**] or [**10**, Chap. 11].

The proof of the Connected Component Bound 4.1 relies on the fundamental bounds on the Betti numbers of real algebraic varieties due to Oleĭnik [**46**], Milnor [**42**], and Thom [**57**]. It will be convenient to denote the sum of all Betti numbers of a semi-algebraic set W by $b(W) := \sum_{k \in \mathbb{N}} b_k(W)$. The following bound on $b(W)$ given by Milnor is particularly useful for us.

MILNOR-THOM BOUND 4.2 (Milnor [**42**], 1964). *Assume W is the zero set of the polynomials $f_1, \ldots, f_p \in \mathbb{R}[X_1, \ldots, X_n]$ of degree at most d. Then we have*

$$b(W) \leq d(2d-1)^{n-1}.$$

The proof ideas of this result are used at various places. For instance, they form the basis for the asymptotically fastest known algorithms for dealing with semi-algebraic sets, compare Grigoriev and Vorobjov [**28**] and the survey by Heintz et al. [**30**]. We remark that we have already used similar ideas in the proof of the Real Degree Bound 3.4.

Because of its importance, we briefly outline the main ideas of the proof of the Milnor-Thom Bound 4.2, for more details the reader may consult the original paper [**42**] or [**7, 3**]. First, the case of a smooth, compact hypersurface given by one equation $f = 0$ of degree $2d$ is settled. One shows that by a linear coordinate transformation, it can be achieved that the system of equations

$$f = 0, \partial_1 f = 0, \ldots, \partial_{n-1} f = 0$$

has only nondegenerate solutions. This means that the restriction to W of the projection $\mathbb{R}^n \to \mathbb{R}$ onto the last coordinate is a Morse function, whose critical points are just the solutions of the above system of equations. By the Bézout Inequality 3.1, the number of critical points is bounded by $2d(2d-1)^{n-1}$. The Morse inequalities (cf. [**41**]) bound the sum $b(W)$ of Betti numbers by the number of critical points and thus provide the estimate in the special case considered (up to a factor of two).

In the general case, we set $f := f_1^2 + \cdots + f_p^2$ and consider the compact sets $K_{\epsilon,r}$ given by the inequality $f(X) + \epsilon^2 ||X||^2 \le \epsilon^2 r^2$ with boundary $W_{\epsilon,r}$ described by the equation $f(X) + \epsilon^2 ||X||^2 = \epsilon^2 r^2$. If $W_{\epsilon,r}$ is a smooth hypersurface, then we can bound its sum of Betti numbers as before. Moreover, Alexander duality [**52**, p. 296] implies for the complement $E_{\epsilon,r}$ of $K_{\epsilon,r}$ in $\mathbb{R}^n$ that $\tilde{H}_k(E_{\epsilon,r}) \simeq H^{n-k-1}(K_{\epsilon,r})$ (the symbol $\tilde{H}_k$ denotes the reduced homology). On the other hand, by Mayer-Vietoris, we have $\tilde{H}_k(W_{\epsilon,r}) \simeq \tilde{H}_k(K_{\epsilon,r}) \oplus \tilde{H}_k(E_{\epsilon,r})$. This implies that $b(W_{\epsilon,r}) = 2\, b(K_{\epsilon,r})$ and we obtain the upper bound $b(K_{\epsilon,r}) \le d(2d-1)^{n-1}$.

We fix now $r > 0$ and use Sard's lemma (cf. [**7**]) to choose monotonically decreasing sequences $\epsilon_i \to 0$ and $r_i \to r$ such that all W_{ϵ_i,r_i} are smooth. The intersection W_r of W with the closed ball of radius r around the origin equals $W_r = \cap_i K_{\epsilon_i,r_i}$. Using standard facts from algebraic topology [**52**, §6.1] and the result that semi-algebraic sets can be triangulated (cf. [**7, 3**]), we see that for each k the cohomology group $H^k(W_r)$ is the direct limit of the $H^k(K_{\epsilon_i,r_i})$ as $i \to \infty$. This implies $b_k(W_r) \le \liminf_i b_k(K_{\epsilon_i,r_i}) \le d(2d-1)^{n-1}$. Finally, it is not hard to see that W_r is a deformation retract of W if r is chosen sufficiently large, which finishes the proof.

COROLLARY 4.3. *Assume that the semi-algebraic set W in $\mathbb{R}^n$ is given by the conditions $f_1 = 0, \ldots, f_p = 0, g_1 > 0, \ldots, g_q > 0$, where f_i, g_j are polynomials of degree at most d. Then we have $b(W) \le d(2d-1)^{n+q-1}$.*

In order to show this, let $\epsilon > 0$ and consider the subset W_ϵ of $\mathbb{R}^n$ given by the conditions $f_i = 0, g_j \ge \epsilon$. This set is homeomorphic to the algebraic subset of $\mathbb{R}^{n+q}$ described by the equations $f_i = 0, g_j = \epsilon + Y_j^2$, to which we can apply the Milnor-Thom Bound 4.2. Finally, we note that $W = \cap_{\epsilon>0} W_\epsilon$ and use the semicontinuity property of the Betti numbers as above.

The proof of the Connected Component Bound 4.1 is now fairly simple. Recall the definition of the leaf set D_v of a computation tree corresponding to a leaf v given at the beginning of Section 3.2. We are going to bound the number of connected components of D_v. Fix a leaf v and consider the corresponding path in the tree. Forgetting for the moment about the test instructions, this path defines a straight-line program with intermediate results $g_{-n+1} = X_1, \ldots, g_0 = X_n, g_1, \ldots, g_r$. Similarly as in the proof of the Degree Bound 3.2, we can describe the graph of $(g_{-n+1}, \ldots, g_r)$ by a system of linear and quadratic equations in the variables $X_1, \ldots, X_n, Y_1, \ldots, Y_r$. We can make this description more concise by eliminating the Y-variables belonging to linear operations. Then the graph is homeomorphic to a subset of $\mathbb{R}^{n+m}$ given by linear and quadratic equations, where m denotes the number of multiplication instructions along the path. By adding the linear inequalities corresponding to the sign tests, we obtain a subset of $\mathbb{R}^{n+m}$, which is homeomorphic to the leaf set D_v. If we assume that there are q sign tests along the path of v, then we obtain from Corollary 4.3 that $b_0(D_v) \le 2 \cdot 3^{n+m+q-1}$

Suppose now that an algebraic computation tree solves the membership problem to the set W in $\mathbb{R}^n$. Then W is the union of all the leaf sets corresponding to 'yes'-leaves and we have the estimate $m + q \leq C(W)$ for each path, since we count only the multiplications and the sign tests. The number of connected components behaves subadditively with respect to the union of sets. Therefore, as there are at most $3^{C(W)}$ leaves, we get

$$b_0(W) \leq \sum_v b_0(D_v) \leq 3^{C(W)} \cdot 2 \cdot 3^{n+C(W)-1} \leq 3^{n+2C(W)},$$

which finishes the proof of the Connected Component Bound 4.1.

4.2. Betti Numbers. Since the Milnor-Thom Bound 4.2 is valid for the sum of all Betti numbers, it is natural to ask whether the Connected Component Bound 4.1 can be extended correspondingly. The difficulty is that the higher Betti numbers do not behave subadditively with respect to the formation of unions (see Figure 1). The key idea is to work with a variant of Betti numbers, which behaves subadditively. This can be achieved by considering cohomology with compact supports. For the following facts from algebraic topology see [**16**, Chap. 8 §6].

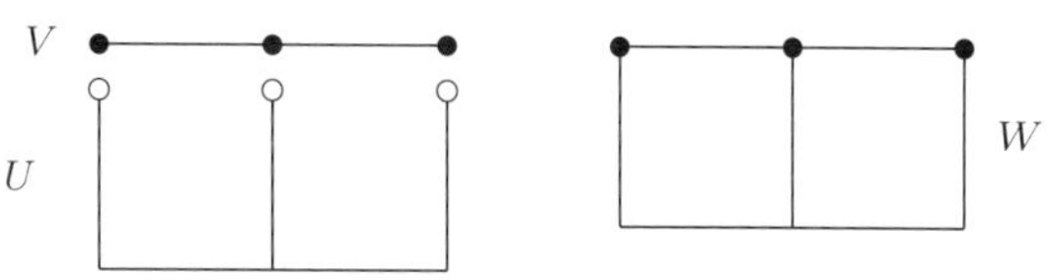

FIGURE 1. Betti numbers are not subadditive.

A locally closed subset of $\mathbb{R}^n$ is defined as the intersection of an open with a closed subset of $\mathbb{R}^n$. Let $V \subseteq W$ be locally closed semi-algebraic subsets of $\mathbb{R}^n$. The *cohomology group with compact supports* $\mathrm{H}_c^k(W, V)$ of the pair (W, V) is defined as the direct limit of the cohomology groups $\mathrm{H}^k(W, V')$ over all subsets V', ordered by reverse inclusion, such that $V \subseteq V' \subseteq W$ and $W \setminus V'$ is compact. By the kth *Betti number with compact supports* $b_{c,k}(W)$ of W we will understand the rank of $\mathrm{H}_c^k(W) := \mathrm{H}_c^k(W, \emptyset)$, and we write $b_c(W) := \sum_{k\in\mathbb{N}} b_{c,k}(W)$ for the sum of these Betti numbers. (We remark that the definition of $\mathrm{H}_c^k(W, V)$ given in [**16**] refers to the Čech-cohomology. However, this coincides with the singular cohomology for semi-algebraic sets V, W, since the latter can be triangulated.)

If V is closed in W and $U := W \setminus V$, one can show that $\mathrm{H}_c^k(W, V) = \mathrm{H}_c^k(U)$ by excision. Under this assumption, we obtain from the long exact sequence

$$\cdots \to \mathrm{H}_c^k(W, V) \to \mathrm{H}_c^k(W) \to \mathrm{H}_c^k(V) \to \mathrm{H}_c^{k+1}(W, V) \to \cdots$$

the long exact sequence

$$\cdots \to \mathrm{H}_c^k(U) \to \mathrm{H}_c^k(W) \to \mathrm{H}_c^k(V) \to \mathrm{H}_c^{k+1}(U) \to \cdots .$$

This implies the following crucial subadditivity property of the Betti numbers with compact supports:

$$b_{c,k}(W) \leq b_{c,k}(U) + b_{c,k}(V) \quad \text{if } V \text{ is closed in } W. \tag{4.1}$$

In order to compute the numbers $b_{c,k}(W)$ for a noncompact locally closed set W, we may use the Alexandrov compactification $\dot{W}$ of W, which compactifies W by adding a "point at infinity" ∞. Since $\dot{W} = W \cup \{\infty\}$ and W is open in $\dot{W}$, we get $\mathrm{H}_c^k(W) = \mathrm{H}^k(\dot{W}, \{\infty\})$ and therefore

$$b_{c,k}(W) := b_k(\dot{W}) \text{ for } k > 0 \text{ and } b_{c,0}(W) = b_0(\dot{W}) - 1.$$

On the other hand, if W is compact, we have $\mathrm{H}_c^k(W) = \mathrm{H}^k(W)$ and thus $b_{c,k}(W) = b_k(W)$. For instance, we obtain $b_c(V) = b(V) = 1$, $b_c(W) = b(W) = 3$, $b_c(U) = 2$ for the example in Figure 1, thus verifying relation (4.1).

We remark that the Betti numbers with compact supports may also be interpreted as the rank of Borel-Moore homology groups ([**8**], [**7**, §11.7]).

The Connected Component Bound 4.1 now generalizes as follows:

BETTI NUMBER BOUND 4.4 (Yao [**63**], 1994). *The decision complexity of a locally closed semi-algebraic set W in $\mathbb{R}^n$ satisfies*

$$C(W) \geq \frac{1}{2}(\log_3 b_c(W) - n - 1).$$

For the proof, we need an extension of Corollary 4.3 to Betti numbers with compact supports.

COROLLARY 4.5. *Assume that the locally closed semi-algebraic set W in $\mathbb{R}^n$ is given by the conditions $f_1 = 0, \dots, f_p = 0, g_1 > 0, \dots, g_q > 0$, where f_i, g_j are polynomials of degree at most d. Then we have $b_c(W) \leq d(2d-1)^{n+q}$.*

In order to show this, assume first that $q = 0$, hence W is closed. We may realize the compactification of W by the inverse of the stereographic projection $S^n \setminus \{(0,\dots,0,1)\} \xrightarrow{\sim} \mathbb{R}^n, x \mapsto y$ given by the equations $y_i = x_i/(1 - x_{n+1})$. Then $\dot{W}$ is described by the equations $(1 - x_{n+1})^d f_i(x_i/(1 - x_{n+1})) = 0$ of degree d and the Milnor-Thom Bound 4.2 implies the assertion. The extension to inequalities can be done as in the proof of Corollary 4.3.

The proof of the Betti Number Bound 4.4 is completely analogous to the one of the Connected Component Bound 4.1. The only remaining issue is to prove that indeed $b_{c,k}(W) \leq \sum_v b_{c,k}(D_v)$ for an algebraic computation tree deciding membership to a locally closed semi-algebraic set W. To a node u of a such a tree we assign the semi-algebraic set W_u consisting of the points $x \in W$ whose path in the tree passes through u. It is obvious that W_u is locally closed. Let L_u denote the set of 'yes'-leaves corresponding to paths passing through u. Then we have $W_u = \cup_{v \in L_u} D_v$. We prove now by reverse induction on the depth of u that

$$b_{c,k}(W_u) \leq \textstyle\sum_{v \in L_u} b_{c,k}(D_v). \tag{4.2}$$

If u is a leaf, then there is nothing to show. Otherwise, let u be a node with the descendents u_-, u_0, u_+ corresponding to the outcome of the sign test with a polynomial f, thus $W_{u_-} = W \cap \{f < 0\}$, $W_{u_0} = W \cap \{f = 0\}$, $W_{u_+} = W \cap \{f > 0\}$. Since W_{u_0} is closed in $W_{u_-} \cup W_{u_0}$ and this set is closed in W_u, we may apply the subadditivity relation (4.1) twice, in order to obtain that $b_{c,k}(W_u) \leq b_{c,k}(W_{u_-}) + b_{c,k}(W_{u_0}) + b_{c,k}(W_{u_+})$. This shows the claim (4.2) and completes the proof of the Betti Number Bound 4.4.

A nice application of the Betti Number Bound 4.4 is the "k-equal problem" considered in Björner et al. [**6, 5**]: Given n real numbers, decide whether k of them are equal ($k \geq 2$). Using a sorting argument, one can show that $O(n \log \frac{n}{k})$ comparisons are sufficient to solve this problem (in the linear decision tree model). The Betti Number Bound 4.4 shows that this is optimal, even in the model of algebraic computation trees. One can compute the Betti numbers of subspace arrangements with the formula of Goresky and MacPherson [**20**], which characterizes the Betti numbers of the complement of a subspace arrangement by its intersection semi-lattice. For more information on this, we refer to the excellent survey by Björner [**4**].

We are currently investigating how the Betti Number Bound 4.4 performs on problems traditionally treated by Degree Bounds. An intriguing open problem is the decision complexity of the discriminant of univariate polynomials (as a function of the coefficients). The Knuth-Schönhage algorithm yields the upper bound $O(M(n)\log n)$ for this problem.

PROBLEM 4.6. Prove a lower bound of order $n \log n$ on the complexity to decide whether a given complex polynomial of degree n has a multiple root!

Degree bounds provide only lower bounds of order $\Omega(n)$ for this problem (compare [**10**, Ex. 8.24] for a related discussion for the resultant). The cohomology of the complement of the complex discriminant variety is intimately linked with the braid group and has been determined by Fuchs [**18**] (see also Vassiliev [**60**]). Unfortunately, the Betti numbers turn out to be too small to answer the above problem.

We remark that an interesting topological method to prove lower bounds was introduced by Smale [**51**] in connection with the problem to approximatively compute the roots of a complex polynomial. He also uses the model of algebraic computation trees over the reals, but considers as complexity measure the number of all branching nodes in the tree (a quantity typically exponentially larger than the complexity we consider in this article). The topology of the complement of the discriminant variety plays a central role in Smale's approach. Vassiliev has considerably improved and extended these results [**58, 59, 60, 61**].

5. Combinatorial Lower Bounds

5.1. Complex Arrangements. We consider an arrangement given by complex hyperplanes $H_1, \dots, H_m$ in $\mathbb{C}^n$ and let W denote the union of these hyperplanes. In Section 4 we have discussed how to bound the decision complexity by topological invariants of W (or its complement). It seems plausible, that also a complicated combinatorial structure leads to high computational complexity, and it turns out that this is indeed the case. By a k-dimensional face of the given arrangement, we understand a k-dimensional intersection of some of the given hyperplanes; a vertex is a 0-dimensional face. Let $N_k(W)$ denote the number of k-dimensional faces of W.

Suppose we have a computation tree deciding membership to W. There is a (unique) leaf set D_v, which is Zariski dense in $\mathbb{C}^n \setminus W$. We call the corresponding path generic, since Zariski almost all inputs follow this path in the tree. Suppose that $f_1, \dots, f_r$ are the nonzero test polynomials computed along the generic path. Then the leaf set D_v is disjoint from W and it is described by the conditions $f_1 \neq 0, \dots, f_r \neq 0$. This implies that on each hyperplane H_i, some of the f_j must vanish, thus the product polynomial $f := f_1 \cdots f_r$ vanishes on W. This means that f is a multiple of $\ell_1 \cdots \ell_m$ if H_i is the zero set of the affine linear polynomial ℓ_i. Note that $L(f) \leq C(W)$ (trading multiplications for branchings).

By Equation (3.1) we know that $L(f) \geq \frac{1}{3}\log_2 \deg(f, \mathrm{grad} f)$. The core of the method consists in the following important lower bound on the degree of $(f, \mathrm{grad} f)$ in terms of the number of vertices of W.

THEOREM 5.1 (Grigoriev [**22**], 1999). *If a nonzero polynomial f vanishes on the union W of a complex hyperplane arrangement in $\mathbb{C}^n$, then we have*

$$\deg(f, \mathrm{grad} f) \geq N_0(W)/2^{2n+1}.$$

This implies a lower bound on $C(W)$ in terms of $N_0(W)$. By intersecting W with a generic $(n-k)$-dimensional hyperplane V and noting that $N_0(W \cap V) = N_k(W)$, we even obtain a lower bound in terms of the number of k-dimensional faces.

COROLLARY 5.2. *The decision complexity of the union W of a complex hyperplane arrangement in $\mathbb{C}^n$ satisfies for all $0 \leq k \leq n$*

$$C(W) \geq \frac{1}{3}(\log_2 N_k(W) - 2(n-k) - 1).$$

It remains to outline the proof of Theorem 5.1. The essential idea behind is the quite intuitive observation that for each vertex x of the arrangement, each direction in $\mathbb{C}^n$ can be obtained as a limit of the directions of $\mathrm{grad} f(x^{(j)})$ for a suitable sequence of points $x^{(j)}$ tending to x. Hereby, a direction means a line in $\mathbb{C}^n$ through the origin, that is, a point in the complex projective space $\mathbb{P}^{n-1}$.

In order to state this observation more succinctly, consider the graph G of $(f, \mathrm{grad} f)$, which is the affine variety in $\mathbb{C}^{n+1} \times \mathbb{C}^n$ defined by

$$G := \{(x_0, x_1, \dots, x_n, y_1, \dots, y_n) \mid x_0 = f(x_1, \dots, x_n), y_i = \partial_i f(x_1, \dots, x_n)\}.$$

Clearly, $\dim G = n$. Let G_1 denote the closure of the image of $G \cap (\mathbb{C}^{n+1} \times (\mathbb{C}^n \setminus 0))$ under the canonical morphism $\mathbb{C}^{n+1} \times (\mathbb{C}^n \setminus 0) \to \mathbb{C}^{n+1} \times \mathbb{P}^{n-1}$. Then the above observation can be stated as follows. We omit the proof (cf. [**22**]):

LEMMA 5.3. *$(0, x_1, \dots, x_n) \times \mathbb{P}^{n-1}$ is contained in G_1 for any vertex $x = (x_1, \dots, x_n)$ of the arrangement.*

Let $\mathcal{H}$ denote the zero set of the coordinate function X_0 in $\mathbb{C}^{n+1} \times \mathbb{P}^{n-1}$. The above lemma implies that $(0, x_1, \dots, x_n) \times \mathbb{P}^{n-1}$ is an irreducible component of $G_1 \cap \mathcal{H}$. We conclude that $G_1 \cap \mathcal{H}$ has at least as many irreducible components as there are vertices in the arrangement.

In order to proceed with a degree estimation, we embed $\mathbb{C}^{n+1} \times (\mathbb{C}^n \setminus 0)$ in a projective space using the Segre embedding:

$$\psi \colon \mathbb{C}^{n+1} \times (\mathbb{C}^n \setminus 0) \hookrightarrow \mathbb{P}^{n+1} \times \mathbb{P}^{n-1} \hookrightarrow \mathbb{P}^{2n^2+2n-1},$$

which maps the point $(x_0, x_1, \dots, x_n, y_1, \dots, y_n)$ to the point with homogeneous coordinates $((x_i y_j)_{0 \leq i \leq n, 1 \leq j \leq n} : y_1 : \dots : y_n)$. Let G_1' and $\mathcal{H}'$ be the images of G_1 and $\mathcal{H}$ under this embedding, respectively. Then $G_1' \cap \mathcal{H}'$ has at least $N_0(W)$ irreducible components, and we conclude that $N_0(W) \leq \deg(G_1' \cap \mathcal{H}') \leq \deg G_1'$ using the Bézout Inequality 3.1 in projective space and $\deg \mathcal{H}' = 1$. By employing the fact that the coordinate functions of ψ are polynomials of degree at most two, it is not hard to see that $\deg G_1 \leq 2^{2n+1} \deg G$ (cf. [**22**]). This implies the assertion of Theorem 5.1.

5.2. Real Arrangements and Polyhedra. We try to extend the previous results to real arrangements given by hyperplanes $H_1, \ldots, H_m$ in $\mathbb{R}^n$. With such an arrangement, we associate the union W of the hyperplanes, as well as the polyhedron W^+ defined as the intersection of the closed halfspaces H_i^+ corresponding to H_i. To be specific: if H_i is the zero set of the affine linear polynomial ℓ_i, then $H_i^+ = \{\ell_i \geq 0\}$. We always assume that W^+ is n-dimensional. Our focus will be on the decision complexity of the polyhedron W^+, but the union W can be treated similarly (even in a simpler way). Note that the Betti Number Bound 4.4 does not provide a meaningful lower bound for the decision complexity of a polyhedron W^+ since it is contractible and thus $b_c(W^+) \leq 1$.

We denote by $N_k(W^+)$ the number of k-dimensional faces of the polyhedron W^+. Yao and Rivest [**64**] proved that at least $\Omega(\log_2 N_k(W^+))$ sign tests are needed in the linear decision tree model for deciding membership to W^+. It is possible to extend this result to the model of algebraic computation trees, answering Problem 11.1 in [**10**]. The Face Bound 5.4 below is a consequence of the methods developed in the work by Grigoriev [**21, 23**] about randomized computation trees. As mentioned in the Introduction, this result has been preceeded by a weaker one due to Grigoriev et al. [**27**]. The proof in that paper is based on completely different concepts and considerably more complicated.

FACE BOUND 5.4. *The decision complexity of a polyhedron W^+ in $\mathbb{R}^n$ given as an intersection of m closed halfspaces satisfies for any $0 \leq k \leq n$*

$$C(W^+) \geq \frac{1}{2}(\log_8 N_k(W^+) - \log_8 m - 2n).$$

This implies for the Element Distinctness problem and for the Knapsack problem lower bounds of the same order of magnitude as the Connected Component Bound 4.1. However, the Face Bound 5.4 has the advantage that it can be extended to the model of randomized computation trees (cf. Section 6).

We will now sketch the proof of the Face Bound 5.4. For simplicity of exposition, we restrict ourselves to the case of vertices ($k = 0$) and remark that the proof technique extends to higher dimensional faces without much difficulty.

Let $H_1, \ldots, H_n$ be hyperplanes in $\mathbb{R}^n$ with $H_1 \cap \ldots \cap H_n = \{0\}$. In order to simplify notation, we assume that $H_i = \{X_i = 0\}$ are the coordinate hyperplanes. We denote by $\mathrm{tc}_n(f)$ the homogeneous part of lowest degree of a nonzero polynomial f in the variables $X_1, \ldots, X_n$. Furthermore, we assign to f nonzero homogeneous polynomials $\mathrm{tc}_i(f) \in \mathbb{R}[X_1, \ldots, X_i]$ and multiplicities $\nu_i(f) \in \mathbb{N}$ by reverse induction on i: we define $\mathrm{tc}_{i-1}(f)$ as the trailing coefficient of $\mathrm{tc}_i(f)$ in the expansion with respect to the variable X_i, that is, the coefficient of the power of X_i of smallest degree in this expansion. We call this smallest degree the ith multiplicity $\nu_i(f)$. In suggestive notation we have

$$\mathrm{tc}_i(f) = \mathrm{tc}_{i-1}(f)\, X_i^{\nu_i(f)} + O(X_i^{\nu_i(f)+1}).$$

We note that $\deg \mathrm{tc}_n(f) = \sum_{i=1}^n \nu_i(f)$ and $\nu_i(fg) = \nu_i(f) + \nu_i(g)$. We call a nonzero polynomial f *strongly singular* at the point 0 iff $\nu_i(f) > 0$ for all $1 \leq i \leq n$. Of course, this notion depends on the chosen hyperplanes H_i as well as on their order. (We remark that the definition in [**23**] is more general, which is necessary for coping with randomized trees.) Any nonzero multiple of a polynomial, which is strongly singular at 0, is strongly singular at 0 as well.

We define now what it means for a polynomial f to be strongly singular at a vertex of an arrangement (or of a polyhedron W^+). To simplify notation, we assume that the vertex is the origin 0. One can show that there exists a system of hyperplanes $H_{i_1}, \dots, H_{i_n}$ of the underlying arrangement such that

$$\dim(W^+ \cap H_{i_\ell} \cap \ldots \cap H_{i_n}) = \ell - 1 \quad \text{for } 1 \le \ell \le n+1. \tag{5.1}$$

We say that the polynomial f is strongly singular at the vertex 0 of the given arrangement, iff it is strongly singular at 0 with respect to such a system of hyperplanes.

LEMMA 5.5. *Assume that an algebraic computation tree T decides membership to the polyhedron W^+. Then for each vertex of W^+ there is a leaf v of the tree such that the product g_v of the (nonzero) test polynomials computed along the path to v is strongly singular at this vertex.*

For the proof it will be convenient to work with infinitesimals. If K is an ordered field, then we may extend this order to the rational function field $K(\epsilon)$ by defining an element $f = \alpha_r \epsilon^r + \alpha_{r+1}\epsilon^{r+1} + \cdots$ of $K[\epsilon]$ to be positive iff the trailing coefficient α_r is positive. As $0 < \epsilon < a$ for all positive $a \in K$, the element ϵ can be interpreted as a positive infinitesimal with respect to K. We set now $K_0 := \mathbb{R}$ and successively define ordered fields $K_i = K_{i-1}(\epsilon_i)$ for $1 \le i \le n+1$ by adjoining infinitesimals ϵ_i. Note that $\epsilon_1 > \cdots > \epsilon_{n+1}$. From this construction of the orders it follows that for $f \in \mathbb{R}[X_1, \dots, X_n]$, for signs $\sigma_j \in \{-1, 1\}$ and $0 \le i \le n$ we have

$$\operatorname{sgn} f(\sigma_1\epsilon_1\epsilon_{n+1}, \dots, \sigma_n\epsilon_n\epsilon_{n+1}) = \sigma_{i+1}^{\nu_{i+1}(f)} \cdots \sigma_n^{\nu_n(f)} \operatorname{sgn} \operatorname{tc}_i(f)(\sigma_1\epsilon_1, \dots, \sigma_i\epsilon_i). \tag{5.2}$$

We note that the polyhedron W^+ can be naturally extended to a polyhedron contained in K_{n+1}^n, which we denote by the same symbol. Also, we may feed points in K_{n+1}^n as inputs to an algebraic computation tree.

We are now going to prove Lemma 5.5 and assume as before that the vertex considered is the origin, and that the hyperplanes H_{i_j} are as in (5.1). Without loss of generality, we assume that $H_{i_j}^+ = \{X_j \ge 0\}$. We have for all $1 \le i \le n$ that

$$E := \epsilon_{n+1}(\epsilon_1, \dots, \epsilon_n) \in W^+, \quad E_i := \epsilon_{n+1}(\epsilon_1, \dots, \epsilon_{i-1}, -\epsilon_i, \epsilon_{i+1}, \dots, \epsilon_n) \notin W^+$$

(in order to show that $E \in W^+$ use condition (5.1)). Let v be the leaf of the computation tree T corresponding to the input E, let $f_1, \dots, f_r$ be the (nonzero) test polynomials computed along this path, and put $g_v := f_1 \cdots f_r$. For each i, the path on input E_i is distinct from the path on input E. Hence there is a test polynomial f_ρ such that the signs of $f_\rho(E)$ and $f_\rho(E_i)$ are distinct. By (5.2), this implies that $\nu_i(f_\rho)$ must be odd, hence positive, hence $\nu_i(g_v) > 0$. As the index i was chosen arbitrarily, we conclude that g_v is strongly singular at the vertex 0 of the arrangement. This proves Lemma 5.5.

One can show the following analogue of Theorem 5.1.

THEOREM 5.6 (Grigoriev [**23**], 1999). *The number $N_0(W, f)$ of vertices of a real arrangement, at which a nonzero, real polynomial f is strongly singular, satisfies*

$$\deg(f, \operatorname{grad} f) \ge \frac{N_0(W, f)}{m\, 16^n}.$$

For the proof, the only thing to check is that Lemma 5.3 remains valid for vertices of the arrangement at which the polynomial f is strongly singular. We omit the details, which can be found in [**23**].

The proof of the Face Bound 5.4 for $k = 0$ is now an easy consequence. Suppose that we have an algebraic computation tree solving the membership problem to W^+. By Lemma 5.5, Theorem 5.6, and Equation (3.1) we obtain

$$N_0(W^+) \leq \sum_v N_0(W, g_v) \leq 3^{C(W^+)} m\, 16^n 8^{C(W^+)} \leq m\, 8^{2n+2C(W^+)},$$

which shows the assertion.

6. Randomized Decision Complexity

Randomization is an important technique for algorithm design and can sometimes speed up computations considerably. One can generalize the model of algebraic computation trees to include randomization for instance by introducing "guess nodes". We say that such a tree decides membership to W if the (two-sided) error probability is less than some positive $\epsilon < 1/2$. We abstain from giving a formal definition here. General results about the simulation of randomized by deterministic computation trees can be found in [**40, 14**].

Here is an example [**11**], which provably shows that randomization can help. The Root Verification problem has been previously shown to be of (deterministic) decision complexity $\Theta(n \log n)$. When allowing for randomization, this problem has complexity $O(n)$, which can be seen as follows: Choose $t \in \mathbb{R}$ uniformly at random in a set of $2n+1$ points. Evaluate the polynomials $f(t) = t^n + \sum_{i=1}^n (-1)^i y_i t^{n-i}$ and $g(t) = \prod_{i=1}^n (t - x_i)$ with $O(n)$ operations, and test whether these values are equal. If $y_i = \sigma_i(x)$ for all $1 \leq i \leq n$, then $f(t) = g(t)$. Otherwise, $f(t) \neq g(t)$ with probability less than $1/2$.

This simple example already shows that the previously discussed lower bounds on decision complexity in terms of degree or Betti numbers in general do not extend to the randomized model! Interestingly, some lower bounds for irreducible hypersurfaces can be extended. For instance, randomized checking of the value of a middle elementary symmetric polynomial, i.e., whether $y_{n/2} = \sigma_{n/2}(x)$, requires $\Omega(n \log n)$ operations [**11**]. In the same paper, it is also proven that randomization does not help to check the value of the determinant (while checking matrix multiplication seems to be easier using randomization [**17**]).

What can be said about the randomized decision complexity of hypersurfaces with many irreducible components, say hyperplane arrangements? Recall that the Element Distinctness and the Knapsack problem are of this type. For the model of randomized linear decision trees, Manber and Tompa [**37**] obtained lower bounds of the same order of magnitude as the deterministic ones. In the model of algebraic computation trees, the situation is considerably more complicated. In a sequence of papers authored by Grigoriev, Karpinski, Meyer auf der Heide, and Smolensky [**25, 24, 26, 22, 23**] this problem was treated, first in the model of randomized algebraic decision trees, and then for randomized algebraic computation trees. So far, the last word was spoken by Grigoriev [**23**], who obtained a bound in terms of the number of faces of a polyhedron, or of an arrangement of real hyperplanes. The proof of this result is an extension of the ideas explained in Section 5, which were in fact developed in order to tackle this question! We do not state Grigoriev's result [**23**], as it is rather technical. As an application, one obtains lower bounds for the Element Distinctness and the Knapsack problem, which are of the same order of magnitude as the deterministic ones.

These lower bounds for hyperplane arrangements in the randomized model in general cannot be extended to subspace arrangements of higher codimension, as the following example shows. Consider the Set Equality problem to decide for two given sequences of reals $x = (x_1, \ldots, x_n)$ and $y = (y_1, \ldots, y_n)$ whether there exists a permutation π such that $y_i = x_{\pi(i)}$. This means that the elementary symmetric polynomials have the same value on x and y. The Connected Component Bound 4.1 shows that the deterministic decision complexity of the Set Equality problem is $\Theta(n \log n)$. However, using randomization, Set Equality can be checked with $O(n)$ operations by the same idea as for the Root Verification problem. (However, the bound $\Omega(n \log n)$ for Set Equality is optimal in the linear decision tree model [**37**].)

Despite the recent impressing progress it seems that the impact of randomization on algebraic decision complexity is still far from being understood.

References

[1] W. Baur and V. Strassen, *The complexity of partial derivatives*, Theoret. Comp. Sci. **22** (1983), 317–330.

[2] M. Ben-Or, *Lower bounds for algebraic computation trees*, Proc. 15th ACM STOC, Boston, 1983, pp. 80–86.

[3] R. Benedetti and J.-J. Risler, *Real algebraic and semi-algebraic sets*, Hermann, 1990.

[4] A. Björner, *Subspace arrangements*, Proc. of 1st European Congress of Mathematics (Paris, 1992), Birkhäuser, 1992, pp. 321–370.

[5] A. Björner and L. Lovász, *Linear decision trees, subspace arrangements and Möbius functions*, J. Amer. Math. Soc. **7** (1994), no. 3, 677–706.

[6] A. Björner, L. Lovász, and A.C. Yao, *Linear decision trees: volume estimates and topological bounds*, Proc. 24th ACM STOC, 1992, pp. 171–177.

[7] J. Bochnak, M. Coste, and M.F. Roy, *Géometrie algébrique réelle*, Ergebnisse der Mathematik und ihrer Grenzgebiete, 3. Folge, vol. 12, Springer Verlag, 1987.

[8] A. Borel and J.C. Moore, *Homology theory for locally compact spaces*, Michigan Math. J. **7** (1960), 137–159.

[9] P. Bürgisser, *Decision complexity of generic complete intersections*, Research Report 8578-CS, Institut für Informatik der Universität Bonn, 1992.

[10] P. Bürgisser, M. Clausen, and M.A. Shokrollahi, *Algebraic complexity theory*, Grundlehren der mathematischen Wissenschaften, vol. 315, Springer Verlag, 1997.

[11] P. Bürgisser, M. Karpinski, and T. Lickteig, *On randomized semialgebraic decision complexity*, J. Compl. **9** (1993), 231–251.

[12] P. Bürgisser and T. Lickteig, *Verification complexity of linear prime ideals*, J. Pure Appl. Alg. **81** (1992), 247–267.

[13] P. Bürgisser, T. Lickteig, and M. Shub, *Test complexity of generic polynomials*, J. Compl. **8** (1992), 203–215.

[14] F. Cucker, M. Karpinski, P. Koiran, T. Lickteig, and K. Werther, *On real Turing machines that toss coins*, Proc. 27th ACM STOC, Las Vegas, 1995, pp. 335–342.

[15] D. Dobkin and R.J. Lipton, *A lower bound of $\frac{1}{2}n^2$ on linear search programs for the knapsack problem*, J. Comp. Syst. Sci. **16** (1978), 413–417.

[16] A. Dold, *Lectures on algebraic topology*, Grundlehren der mathematischen Wissenschaften, vol. 200, Springer Verlag, 1972.

[17] R. Freivalds, *Fast probabilistic algorithms*, LNCS, no. 74, Springer Verlag, 1979, pp. 57–69.

[18] D.B. Fuchs, *Cohomology of the braid group mod 2*, Funct. Anal. Appl. **4(2)** (1970), 46–59.

[19] J. von zur Gathen and J. Gerhard, *Modern computer algebra*, Cambridge University Press, 1999.

[20] M. Goresky and R.D. MacPherson, *Stratified Morse theory*, Ergebnisse der Mathematik und ihrer Grenzgebiete, 3. Folge, vol. 14, Springer Verlag, 1988.

[21] D.Yu. Grigoriev, *Randomized complexity lower bounds*, Proc. 30th ACM STOC, 1998, pp. 219–223.

[22] ———, *Complexity lower bounds for randomized computation trees over zero characteristic fields*, Comp. Compl. **8** (1999), 316–329.

[23] ———, *Randomized complexity lower bound for arrangements and polyhedra*, Discrete Comput. Geom. **21(3)** (1999), 329–344.

[24] D.Yu. Grigoriev and M. Karpinski, *Randomized quadratic lower bound for knapsack*, Proc. 29th ACM STOC, 1997, pp. 76–85.

[25] D.Yu. Grigoriev, M. Karpinski, F. Meyer auf der Heide, and R. Smolensky, *A lower bound for randomized algebraic decision trees*, Proc. 37th STOC, 1996, pp. 612–619.

[26] D.Yu. Grigoriev, M. Karpinski, and R. Smolensky, *Randomization and the computational power of analytic and algebraic decision trees*, Comp. Compl. **6** (1996/97), no. 4, 376–388.

[27] D.Yu. Grigoriev, M. Karpinski, and N. Vorobjov, *Lower bound on testing membership to a polyhedron by algebraic decision and computation trees*, Discrete Comput. Geom. **17** (1997), 191–215.

[28] D.Yu. Grigoriev and N. Vorobjov, *Solving systems of polynomial inequalities in subexponential time*, J. Symb. Comp. **5** (1988), 37–64.

[29] R. Hartshorne, *Algebraic geometry*, GTM, Springer Verlag, 1977.

[30] J. Heintz, T. Recio, and M.F. Roy, *Algorithms in real algebraic geometry and applications to computational geometry*, DIMACS Series in Discrete Mathematics and Theoretical Computer Science **6** (1991), 137–163.

[31] R.M. Karp, *Reducibility among combinatorial problems*, Complexity of Computer Computations (R.E. Miller and J.W. Thatcher, eds.), New York, 1972, pp. 85–104.

[32] D.E. Knuth, *The analysis of algorithms*, Actes du congrès international des Mathématiciens, Nice, vol. 3, 1970, pp. 269–274.

[33] D.H. Lehmer, *Euclid's algorithm for large numbers*, Amer. Math. Monthly **45** (1938), 227–233.

[34] T. Lickteig, *On semialgebraic decision complexity*, Tech. Report TR-90-052, Int. Comp. Sc. Inst., Berkeley, 1990, Habilitationsschrift, Universität Tübingen.

[35] ———, *Semi-algebraic decision complexity, the real spectrum, and degree*, J. Pure Appl. Alg. **110** (1996), 131–184.

[36] T. Lickteig and M.F. Roy, *Semi-algebraic complexity of quotients and sign determination of remainders*, J. Compl. **12** (1996), 545–571.

[37] U. Manber and M. Tompa, *Probabilistic, nondeterministic, and alternating decision trees*, Proc. 14th ACM STOC, 1982, pp. 234–244.

[38] S. Meiser, *Point location in arrangements of hyperplanes*, Information and Computation **106** (1993), 286–303.

[39] F. Meyer auf der Heide, *A polynomial linear search algorithm for the n-dimensional knapsack problem*, J. ACM **31** (1984), 668–676.

[40] ———, *Simulating probabilistic by deterministic algebraic computation trees*, Theoret. Comp. Sci. **41** (1985), 325–330.

[41] J. Milnor, *Morse theory*, Annals of Math. Studies, no. 51, Princeton University Press, 1963.

[42] ———, *On the Betti numbers of real varieties*, Proc. AMS, vol. 15, 1964, pp. 275–280.

[43] R.T. Moenck, *Fast computation of GCDs*, Proc. 5th ACM STOC, 1973, pp. 142–151.

[44] J.L. Montaña, J.E. Morais, and L.M. Pardo, *Lower bounds for arithmetic networks II: Sum of Betti numbers*, AAECC **7** (1996), 41–51.

[45] D. Mumford, *Algebraic geometry I: Complex projective varieties*, Springer Verlag, 1976.

[46] O.A. Oleĭnik, *Estimates of the Betti numbers of real algebraic hypersurfaces*, Math. Sb. (N.S.) **28 (70)** (1951), 635–640.

[47] F.P. Preparata and M.I. Shamos, *Computational geometry, an introduction*, Texts and Monographs in Computer Science, Springer Verlag, 1985.

[48] A. Schönhage, *Schnelle Berechnung von Kettenbruchentwicklungen*, Act. Inf. **1** (1971), 139–144.

[49] P. Schuster, *Interpolation und Kettenbruchentwicklung. Die Komplexität einiger Berechnungsaufgaben*, Ph.D. thesis, Univ. Zürich, 1980.

[50] I.R. Shafarevich, *Basic algebraic geometry*, Springer Verlag, 1974.

[51] S. Smale, *On the topology of root finding*, J. Compl. **3** (1987), 81–89.

[52] E. Spanier, *Algebraic topology*, MacGraw-Hill, 1966.

[53] J.M. Steele and A.C. Yao, *Lower bounds of algebraic decision trees*, J. Algorithms **3** (1982), 1–8.

[54] V. Strassen, *Die Berechnungskomplexität von elementarsymmetrischen Funktionen und von Interpolationskoeffizienten*, Num. Math. **20** (1973), 238–251.

[55] ———, *The computational complexity of continued fractions*, SIAM J. Comp. **12** (1983), 1–27.

[56] ———, *Algebraic complexity theory*, Handbook of Theoretical Computer Science (J. van Leeuwen, ed.), vol. A, Elsevier Science Publishers B. V., Amsterdam, 1990, pp. 634–672.

[57] R. Thom, *Sur l'homologie des variétés algébriques réelles*, Differential and Combinatorial Topology (S.S. Cairns, ed.), Princeton Univ. Press, 1965, pp. 255–265.

[58] V.A. Vassiliev, *Cohomology of braid groups and complexity of algorithms*, Funct. Anal. Appl. **22** (1988), 15–24.

[59] ———, *Topological complexity of algorithms of approximate solving the systems of polynomial equations*, Algebra Anal. **1** (1989), 198–213.

[60] ———, *Complements of discriminants of smooth maps: Topology and applications*, Transl. of Math. Monographs, no. 98, AMS, 1992.

[61] ———, *Cohomology of braid groups and complexity*, From Topology to Computation: Proceedings of the SMALEFEST, Springer Verlag, 1993, pp. 359–367.

[62] A.C. Yao, *Algebraic decision trees and Euler characteristic*, Proc. 33rd FOCS, 1992.

[63] ———, *Decision tree complexity and Betti numbers*, Proc. 26th ACM STOC, 1994.

[64] A.C. Yao and R.L. Rivest, *On the polyhedral decision problem*, SIAM J. Comp. **9** (1980), 343–347.

Dept. of Mathematics and Computer Science, University of Paderborn, D-33095 Paderborn, Germany

E-mail address: buergisser@upb.de

DIMACS Series in Discrete Mathematics
and Theoretical Computer Science
Volume **60**, 2003

The Viro method applied with quadratic transforms

Benoit Chevallier

ABSTRACT. The problem of deciding the existence of a set X_d, defined as the real zero set of a real homogeneous polynomial in $q+1$ variables, isotopic to a given smooth submanifold of codimension 1 in $\mathbb{RP}^q$ is considered.

Fix positive integers d and q. Let Y be a smooth submanifold of codimension 1 in $\mathbb{RP}^q$ (resp. in $\mathbb{R}^q$). Note X_d the real zero set of a real homogeneous polynomial in $q+1$ variables (resp. a real polynomial in q variables) of degree d. This paper is devoted to the study of the question: *Does it exist a set X_d isotopic to a given Y?*

We shall write this question in the form of the following equation $Y = X_d$.

A basic computational problem is to find roots of a given polynomial. Here on the contrary we are given a set and we want a polynomial of degree d having this set as roots, up to ambient isotopy.

The powerful Viro gluing theorem allows to construct a polynomial zero set from gluing pieces of more simple ones. See [**8**] or [**10**]. It allows to study smoothings of isolated singular points of hypersurfaces of given degree. However this theorem is based on hypotheses of genericity on the coefficients (genericity with respect to the Newton diagram, see e.g. [**8**]). Removing these hypotheses has interesting consequences (see e.g. [**7**]).

We discuss in this paper the smoothing of the singularity of hypersufaces of some special class ($\{f(X) = 0\}$ where f is defined in the beginning of section 3). This singularity is degenerate with respect to its Newton diagram, thus one needs to introduce some refinements around the Viro method. Lemma 2.1 is a general statement on smoothing of singularities of this type for all dimensions when the degree of the hypersurface is even. An analogous statement for odd degrees can be proved similarly. The applications presented are the followings:

Four new isotopy types of plane curves of degree 8 (with the maximal possible number of ovals) are constructed (Theorem 2.2, a complete proof of this result is published in [**1**]). Many isotopy types of spatial surfaces of degree four ($q = 3$, $d = 4$) are realized (section 4.3 and 4.4). In the latter case it is probable that all the isotopy types can be realized using our construction.

These applications show new opportunities for the use of the *Viro method* and in particular its most combinatorial version known as T-construction or patchworking

---

1991 *Mathematics Subject Classification.* 14B07, 14B12, 14H20, 14P05.
Key words and phrases. Viro's method, M–curves, smooth surfaces.

simplices. T-construction gives the best suited algorithm for finding various types of zero sets through systematic computations (see [**5**]).

Let us recall a few facts from the (elementary) classification of isotopic smooth plane projective curves of degree 4 in ($\mathbb{RP}^2$). Such curves are empty or the union of embedded circles called *ovals*. Those components carry a natural notion of interior (resp. exterior), namely the part of the complement homeomorphic to a disk (resp. a Moebius band). All connected components of a smooth non empty real algebraic plane curve in $\mathbb{RP}^2$ are ovals, except one component homotopic to a straight line if the degree is odd. The intersection of a curve of degree 4 with straight lines or conics (Bézout theorem) allows only the following five non empty zero sets up to isotopy:

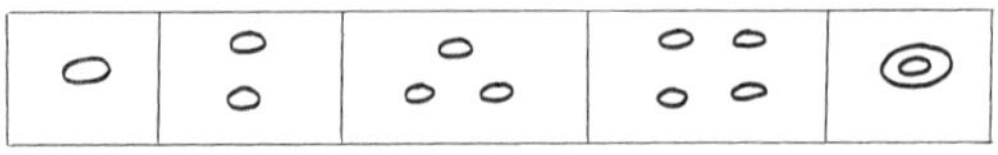

Figure 1

It is easy to check that they are indeed realizable: consider equations $f_1(x,y)\cdot f_2(x,y)+t=0$ where $\{f_1(x,y)=0\}$, $\{f_2(x,y)=0\}$ are two ellipses in some intersecting positions and t is a small enough parameter. The effect of the condition $|t| \ll 1$ is to keep the curve $\{f_1f_2+t=0\}$ close to the smooth part of the union of the ellipses and simultaneously eliminate the singular intersection points according to the signs of f_1, f_2, t.

Those simple arguments fail to solve the equation $Y = X_d$ for all Y when $q=2$, $d \geq 6$ or $q=3$, $d \geq 4$. D. Hilbert explicitely pointed out these cases in the list of his 23 problems at the Paris international congress of mathematicians in 1900 ([**4**]). See the reviews [**3**], [**9**] and [**2**].

Note that real plane curves of degree d with the maximal number $(d-1)(d-2)/2+1$ of connected components (see[**3**]) are of particular interest because they are in some sense generators (the vanishing of most ovals through Morse points can be realized with all known methods of construction). They are necessarily smooth by the genus formula and were called *M-curves* by I.G. Petrovskii. According to [**5**], using in particular a T-construction of B. Haas, the number of isotopy types of M-curves of degree d is asymptotically between $a^{d^{3/2}}$ and b^{d^2} when d goes to infinity.

Two general questions beyond the scope of this paper arise naturally from a large collection of solved equations $Y = X_d$, i.e. a large collection of realized isotopy types:

(1) *find non trivial sets of submanifolds Y for which there could be no solution to $Y = X_d$.* In general such sets occur when the Betti numbers of X_d are forced to verify some extremal conditions. There, new statements about restrictions on isotopy types are likely to exist. A famous example is the Gudkov-Arnol'd-Rokhlin-Guillou-Marin congruence modulo 8 about M-curves of even degrees (see [**3**],[**2**]).

(2) *Could the set of isotopy types realizable by the real zero set of a real polynomial of given degree be a useful information for questions coming from other fields?* For example O.Viro working with isotopy types of surfaces has proved a conjecture of V.I.Arnold about the eigenfunctions of the Laplace operator on the sphere S^3 (see [**11**]; according to Arnold, this result disproves a theorem of Courant-Herman).

The three sections below are organized as follows. In section 2 the non familiar reader shall have a glimpse without proofs on the Viro method as an approach to dissipate elaborate real plane curve singularities. The new constructions (section 3) incorporate this method.

In section 3 is explained how the Viro method combined with quadratic transforms remains applicable in any dimension (general notations, Lemma 2.1, Conjecture 2.3) and particularly in dimension 2: a recent result about $Y = X_8$ when Y is a plane curve is recalled (Theorem 2.2). New quantitative results also exist for $Y = X_9$ and $Y = X_{10}$ (author's unpublished).

In section 4 condensed proofs are given as well as examples of solutions of $Y = X_4$ with two historically important (isotopy type of) spatial surfaces Y. Conjecture 2.3 gives realizations of surfaces of degree 4 relying on E. Shustin's results about smoothings of singular plane curves of degree 8, which are not under a completed form yet. See [**2**] for previous works on surfaces of degree 4. As a rule the equation $Y = X_d$ is more delicate to solve when Y is a surface than if Y is a curve.

The author expresses his thanks to the referees for their remarks.

1. A sketch of how to perturb a singular real algebraic plane curves, keeping the degree fixed, according to the Viro method

Let

$$f(X) = \sum_{i+j\leq d} a_{i,j}x^iy^j \qquad N_f = \{(i,j) \in \mathbb{N}^2; a_{i,j} \neq 0\}$$

and $\triangle_f$ be the convex hull of N_f. This convex hull is called the Newton polygon of f or $\{f = 0\}$. Suppose:

(1) the Newton polygon $\triangle_f$ touches the two axis of $\mathbb{R}^2$ (i.e. the variables x and y do not factor in f),
(2) the origin is a singular point of $\{f = 0\}$,
(3) the restriction of f to each face of $\triangle_f$ seen from the origin of $\mathbb{N}^2$ has no other multiple factors than powers of x or y.

Then (under certain natural additional conditions) there exists a curve $\{F = 0\}$ of degree d realizing a "gluing" of algebraic curves $\{f_k = 0\}$ understood as fragments, between each others, and of $\{f = 0\}$ itself, such that the associated Newton polygons $\triangle_{f_k}$ make altogether a partition (or a patchwork) of a given region of $\mathbb{R}^2_{\geq 0}$ under $\triangle_f$. Clearly the restrictions of the f_k's on the $\partial\triangle_{f_k}$'s must be compatible. We get a smoothing or a lowering of the Milnor number of the initial curve $\{f = 0\}$ at the origin. The pattern of this smoothing is deduced from the patterns of the fragments. The Viro polynomial F is:

$$F(x,y;t) = \sum_{(i,j)\in\triangle_f \bigcup \triangle_{f_1} \bigcup \cdots \bigcup \triangle_{f_k} \bigcup \cdots} b_{i,j}x^iy^jt^{\nu(i,j)}$$

where t is a small enough real parameter and ν is a continuous arbitrary function from $\mathbb{R}^2$ to $\mathbb{R}$, sending $\mathbb{Z}^2$ to $\mathbb{Z}$, affine over each $\triangle_{f_k}$ but non affine over $\triangle_{f_k} \bigcup \triangle_{f_l}$ if $\triangle_{f_k}$ and $\triangle_{f_l}$ are adjacent. The coefficients $b_{i,j}$ are those of the polynomials f, $f_1, \ldots, f_k, \ldots$, following the position of (i,j) in $\mathbb{R}^2$. General statements with old and new detailed proofs are on O.Viro's web site: http://www.math.uu.se/~oleg

The first interesting application of the Viro method is a unified simple realization of the three isotopy types of M-curves of degree 6 ([**12**], one of them had been conjectured by D. Hilbert not to exist [**4**]). The initial singular curve chosen by Viro is the union of three ellipses with two points of contact of order 2 (Figure 5a). The two points are singular points of type J_{10}. In appropriate coordinates, the Newton polygon of such a curve is the segment $[(0,3),(6,0)]$. Each such singular point (at the origin and at infinity in these coordinates) can be independently smoothed out from gluing preconstructed smooth fragments. In this case it is even possible to use as fragments only curves $\{f_k = 0\}$ isotopic to segments and such that each polygon $\triangle_{f_k}$ is a triangle whose vertices are its only integer points. The three isotopy types are well known (see Figure 3 for the codes): $9 \sqcup 1\langle 1 \rangle$, $5 \sqcup 1\langle 5 \rangle$ and $1 \sqcup 1\langle 9 \rangle$.

In [**12**] Viro also solves all the equations $Y = X_7$ in the sense of section 1 (in particular, there are 14 isotopy types of M-curves). However, up to today, all equations $Y = X_d$ are not solved if $d \geq 8$. When $d = 8$ Theorem 2.2 lowers the number of remaining unknown cases from 13 to 9 ([**1**]). Note also that for surfaces of degree d the maximal number of connected components is not known if $d > 4$. It is 10 (resp. 2) when $d = 4$ (resp. $d = 3$) ([**2**]).

2. Notations and results

Consider the positive integers d, q, with d even, $q \geq 2$, the affine coordinates $x_1, \dots, x_q$, homogeneous polynomials $h_{d-2i}(x_1, \dots, x_q)$ of degree $d - 2i$, some real numbers c, $\alpha_1, \dots \alpha_{d/2}$ and the polynomials

$$f(X) = \sum_{i=0}^{d/2} \alpha_i (x_1 - c x_q^2)^i h_{d-2i}(x_1, \dots, x_{q-1}),$$

$$\tilde{f}(U) = \sum_{i=0}^{d/2} \alpha_i u_1^i h_{d-2i}(u_1 + c u_q^2, u_2 \dots, u_{q-1}),$$

obtained from one another by the invertible change of variables

$$u_1 = x_1 - c x_q^2, \qquad u_j = x_j, \quad 2 \leq j \leq q$$
$$x_1 = u_1 + c u_q^2, \qquad x_j = u_j, \quad 2 \leq j \leq q$$

The Newton polyhedra of $f(X)$ and $\tilde{f}(U)$ are represented on Figures 2.a, 2.b, when $q = 3$, as the convex hulls of the points G, A_1, A_2, A_3 and G, B_1, B_2, B_3 respectively.

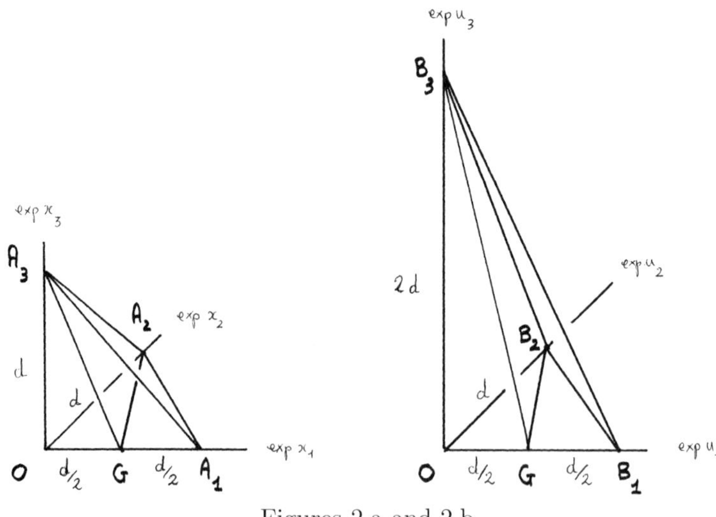

Figures 2.a and 2.b

Let us analyze the case $q = 2$ in detail (the context of Theorem 2.2). To have this case as a direct specialization of $q = 3$ (in particular, to be able to refer to Figure 2) we label our two coordinates x_1, x_3 and u_1, u_3 instead of x_1, x_2 and u_1, u_2. We have (assuming $\alpha_{d/2} = 1$ which is no loss of generality):

$$f(X) = \sum_{i=0}^{d/2} \alpha_i (x_1 - cx_3^2)^i (x_1^2)^{d/2-i} = \prod_{i=0}^{d/2} (x_1 - cx_3^2 - a_i x_1^2)$$

$$\tilde{f}(U) = \prod_{i=0}^{d/2} (u_1 - a_i(u_1 + cu_3^2)^2).$$

Suppose the α_i's are such that the a_i's are real and distinct. The curve $\{f(X) = 0\}$ is the union of $d/2$ non-singular conics with a unique intersection point (at the origin). The truncation of f (resp. $\tilde{f}$) to its Newton diagram $[(d/2, 0), (0, d)]$ (resp. $[(d/2, 0), (0, 2d)]$) shows multiple factors (resp. no multiple factors): the Viro hypotheses are not (resp. are) verified. Theorem 2.2 arises from perturbations of $\tilde{f}$ transformed into perturbations of f.

Let us go back to the the general case. Consider the indices $I = (i_1, \dots, i_q)$, $R = (r_1, \dots, r_q)$, monomials

$$b_I U^I = b_{i_1, \dots i_q} u_1^{i_1} \dots u_q^{i_q},$$

$$\varepsilon_R X^R = \varepsilon_{r_1, \dots, r_q} x_1^{r_1} \dots x_q^{r_q},$$

the sums of indices

$$K = 2d - (2(i_1 + \dots + i_{q-1}) + i_q),$$
$$E = 2d - (4i_1 + 2i_2 + \dots + 2i_{q-1} + i_q),$$
$$s(R) = r_1 + \dots + r_q,$$

and the polynomials

$$\tilde{g}(U) = \sum_{E \geq 0} b_I U^I, \qquad \tilde{g}_t(U) = \sum_{E \geq 0} b_I(t) U^I$$

with $b_I(t) = t^E b_I$.

Assume $\tilde{g}(U)$ and $\tilde{g}_t(U)$ have a truncation equal to the truncation of $\tilde{f}(U)$ on the Newton diagram (which corresponds to the condition $E = 0$, i.e. the convex hull of the integer points $I = (p_1, \dots, p_{q-1}, 2d - (4p_1 + 2p_2 + \dots + 2p_{q-1})$, i.e. the face G, B_2, B_3 of Figure 2.b when $q = 3$). The polynomial $\tilde{g}_t(U)$ parametrized by t is considered by Viro for gluing the hypersurface$\{\tilde{g}(U) = 0\}$ in this context

Consider the polynomials

$$\varphi(X) = \sum_{s(R) \leq d} \varepsilon_R X^R,$$

$$\tilde{\varphi}(U) = \varphi(u_1 + cu_q^2, u_2, \dots, u_{q-1}, u_q) = \sum_{K \geq 0} \eta_I(\vec{\varepsilon}) U^I,$$

where φ and $\tilde{\varphi}$ are obtained from one another like f and $\tilde{f}$, i.e. by the invertible change of variables introduced above. The components of the vector $\vec{\varepsilon}$ are the ε_R's and the $\eta_I(\vec{\varepsilon})$'s are the linear functions of $\vec{\varepsilon}$ deduced from the change of coordinates. If $q = 3$ (resp. $q = 2$) the Newton polyhedron of $\tilde{\varphi}(U)$is the convex hull of O, B_1, B_2, B_3, (resp. O, B_1, B_3) on Figure 2.b.

Now consider the system (**s**):

$$\begin{cases} \eta_I(\vec{\varepsilon}) = b_I(t) & \text{when } E > 0 \\ \eta_I(\vec{\varepsilon}) = 0 & \text{when } E = 0 \end{cases}$$

If $q = 3$ (resp. $q = 2$) the condition $E > 0$ defines the convex hull of O, G, B_2, B_3, (resp. O, G, B_3). The system (**s**) is a set of p triangular and non degenerate subsystems. Each of the subsystems corresponds to an element of the set

$$M = \{(m_2, \dots, m_q) \in \mathbf{N}^{q-1}; 2(m_2 + \cdots + m_{q-1}) + m_q \leq 2d\}$$

and in total $p = \operatorname{card} M$. More precisely for each $(m_2, \dots, m_q) \in M$ the triangular system involves the indices $(i_1, \dots, i_q)$ such that $2i_1 + i_q = m_q$ and $i_2 = m_2, \dots, i_{q-1} = m_{q-1}$.

Denote $\varphi_t(X)$ and $\tilde{\varphi}_t(U)$ the polynomials defined by substituting $\vec{\varepsilon}(t)$ to $\vec{\varepsilon}$ in $\varphi(X)$ and $\tilde{\varphi}(U)$, where $\vec{\varepsilon}(t)$ is the unique solution of (**s**).

Introduce the polynomials $f_t(X) = f(X) + \varphi_t(X)$ and $\tilde{f}_t(U) = \tilde{f}(U) + \tilde{\varphi}_t(U)$.

LEMMA 2.1. *There exists $t_0 > 0$ such that for $t \in]0, t_0]$ the hypersurface $\{\tilde{f}_t(U) = 0\}$ is obtained from $\{\tilde{f}(U) = 0\}$ by gluing the pattern $\{\tilde{g}(U) = 0\}$ in the place of the singular point at the origin.*

The hypersurface $\{f_t(X) = 0\}$ is a perturbation of $\{f(X) = 0\}$, with constant degree, at the origin which is a degenerate singular point. This perturbation is deduced from $\{f_t(U) = 0\}$ through the change of variables

$$\begin{cases} x_1 = u_1 + cu_q^2 & \\ x_j = u_j & 2 \leq j \leq q \end{cases}$$

which decomposes in

$$\begin{cases} x_1 = x'_1 x'_q \\ x_j = x'_j x'_q \\ x_q = x'_q \end{cases} \quad \begin{cases} x'_1 = x''_1 + cx''_q \\ x'_j = x''_j \\ x'_q = x''_q \end{cases} \quad \begin{cases} x''_1 = u_1/u_q \\ x''_j = u_j/u_q & 2 \leq j \leq q-1 \\ x''_q = u_q \end{cases}$$

Applying the Viro method in the U-coordinates and these above transformations to come back to the X-coordinates, it comes:

THEOREM 2.2. *If $q = 2$, $d = 8$, the following new isotopy types of M-curves can be realized as $\{f_t(X) = 0\}$ when $|\, t \,|$ is small enough:*

$16 \sqcup 1\langle 2 \sqcup 1\langle 2 \rangle\rangle$, $2 \sqcup 1\langle 2 \sqcup 1\langle 16 \rangle\rangle$, $13 \sqcup 1\langle 2 \sqcup 1\langle 5 \rangle\rangle$, $5 \sqcup 1\langle 2 \sqcup 1\langle 13 \rangle\rangle$,

as well as most of the known ones. (The notation $a \sqcup 1\langle b \sqcup 1\langle c \rangle\rangle$ is explained in Figure 3.)

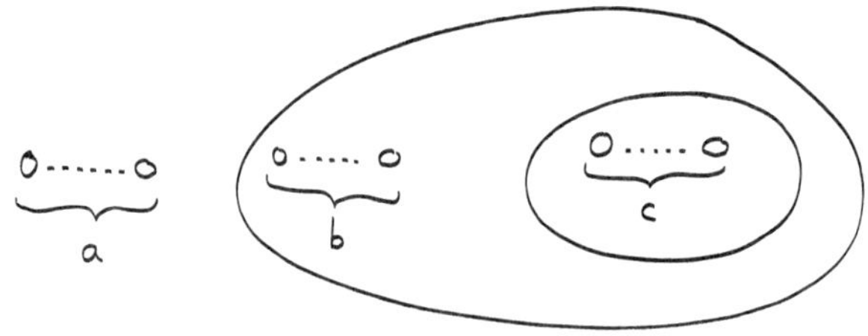

Figure 3

CONJECTURE 2.3. *If $q = 3$, $d = 4$, all the isotopy types of smooth surfaces homotopic or not homotopic to a point in $\mathbb{RP}^3$, can be realized as $\{f_t(X) = 0\}$ when $|\, t \,|$ is small enough.*

As mentioned in section I, conjecture 2.3 would be the first application of Lemma 2.1 to surfaces. More precisely: surfaces of degree 4 are realized in 4.3 from smoothing a singularity of type X_{21} on curves of degree 8.

3. Proofs

3.1. Proof of Lemma 2.1. The polynomial $\tilde{f}_t(U)=\tilde{f}(U)+\tilde{\varphi}_t(U)$ can be written as:

$$\tilde{f}_t(U) = \tilde{f}(U) + \sum_{E\geq 0} b_I(t)U^I + \sum_{E<0, J\leq 2d} c_I(t)U^I$$

with E as above, $J = 2(i_1 + i_2 + \cdots + i_{q-1}) + i_q$, and $\lim_{t\to 0} c_I(t) = 0$.

Note that when $| t | \ll 1$ the term $\sum_{E<0,J\leq 2d} c_I(t)U^I$ does not perturb the genericity of $\tilde{f}(U)$, and when $q = 3$ the Newton polyhedra of $\tilde{f}(U)$; $\tilde{g}(U)$ or $\sum_{E\geq 0} b_I(t)U^I$; $\sum_{E<0,J\leq 2d} c_I(t)U^I$ appear on Figure 2.b as the respective convex hulls of G, B_1, B_2, B_3; O, G, B_2, B_3; G, B_1, B_2, B_3.

The lemma follows from the same arguments as in the proof of Viro's gluing theorem, or from more general statements and proofs on singularity smoothings given in [**7**].

3.2. Proof of Theorem 2.2. We use here the codes of singularities and the catalogs of smoothings reviewed in ([**9**], 4.1, 4.2 and 4.3). Consider the Newton polygon of $\tilde{g}(U)$ partitioned as in Figure 4, and the following change of coordinates (the index 3 is not a mistake, see section 3)

$$\begin{cases} u_1 = u_1'u_3' \\ u_3 = u_3' \end{cases} \qquad \begin{cases} u_1' = u_1''u_3'' \\ u_3' = u_3'' \end{cases} \qquad \begin{cases} u_1'' = u_1'''u_3''' \\ u_3'' = u_3''' \end{cases}$$

We have four sets of coordinates, let say U, U', U'', U''' and four Newton polygons whose integer points are in one to one correspondences. It is natural to keep the same label for an integer point and for its images in the various Newton polygons.

The curve (resp. curves) drawn in Figure 5.a (resp. Figure 5.b) are in the (projective) coordinates U''' and the associated Newton polygon is the convex hull of F, B_3 (resp. F, G, B_3, D), see Figure 4. It is a union of three real conics (resp. they are perturbations of the two singular points J_{10} on the union of the three conics. Such perturbations can be realized as gluings of fragments, which are themselves smoothings of singularities D_4 and E_6 in appropriate coordinates (see [**9**]). The curves corresponding to Figure 5.b have a singularity Z_{15} at the origin in the U' coordinates. The smoothings of these singular curves lead to curves whose Newton polygon is the convex hull of E, G, B_3, K (see Figure 4). They have themselves a singularity J_{10} at the origin in the U coordinates. Smoothing again gives curves $\{\tilde{g}(U) = 0\}$ whose patterns are the good ones for the realizations of the four isotopy types announced in Theorem 2.2. The gluings to $\{\tilde{f}(U) = 0\}$ are straightforward since the coefficients a_i introduced for f in section 3 can be chosen arbitrarily. The intersection with the line at infinity in Figures 5.a, 5.b shows that here we have the inequalities $a_1 < a_2 < a_3 < 0 < a_4$. Note that all the isotopy types of smoothings of a curve $\{\psi = 0\}$ having one of the Newton polygons and one of the nondegenerate singularities mentionned above, are known to be independent of the values of the coefficients of the truncation of ψ on each face of the Newton diagram, as long as a double root does not appear (see [**9**]). This means that there is no problem for the use of the gluing theorem.

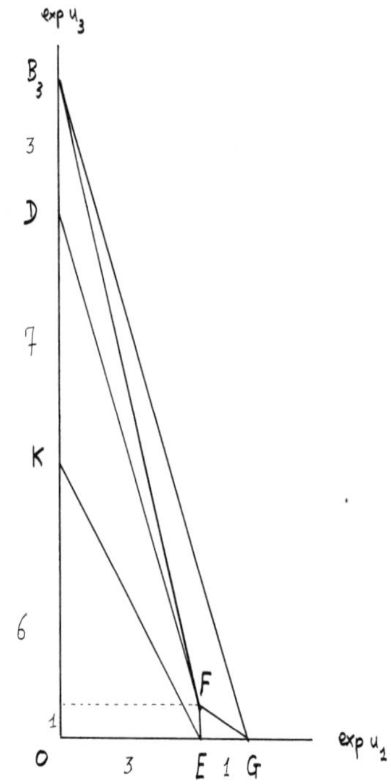

Figure 4

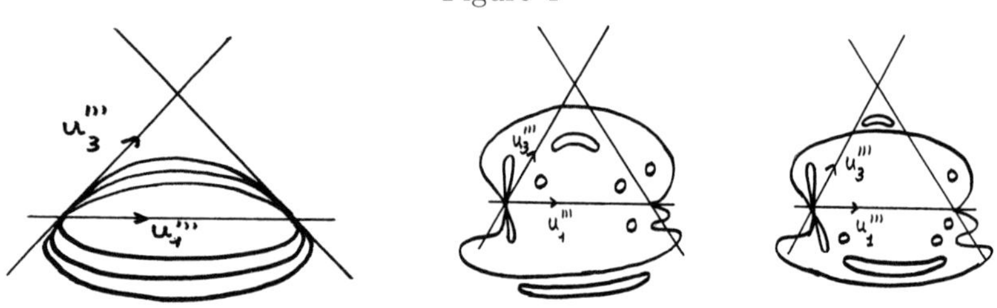

Figures 5.a (the one to the left) and 5.b

3.3. About the Conjecture 2.3. It is not difficult to represent a singular surface $\{f(X) = 0\}$ introduced in section 3. Homogenize $f(X)$ with y, then dehomogenize with x_1 and set $y' = y - cx_3^2$. From Figures 6.a and 6.b representing zero sets of $\alpha_2 {y'}^2 + \alpha_1 y' h_2(1, x_2) + h_4(1, x_2)$ in projective coordinates, follows the artist view of surfaces $\{f(X) = 0\}$ in Figures 7.a, 7.b, also in projective coordinates. Locally at the origin, these surfaces are branched double coverings over the (x_2, x_3)-plane. The local projections of the surfaces onto this plane are given in Figures 7.a, 7.b as well (shaded regions).

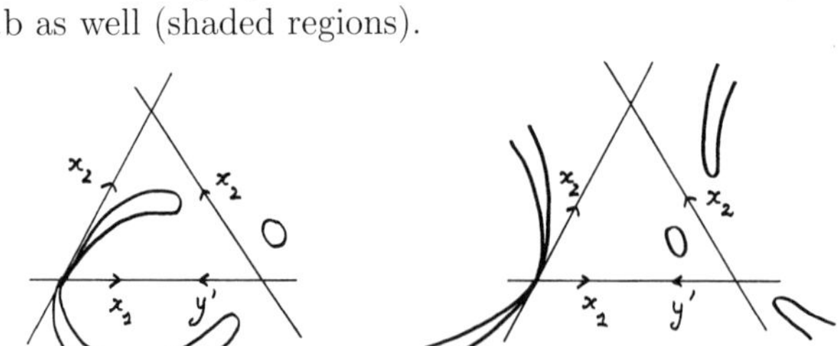

Figures 6.a and 6.b

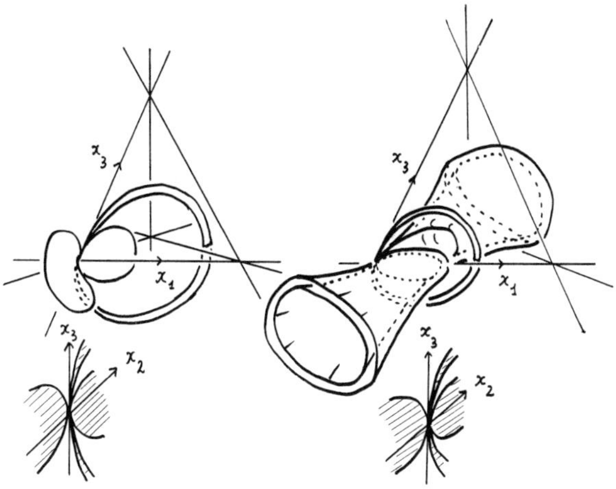

Figures 7.a and 7.b

Call Γ the boundary curve of the projection. Following the change of coordinates

$$\begin{cases} x_1 = u_1 + cu_3^2 \\ x_j = u_j & 2 \leq j \leq 3 \end{cases}$$

the local projection of the surface $\{\tilde{f}(U) = 0\}$ on the (u_2, u_3)-plane has a boundary curve locally diffeomorphic to Γ and is locally diffeomorphic to a curve of degree 8 having the segment B_2, B_3 on Figure 2.b as Newton polygon (the diffeomorphism preserves the sign of curvature of each local branch at the origin). Consequently this last curve has a singularity X_{21} at the origin (see ([**9**], 4.7). The catalog of its smoothings are reviewed in more detail in [**6**]. These smoothings are exactly the projections on the (u_2, u_3)-plane of the ramification curves of the double covering $\{\tilde{g}(U) = 0\}$ to be glued to $\{\tilde{f}(U) = 0\}$, according to Lemma 2.1.

E. Shustin informed me (unpublished) that isotopy types of smoothings of X_{21} known to be independent enough of the curvatures at the initial singular point (i.e. usable for gluing purposes), are in particular those of the family B, and probably those of the family C, where B and C are defined in the catalogs of [**6**].

The realization in the X-coordinates of all isotopy types of real algebraic surfaces of degree 4, homotopic or not homotopic to a point, basically follows from those two families and from oval contractions through a Morse point (the possible exceptions, like for nested components, are easily worked out from other choices of the singular surface $\{f(X) = 0\}$).

3.4. Examples. The smoothing shown in Figure 8 falls in the family B of [**6**]. It is possible to glue according to E. Shustin because the two middle local branches can be made arbitrarily close to each other (contract the small oval on Figures 6.a, 6.b onto an interior point according to classical constructions). The resulting surfaces are drawn on Figures 9.a, 9.b in the X-coordinates. The surface in Figure 9.a (resp. 9.b) is the union of 10 embedded spheres (resp. 9 embedded spheres and a connected component of genus 2). It has the maximal number of connected components for a surface of degree 4 in $\mathbb{RP}^3$ homotopic (resp. non homotopic) to a point. The existence of such an isotopy types for the degree four was first proved by K. Rohn in 1911 (resp. by V.M. Kharlamov in 1974) with different methods (see [**2**]). The knowledge of various approaches seems necessary in view of solving

$Y = X_5$ or $Y = X_6$ with a reasonable success. The generalization from d to $d+1$ is not straightforward.

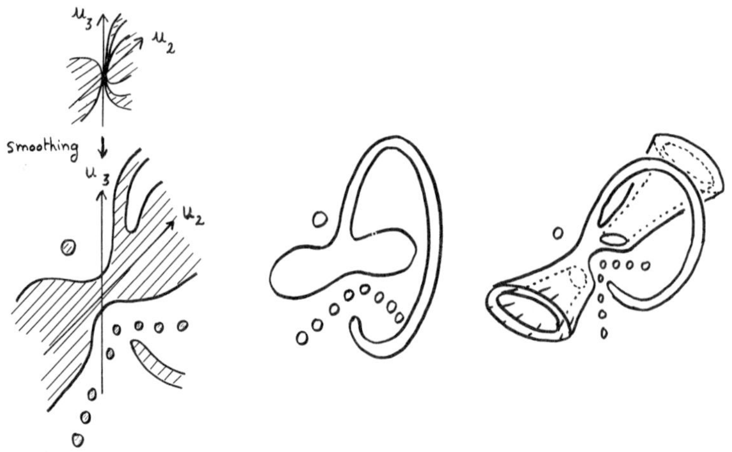

Figures 8 (the one to the left), 9.a and 9.b

References

[1] B. Chevallier, *Four M-curves of degree 8*, Functional Anal.Appl. 36, 2002, 76–78.

[2] A. Degtyarev, V.M. Kharlamov, *Topological properties of real algebraic varieties: Rokhlin's way*, Russ. Math. Surveys 55, 2000, 735–814.

[3] D. A. Gudkov, *The topology of real projective algebraic varieties*, Russ. Math. Surveys 29, 1974, 1–79.

[4] D. Hilbert, *Sur les problémes futurs des mathématiques*, Arkiv für Math. und Phis., Ser. IV, 1, 1901, 213–237.

[5] V. M. Kharlamov, S. Yu. Orevkov, *Asymptotic growth of the number of classes of real plane algebraic curves when the degree increases*, Zapiski Nauch. Seminarov POMI, 266, 2000, 218–233.

[6] G. M. Polotovskii, *On the classification of non singular curves of degree 8*, Topology and geometry Rokhlin seminar, Lecture Notes in Math. 1346, 1989, 455–485.

[7] E. I. Shustin, *Lower deformations of isolated hypersurface singularities*, Algebra i Analiz 10, 5, 1999, 221–249. English translation: St. Petersbourg Math. J. 11, 5, 2000, 883–908.

[8] O. Y. Viro, *Constructing real algebraic varieties with prescribed topology*, D. Sc. Thesis. English translation of Chapter I: *Patchworking real algebraic varieties* (available at http://www.uu.se/~oleg).

[9] O.Y. Viro, *Real algebraic curves: construction with controlled topology*, Algebra i Analiz 1, 5, 1989, 1–73. English translation: Leningrad Math. J. 1, 5, 1990, 1059–1134.

[10] O. Y. Viro, *Gluing of algebraic hypersurfaces, smoothing of singularities and construction of curves*, Proc. Leningrad topological conf., Nauka Leningrad, 1983, 149 - 197 (Russian).

[11] O. Y. Viro, *Construction of multicomponent real algebraic surfaces*, Soviet Math. Doklady 20, 5, 1979, 991–995.

[12] O.Y. Viro, *Gluing of plane algebraic curves and construction of curves of degree* 6 *and* 7, Lecture Notes in Math. 1060, 1984, 187–200.

Département de Mathématiques, Université Toulouse II, 5 allées Antonio Machado, 31058 Toulouse, France

Current address: Laboratoire E. Picard, CNRS UMR 5580, Université Toulouse IV, 118 route de Narbonne, 31062 Toulouse, France

E-mail address: `chevalli@univ-tlse2.fr`

DIMACS Series in Discrete Mathematics
and Theoretical Computer Science
Volume **60**, 2003

On the Number of Connected Components of the Relative Closure of a Semi-Pfaffian Family

Andrei Gabrielov and Thierry Zell

ABSTRACT. The notion of relative closure $(X,Y)_0$ of a semi-Pfaffian couple (X,Y) was introduced by Gabrielov to give a description of the o-minimal structure generated by Pfaffian functions. In this paper, an effective bound is given for the number of connected components of $(X,Y)_0$ in terms of the Pfaffian complexity of X and Y.

Introduction

Pfaffian functions are real analytic functions that are solutions to certain triangular systems of polynomial partial differential equations (see Definition 1). They were introduced by Khovanskiĭ [**12**], who showed that these transcendental functions exhibit, in the real domain, global finiteness properties similar to the properties of polynomials. It follows in turn that the geometrical and topological characteristics of sets defined using Pfaffian functions are also well controlled. Effective upper bounds for those geometric properties can be found in [**3, 5, 7, 8, 18, 22**].

O-minimality is a natural framework for the study of Pfaffian functions. (The reader can refer to van den Dries [**2**] for definitions.) Wilkie proved in [**21**] that the structure generated by Pfaffian functions is o-minimal (see also [**9, 19**] for generalizations of this result.)

In [**6**], Gabrielov introduced the notion of *relative closure* of a semi-Pfaffian couple, as an alternative to Wilkie's construction. In this way, he obtained an o-minimal structure in which definable sets (called *limit sets*) have a simple presentation. As a result, this structure supports a notion of complexity which naturally extends the usual Pfaffian complexity, and allows one to estimate the complexity of Boolean operations in that structure.

In this paper, we use this notion of complexity to give an explicit bound on the number of connected components of a limit set. In the first part, we recall the usual definitions related to Pfaffian functions, then we cover the essential material we will use from [**6**]. In the second part, we give explicit bounds for the smooth case (Theorem 15) and for the general case (Theorem 17). Section 3 is devoted to two applications to the fewnomial case: Theorem 19 and Theorem 20.

1991 *Mathematics Subject Classification.* Primary 32B20; Secondary 03C64, 14P15.
Key words and phrases. Pfaffian functions, fewnomials, o-minimality.

Notations. If X is a subset of $\mathbb{R}^n$, we'll denote by $\overline{X}$ its closure, and $\partial X = \overline{X} \setminus X$ its frontier. Unless otherwise noted, all subsets X are assumed to be relatively compact.

1. Semi-Pfaffian sets and relative closure

We will recall in this section the usual definitions of Pfaffian functions and semi-Pfaffian sets as given by Khovanskiĭ. Then, we will define the relative closure of a semi-Pfaffian couple that appears in [**6**].

1.1. Pfaffian functions. Let $\mathscr{U} \subseteq \mathbb{R}^n$ be an open domain. The following definitions are due to Khovanskiĭ [**12**] (see also [**10, 11**]).

DEFINITION 1. *A sequence $(f_1, \ldots, f_\ell)$ of functions that are defined and analytic in $\mathscr{U}$ is called a* Pfaffian chain *if it satisfies on $\mathscr{U}$ a differential system of the form:*

$$df_i = \sum_{j=1}^{n} P_{i,j}(x, f_1(x), \ldots, f_i(x))dx_j, \tag{1}$$

where each $P_{i,j}$ is a polynomial in $x, f_1, \ldots, f_i$, and the following holds.

(P1) *The graph Γ_i of f_i is contained in a domain Ω_i defined by polynomial inequalities in $(x, f_1(x), \ldots, f_{i-1}(x), t)$, and such that $\partial\Gamma_i \subseteq \partial\Omega_i$.*

(P2) Γ_i *is a* separating submanifold *in* Ω_i, i.e. $\Omega_i \setminus \Gamma_i$ *is a disjoint union of two open sets* Ω_i^+ *and* Ω_i^-. *(See* [**12**, p. 38]. *This is also called the* Rolle leaf *condition in the terminology of* [**14, 15**].)

If $\alpha \in \mathbb{N}$ is a bound on the degrees of the polynomials $P_{i,j}$, we say that the degree *of the chain is (at most) α. The integer ℓ is called the* length *of the chain.*

A Pfaffian function *$q(x)$ with the chain $(f_1, \ldots, f_\ell)$ is any function that can be written as*

$$q(x) = Q(x, f_1(x), \ldots, f_\ell(x)),$$

for some polynomial $Q(x, y_1, \ldots, y_\ell)$. If the degree of Q is β, we say that the degree of the function $q(x)$ is (at most) β.

EXAMPLE 2. *For any $a = (a_1, \ldots, a_n) \in \mathbb{R}^n$ the function $e_a : x \mapsto \exp(a \cdot x)$ is a Pfaffian function on $\mathbb{R}^n$, since*

$$de_a(x) = \sum_{j=1}^{n} a_j\, e_a(x)dx_j.$$

Pfaffian functions form a rather large class of functions. Note in particular that elementary functions can be seen as Pfaffian functions on appropriate subsets of their domain of definition (see the Chapter 1 of [**12**] or [**7**] for more precise statements).

In sections 2 and 3, we will take $\mathscr{U}$ to be of the form $\mathbb{R}^n$, $\mathbb{R}^{n-1} \times \mathbb{R}_+$, or $(\mathbb{R}_+)^n$. In that case, we will use the following result which is due to Khovanskiĭ [**12**, p. 79]. The reader can also refer to the revised edition (in Russian) [**13**].

THEOREM 3. *Let $(f_1, \ldots, f_\ell)$ be a Pfaffian chain on $\mathscr{U}$, where $\mathscr{U}$ is one of $\mathbb{R}^n$, $\mathbb{R}^{n-1} \times \mathbb{R}_+$, or $(\mathbb{R}_+)^n$. If $q_1, \ldots, q_n$ are Pfaffian functions of respective degrees β_i, the number of solutions of the system*

$$q_1(x) = \cdots = q_n(x) = 0,$$

for which the Jacobian determinant is not zero is bounded by

$$2^{\ell(\ell-1)/2}\beta_1 \cdots \beta_n \left[\beta_1 + \cdots + \beta_n - n + \min(n,\ell)\alpha + 1\right]^{\ell}. \tag{2}$$

Remark. It is possible to give effective estimates when the domain $\mathscr{U}$ is not of the above form, but one has to modify the estimates to take into account the complexity of the domain.

1.2. Semi-Pfaffian sets. Let $(f_1, \ldots, f_\ell)$ be a Pfaffian chain of degree α, defined on a domain $\mathscr{U} \subseteq \mathbb{R}^n$.

DEFINITION 4. *A* basic semi-Pfaffian set *is a set X of the form:*

$$X = \{x \in \mathscr{U} \mid \varphi_i(x) = 0,\ \psi_j(x) > 0,\ \textit{for } i = 1,..,I;\ j = 1,..,J\}, \tag{3}$$

where all the functions above are Pfaffian with the chain $(f_1, \ldots, f_\ell)$. If all these functions have degree at most β, the set X is said to have format $(I, J, n, \ell, \alpha, \beta)$.

A semi-Pfaffian set *is any finite union of basic semi-Pfaffian sets. A semi-Pfaffian set has format $(N, I, J, n, \ell, \alpha, \beta)$ if it is the union of at most N basic semi-Pfaffian sets having formats component-wise not exceeding $(I, J, n, \ell, \alpha, \beta)$.*

We say that a semi-Pfaffian set X is restricted *if it is relatively compact in $\mathscr{U}$.*

DEFINITION 5. *A basic semi-Pfaffian set is called* (effectively) non-singular *if for all $x \in X$, the functions φ_i appearing in* (3) *verify*

$$d\varphi_1(x) \wedge \cdots \wedge d\varphi_I(x) \neq 0.$$

1.3. Effective bounds on the Betti numbers. Following ideas of Oleĭnik-Petrovskiĭ [**17**], Milnor [**16**] and Thom [**20**] for algebraic varieties, we can use the bound appearing in Theorem 3 to estimate the sum of the Betti numbers of a semi-Pfaffian set [**12, 22**].

Assume X is a compact semi-Pfaffian set defined only by equations,

$$X = \{x \in \mathbb{R}^n \mid p_1(x) = \cdots = p_r(x) = 0\}, \tag{4}$$

and let $p = p_1^2 + \cdots + p_r^2$. Then, it can be shown that the sum of the Betti numbers of X is exactly one-half the sum of the Betti numbers of the compact components of the set $\{p = \varepsilon\}$, where ε is a small positive real number.

Counting the number of critical points of a generic projection, we obtain the following bound.

COROLLARY 6. *Assume $X \subseteq \mathbb{R}^n$ is of the form* (4)*, where $p_1, \ldots, p_r$ are of degree at most β in a Pfaffian chain $(f_1, \ldots, f_\ell)$ defined in $\mathscr{U} = \mathbb{R}^n$ with a length ℓ and degree α. Then, the sum of the Betti numbers of X is bounded by*

$$2^{\ell(\ell-1)/2}\beta(\alpha + 2\beta - 1)^{n-1}[(2n-1)\beta + (n-1)(\alpha-2) + \min(n,\ell)\alpha]^{\ell}. \tag{5}$$

1.4. Relative closure. ¿From now on, we consider semi-Pfaffian subsets of $\mathbb{R}^n \times \mathbb{R}_+$ with a fixed Pfaffian chain $(f_1, \ldots, f_\ell)$ in a domain $\mathscr{U}$. We write $(x_1, \ldots, x_n)$ for the coordinates in $\mathbb{R}^n$ and λ for the last coordinate (which we think of as a parameter.) If X is such a subset, we let $X_\lambda = \{x \mid (x, \lambda) \in X\} \subseteq \mathbb{R}^n$ and consider X as the family of its fibers X_λ.

We let $X_+ = X \cap \{\lambda > 0\}$ and $\check{X} = \overline{X_+} \cap \{\lambda = 0\}$. The following definitions appear in [**6**].

DEFINITION 7. *Let X be a relatively compact semi-Pfaffian subset of $\mathbb{R}^n \times \mathbb{R}_+$. The family X_λ is said to be a* semi-Pfaffian family *if for any $\varepsilon > 0$, the set $X \cap \{\lambda > \varepsilon\}$ is restricted. (See Definition)*

The format *of the family X is the format of the fiber X_λ for a small $\lambda > 0$.*

Remark. Note that the format of X as a semi-Pfaffian set is different from its format as a semi-Pfaffian family.

DEFINITION 8. *Let X and Y be semi-Pfaffian families in $\mathscr{U}$ with a common chain $(f_1, \dots, f_\ell)$. They form a* semi-Pfaffian couple *if the following properties are verified:*

- $(\overline{Y})_+ = Y_+$;
- $(\partial X)_+ \subseteq Y$.

Then, the format *of the couple (X, Y) is the component-wise maximum of the formats of the families X and Y.*

DEFINITION 9. *Let (X, Y) be a semi-Pfaffian couple in $\mathscr{U}$. We define the* relative closure *of (X, Y) at $\lambda = 0$ by*

$$(X, Y)_0 = \check{X} \setminus \check{Y} \subseteq \check{\mathscr{U}}. \tag{6}$$

DEFINITION 10. *Let $\Omega \subseteq \mathbb{R}^n$ be an open domain. A* limit set *in Ω is a set of the form $(X_1, Y_1)_0 \cup \cdots \cup (X_K, Y_K)_0$, where (X_i, Y_i) are semi-Pfaffian couples respectively defined in domains $\mathscr{U}_i \subseteq \mathbb{R}^n \times \mathbb{R}_+$, such that $\check{\mathscr{U}}_i = \Omega$ for $1 \leq i \leq K$. If the formats of the couples (X_i, Y_i) are bounded component-wise by $(N, I, J, n, \ell, \alpha, \beta)$, we say that the* format *of the limit set is $(K, N, I, J, n, \ell, \alpha, \beta)$.*

EXAMPLE 11. *Any (not necessarily restricted) semi-Pfaffian set X is a limit set.*

PROOF. It is enough to prove the result for a basic set $X \subseteq \mathscr{U}$ of the form (3). Assume $\mathscr{U} = \{x \in \mathbb{R}^n \mid g_i(x) > 0,\ 1 \leq i \leq r\}$. Let $\psi = \psi_1 \cdots \psi_J$ and $g = g_1 \cdots g_r$. Define the sets

$$W = \{(x, \lambda) \in X \times \Lambda \mid g(x) > \lambda,\ |x| < \lambda^{-1}\};$$
$$Y_1 = \{(x, \lambda) \in \mathscr{U} \times \Lambda \mid \varphi_1(x) = \cdots = \varphi_I(x) = 0,\ \psi(x) = 0,\ g(x) \geq \lambda,\ |x| \leq \lambda^{-1}\};$$
$$Y_2 = \{(x, \lambda) \in \mathscr{U} \times \Lambda \mid \varphi_1(x) = \cdots = \varphi_I(x) = 0,\ g(x) = \lambda,\ |x| \leq \lambda^{-1}\};$$
$$Y_3 = \{(x, \lambda) \in \mathscr{U} \times \Lambda \mid \varphi_1(x) = \cdots = \varphi_I(x) = 0,\ g(x) \geq \lambda,\ |x| = \lambda^{-1}\};$$

where $\Lambda = (0, 1]$. If $Y = Y_1 \cup Y_2 \cup Y_3$, it is clear that (W, Y) satisfies the requirements of Definition 8. Thus (W, Y) is a semi-Pfaffian couple; its relative closure is X. □

For all $n \in \mathbb{N}$ we let $\mathcal{S}_n$ be the collection of limit sets in $\mathbb{R}^n$, and $\mathcal{S} = \cup_{n \in \mathbb{N}} \mathcal{S}_n$. The following theorem sums up the results in [**6**, Theorems 2.9 and 5.1].

THEOREM 12. *The structure $\mathcal{S}$ is o-minimal. Moreover, if X is a definable set defined by a formula involving the limit sets $L_1, \dots, L_N$, X can be presented as a limit set whose format is bounded by an effective function of the formats of $L_1, \dots, L_N$.*

2. Connected components of a limit set

In this section, we establish effective bounds on the number of connected components of the relative closure of a semi-Pfaffian couple (X, Y). In what follows, we will always assume that Y is not empty. Note that if Y is empty, it follows from Definition 8 that $(\partial X)_+$ must be empty too; then, X_λ is compact and the number of connected components of X_0 is at most the number of connected components of X_λ. Effective bounds for this case appear in [**12, 22**].

As announced in the introduction, we assume as in [**6**] that X and Y are relatively compact. To keep the estimates simple, we'll also assume that the Pfaffian chain $(f_1, \ldots, f_\ell)$ is defined over the whole of $\mathbb{R}^n \times \mathbb{R}_+$, so that the bound given in Theorem 3 is applicable.

2.1. General considerations. We show here how to reduce the problem of counting the number of connected components of a limit set to a problem in the semi-Pfaffian setting.

Let Φ be the (squared) distance function on $\mathbb{R}^n \times \mathbb{R}^n$:

$$\Phi : \mathbb{R}^n \times \mathbb{R}^n \longrightarrow \mathbb{R} \qquad (x,y) \mapsto |x-y|^2 \tag{7}$$

For any $\lambda > 0$, we can define the distance to Y_λ, Ψ_λ on X_λ by:

$$\Psi_\lambda(x) = \min_{y \in Y_\lambda} \Phi(x, y). \tag{8}$$

Define similarly for $x \in \check{X}$:

$$\Psi(x) = \min_{y \in \check{Y}} \Phi(x, y). \tag{9}$$

The following result appears in [**6**, Theorem 2.12]. The proof is presented here for the sake of self-containment.

THEOREM 13. *Let (X, Y) be a semi-Pfaffian couple. Then, there exists $\lambda \ll 1$ such that for every connected component C of $(X, Y)_0$, we can find a connected component D_λ of the set of local maxima of Ψ_λ such that D_λ is arbitrarily close to C.*

PROOF. Let C be a connected component of $(X, Y)_0$. Note that by definition of the relative closure, if x is in C, it cannot be in $\check{Y}$. So we must have $\Phi(x, y) > 0$ for all $y \in \check{Y}$, and since $\check{Y}$ is compact, we must have $\Psi(x) > 0$. Also, any point in ∂C must be in $\check{X}$, but not in $(X, Y)_0$. So we must have $\partial C \subseteq \check{Y}$, hence $\Psi|_{\partial C} \equiv 0$. This means that the restriction of Ψ to C takes its maximum inside of C.

Choose $x_0 \in C$, and let $c = \Psi(x_0) > 0$. For a small λ, there is a point $x_\lambda \in X_\lambda$ close to x_0 such that $c_\lambda = \Psi_\lambda(x_\lambda)$ is close to c, and is greater than the maximum of the values of Ψ_λ over points of X_λ close to ∂C. Hence the set $\{x \in X_\lambda \mid \Psi_\lambda(x) \geq c_\lambda\}$ is nonempty, and the connected component A_λ of this set that contains x_λ is close to C. There exists a local maximum $x^*_\lambda \in A_\lambda$ of Ψ_λ. If D_λ is the connected component in the set of local maxima of Ψ_λ, it is contained in Z_λ and is close to C. □

2.2. A bound for the smooth case. We will now show how the number of connected components of the set of local maxima of Ψ_λ that appear in Theorem 13 can be estimated when the sets X_λ and Y_λ are smooth.

Define for all p,

$$Z_\lambda^p = \{(x, y_0, \dots, y_p) \in W_\lambda^p \mid \Phi(x, y_0) = \dots = \Phi(x, y_p)\}, \tag{10}$$

where

$$W_\lambda^p = \{(x, y_0, \dots, y_p) \in X_\lambda \times (Y_\lambda)^{p+1} \mid y_i \neq y_j, 0 \leq i < j \leq p\}. \tag{11}$$

LEMMA 14. *Assume (X, Y) is a Pfaffian couple such that X_λ and Y_λ are smooth for all $\lambda > 0$. For a given $\lambda > 0$, let x^* be a local maximum of $\Psi_\lambda(x)$. Then, there exists $0 \leq p \leq \dim(X_\lambda)$ and a point $z^* = (x^*, y_0^*, \dots, y_p^*) \in Z_\lambda^p$ such that Z_λ^p is smooth at z^*, and z^* is a critical point of $\Phi(x, y_0)$ on Z_λ^p.*

PROOF. Since x^* is a local maximum of $\Psi_\lambda(x)$, there exists a point $y_0^* \in Y_\lambda$ such that $\Phi(x^*, y_0^*) = \min_{y \in Y_\lambda} \Phi(x^*, y) = \Psi_\lambda(x^*)$. In particular, $d_y\Phi(x, y) = 0$ at $(x, y) = (x^*, y_0^*)$. If (x^*, y_0^*) is a critical point of $\Phi(x, y)$ (this is always the case when $\dim(X_\lambda) = 0$) the statement holds for $p = 0$. Otherwise $d_x\Phi(x, y_0^*) \neq 0$ at $x = x^*$. Let ξ be a tangent vector to X at x^* such that $d_x\Phi(x^*, y_0^*)(\xi) > 0$.

Assume that for all $y \in Y_\lambda$ such that $\Phi(x^*, y) = \Psi_\lambda(x^*)$, we have $d_x\Phi(x, y)(\xi) > 0$ when $x = x^*$. Let $\gamma(t)$ be a curve on X_λ such that $\gamma(0) = x^*$ and $\dot\gamma(0) = \xi$. For all $y \in Y_\lambda$, there exists T_y such that for all $0 < t < T_y$, the inequality $\Phi(\gamma(t), y) > \Phi(x^*, y)$ holds. By compactness of Y_λ, this means we can find some t such that that inequality holds for all $y \in Y_\lambda$. Hence, $\Psi_\lambda(\gamma(t)) > \Psi_\lambda(x^*)$, which contradicts the hypothesis that Ψ_λ has a local maximum at x^*.

Since x^* is a local maximum of $\Psi_\lambda(x)$, there exists a point $y_1^* \in Y_\lambda$ such that $d_x\Phi(x, y_1^*)(\xi) \leq 0$ at $x = x^*$ and $\Phi(x^*, y_1^*) = \Psi_\lambda(x^*)$. In particular, $y_1^* \neq y_0^*$, $d_y\Phi(x^*, y) = 0$ at $y = y_1^*$, and $d_x\Phi(x, y_1^*) \neq d_x\Phi(x, y_0^*)$ at $x = x^*$. This implies that $(x^*, y_0^*, y_1^*) \in Z_\lambda^1$, and Z_λ^1 is smooth at (x^*, y_0^*, y_1^*). If (x^*, y_0^*, y_1^*) is a critical point of $\Phi(x, y_0)$ on Z_λ^1 (this is always the case when $\dim(X_\lambda) = 1$) the statement holds for $p = 1$. Otherwise $d_x\Phi(x, y_0^*)$ and $d_x\Phi(x, y_1^*)$ are linearly independent at $x = x^*$. Since $\dim(X_\lambda) \geq 2$, there exists a tangent vector ξ to X_λ at x^* such that $d_x\Phi(x^*, y_0^*)(\xi) > 0$ and $d_x\Phi(x^*, y_0^*)(\xi) > 0$. Since x^* is a local maximum of $\Psi_\lambda(x)$, there exists a point $y_2^* \in Y_\lambda$ such that $d_x\Phi(x, y_2^*)(\xi) \leq 0$ at $x = x^*$ and $\Phi(x^*, y_2^*) = \Psi_\lambda(x^*)$. This implies that $(x^*, y_0^*, y_1^*, y_2^*) \in Z_\lambda^2$, and Z_λ^2 is smooth at $(x^*, y_0^*, y_1^*, y_2^*)$. The above arguments can be repeated now for Z_λ^2, Z_λ^3, etc., to prove the statement for all $p \leq \dim(X_\lambda)$. □

Assume now that X_λ and Y_λ are *effectively* non-singular, *i.e.* they are of the following form:

$$\begin{aligned} X_\lambda &= \{x \in \mathbb{R}^n \mid p_1(x, \lambda) = \dots = p_{n-d}(x, \lambda) = 0\}; \\ Y_\lambda &= \{y \in \mathbb{R}^n \mid q_1(y, \lambda) = \dots = q_{n-k}(y, \lambda) = 0\}; \end{aligned} \tag{12}$$

where, for all $\lambda > 0$, we assume that $d_xp_1 \wedge \dots \wedge d_xp_{n-d} \neq 0$ on X_λ and that $d_yq_1 \wedge \dots \wedge d_yq_{n-k} \neq 0$ on Y_λ. In particular, we have $\dim(X_\lambda) = d$ and $\dim(Y_\lambda) = k$.

Remark. Note that we assume that no inequalities appear in (12). We can clearly make that assumption for Y_λ, since that set has to be closed for all $\lambda > 0$. For X_λ, we observe the following: if C is a connected component of C_λ^p, the critical set of

$\Phi|_{Z_\lambda^p}$, the function Φ is constant on C. If C contains a local maximum for Ψ_λ, it cannot meet ∂X_λ because $\partial X_\lambda \subseteq Y_\lambda$. Hence, we do not need to take into account the inequalities appearing in the definition of X_λ.

Let us now define for all p,

$$\theta_p : (y_0, \dots, y_p) \in (Y_\lambda)^{p+1} \mapsto \sum_{0\le i<j\le p} |y_i - y_j|^2. \tag{13}$$

Then, for X_λ and Y_λ as in (12), the sets Z_λ^p are defined for all p by the following conditions.

$$\begin{cases} p_1(x,\lambda) = \cdots = p_{n-d}(x,\lambda) = 0; \\ q_1(y_i,\lambda) = \cdots = q_{n-k}(y_i,\lambda) = 0, \quad 0 \le i \le p; \\ \Phi(x,y_i) - \Phi(x,y_j) = 0, \quad 0 \le i < j \le p; \end{cases} \tag{14}$$

and the inequality

$$\theta_p(y_0, \dots, y_p) > 0. \tag{15}$$

Under these hypotheses, we obtain the following bound.

THEOREM 15. *Let (X, Y) be a semi-Pfaffian couple such that for all small $\lambda > 0$, X_λ and Y_λ are effectively non-singular basic sets of dimension respectively d and k. If the format of (X, Y) is $(1, I, J, n, \ell, \alpha, \beta)$, the number of connected components of $(X, Y)_0$ is bounded by*

$$2\sum_{p=0}^{d} 2^{q\ell(q\ell-1)/2}\beta_p(\alpha+2\beta_p)^{nq-1}[nq(2\beta_p+\alpha-2)+q\min(n,\ell)\alpha]^{q\ell}, \tag{16}$$

where $q = p+2$ and $\beta_p = \max\{1+(n-k)(\alpha+\beta-1), 1+(n-d+p)(\alpha+\beta-1)\}$.

PROOF. According to Theorem 13, we can chose $\lambda > 0$ such that for any connected component C of $(X, Y)_0$, we can find a connected component D_λ of the set of local maxima of Ψ_λ such that D_λ is close to C. We see that for λ small enough, two connected components C and C' of $(X, Y)_0$ cannot share the same connected component D_λ, since D_λ cannot meet $\check{Y}$ for λ small enough. Indeed, the distance from D_λ to $\check{Y}$ is bounded from below by the distance from D_λ to Y_λ, – which is at least c_λ, – minus the distance between Y_λ and $\check{Y}$. But the latter distance goes to zero, whereas the former goes to a positive constant c when λ goes to zero.

Once that λ is fixed, all we need to do is estimate the number of connected components of the set of local maxima of Ψ_λ. According to Lemma 14, we can reduce to estimating the number of connected components of the critical sets C_λ^p of the restriction $\Phi|_{Z_\lambda^p}$ for $0 \le \lambda \le d$.

For the sake of concision, we will drop λ from the notations in this proof, writing Z^p for Z_λ^p, $p_i(x)$ for $p_i(x, \lambda)$, *etc...*

A point $z = (x, y_0, \dots, y_p) \in Z^p$ is in C^p if and only if the following conditions are satisfied:

$$\begin{cases} d_y\Phi(x,y_j) = 0, \quad 0 \le j \le p; \\ \operatorname{rank}(d_x\Phi(x,y_0), \dots, d_x\Phi(x,y_p)) \le p. \end{cases} \tag{17}$$

For X_λ and Y_λ as in (12), those conditions translate into:

(18)
$$\begin{cases} \operatorname{rank}\{\nabla_y q_1(y_i), \dots, \nabla_y q_{n-k}(y_i), \nabla_y \Phi(x, y_i)\} \leq n-k, \quad 0 \leq i \leq p; \\ \operatorname{rank}\{\nabla_x p_1(x), \dots, \nabla_x p_{n-d}(x), \nabla_x \Phi(x, y_0), \dots, \nabla_x \Phi(x, y_0)\} \leq n-d+p. \end{cases}$$

Those conditions translate into all the maximal minors of the corresponding matrices vanishing. These minors are Pfaffian functions in the chain used to define X and Y. Their degrees are respectively $1+(n-k)(\alpha+\beta-1)$ and $1+(n-d+p)(\alpha+\beta-1)$.

The number of connected components of C^p is bounded by the number of connected components of the set D^p defined by the conditions in (14) and (15), and the vanishing of the maximal minors corresponding to the conditions in (18).

Let E^p be the set defined by the equations (14) and (18), so that $D^p = E^p \cap \{\theta_p > 0\}$. Then, the number of connected components of D^p is bounded by the number of connected components of E^p plus the number of connected components of $E^p \cap \{\theta_p = \varepsilon\}$ for a choice of $\varepsilon > 0$ small enough.

Hence, we're reduced to the problem of estimating the number of connected components of two semi-Pfaffian sets in $\mathbb{R}^{n(p+2)}$ that are defined without inequalities using a Pfaffian chain in $n(p+2)$ variables of degree α and length $(p+2)\ell$. Using the bounds on the Betti numbers from Corollary 6, we obtain (16). □

2.3. Bounds for the singular case. Let's consider now the case where X_λ and Y_λ may be singular. We can use deformation techniques to reduce to the smooth case. First, the following lemma shows we can reduce to the case where X_λ is a basic set.

LEMMA 16. *Let X_1, X_2 and Y be semi-Pfaffian sets such that (X_1, Y) and (X_2, Y) are Pfaffian families. Then, $(X_1 \cup X_2, Y)_0 = (X_1, Y)_0 \cup (X_2, Y)_0$.*

The proof follows from the definition of the relative closure.

THEOREM 17. *Let (X, Y) be a semi-Pfaffian couple. Assume X_λ and Y_λ are unions of basic sets having a format of the form $(I, J, n, \ell, \alpha, \beta)$. If the number of basic sets in X_λ is M and the number of basic sets in Y_λ is N, then the number of connected components of $(X, Y)_0$ is bounded by*

$$2\,MN \sum_{p=0}^{n-1} 2^{q\ell(q\ell-1)/2} \gamma_p (\alpha + 2\gamma_p)^{nq-1} [nq(2\gamma_p + \alpha - 2) + q \min(n, \ell)\alpha]^{q\ell}, \tag{19}$$

where $q = p+2$ and $\gamma_p = 1 + (p+1)(\alpha + 2\beta - 1)$.

PROOF. Again, we want to estimate the number of local maxima of the function Ψ_λ defined in (8).

By Lemma 16, we can restrict ourselves to the case where X is basic. Let $Y = Y_1 \cup \dots \cup Y_N$, where all the sets Y_i are basic. For each basic set, we take the sum of squares of the equations defining it: the corresponding positive functions, which we denote by p and $q_1, \dots, q_N$, have degree 2β in the chain. Fix $\varepsilon_i > 0$, for $0 \leq i \leq N$, and $\lambda > 0$, and let $\mathcal{X} = \{p(x, \lambda) = \varepsilon_0\}$ and for all $1 \leq i \leq N$, let $\mathcal{Y}_i = \{q_i(x, \lambda) = \varepsilon_i\}$.

Since Y_λ is compact, if x^* is a point in X_λ such that Ψ_λ has a local maximum at $x = x^*$, there is a point y^* in some $(Y_i)_\lambda$ such that $\Phi(x, y) = \Psi_\lambda(x)$. Then, we can find a couple $(x', y') \in \mathcal{X}_\lambda \times (\mathcal{Y}_i)_\lambda$ close to (x^*, y^*) such that $\Phi(x', y')$ is a local maximum of the distance (measured by Φ) from $\mathcal{X}_\lambda$ to $(\mathcal{Y}_i)_\lambda$.

Since for small enough $\varepsilon_0, \ldots, \varepsilon_N$, the sets $\mathcal{X}_\lambda$ and $(\mathcal{Y}_i)_\lambda$ are effectively non-singular hypersurfaces, the number of local maxima of the distance of $\mathcal{X}_\lambda$ to $(\mathcal{Y}_i)_\lambda$ can be bounded by (16), for appropriate values of the parameters. The estimate (19) follows. □

3. Application to fewnomials

In this section, we will apply our previous results to the case where the Pfaffian functions we consider are fewnomials.

3.1. Fewnomials and low additive complexity. Recall that we can consider the restriction of any polynomial q to an orthant as a Pfaffian function whose complexity depends only on the number of non zero monomials in q. Fix $\mathcal{K} = \{m_1, \ldots, m_r\} \in \mathbb{N}^n$ a set of exponents.

DEFINITION 18. *The polynomial q is a $\mathcal{K}$-*fewnomial *if it is of the form:*

$$q(x) = a_0 + a_1 x^{m_1} + \cdots a_r x^{m_r}, \quad \textit{where} \quad a_0, \ldots, a_r \in \mathbb{R}.$$

Let $\ell = n + r$, and $(f_1, \ldots, f_\ell)$ be the functions defined by:

$$f_i(x) = \begin{cases} x_i^{-1} \text{ if } 1 \le i \le n, \\ x^{m_{i-n}} \text{ if } i > n. \end{cases} \tag{20}$$

It is easy to see that $(f_1, \ldots, f_\ell)$ is a Pfaffian chain of length ℓ and degree $\alpha = 2$ in the positive orthant $\mathcal{U} = (\mathbb{R}_+)^n$, since we have:

$$\frac{\partial f_i}{\partial x_j} = \begin{cases} -f_i^2 \text{ if } i = j \le n, \\ f_j f_i \text{ if } i > n. \end{cases}$$

Then, a $\mathcal{K}$-fewnomial as defined above is just a Pfaffian function in the chain $(f_1, \ldots, f_\ell)$ of degree $\beta = 1$, involving only the last r functions of the chain. In general, if $q(x)$ is a polynomial, and if $q(x)$ can be expressed as a Pfaffian function of degree at most β in the chain (20), we say that q has a *low additive complexity*, and call β the *pseudo-degree* of q. (For more on additive complexity, see [**1, 12**].)

Example. To make the distinction between fewnomials and polynomials with low additive complexity clearer, if a and b are any integers, the univariate polynomial $(x^a + 1)^b$ has pseudo degree b in the obvious Pfaffian chain. But under our terminology, we wouldn't call it a fewnomial, and it has indeed many monomials if expanded.

Let now $S \subseteq (\mathbb{R}_+)^n$ be a semi-algebraic set defined by $\mathcal{K}$-fewnomials, and assume S is bounded. We can define from S a semi-Pfaffian family $X \subseteq \mathbb{R}^n \times \mathbb{R}$ by:

$$X = \{(x, \lambda) \in \mathcal{U} \times \mathbb{R}_+ \mid x \in S,\ x_1 > \lambda, \ldots, x_n > \lambda\}. \tag{21}$$

We can apply the results from Theorems 15 and 17 to X, to obtain a bound on the number of connected components of $\overline{S} \cap \partial\mathcal{U}$. Note that this is indeed an important case, since an example is given in [**4**] of a fewnomial semialgebraic set for which the closure cannot be described independently of the degrees of the defining polynomials.

3.2. Bounds for fewnomial couples. We now give explicit bounds in the case where the fibers in the semi-Pfaffian couple (X, Y) can both be defined $\mathcal{K}$-fewnomials, *i.e.* by degree 1 Pfaffian function in a chain of the type (20). In particular, if S is a semi-algebriac set such that $\partial S \cap \mathscr{U} = \emptyset$, these bounds will apply to X as in (21) and to $Y = \{x_1 = \lambda\} \cup \cdots \cup \{x_n = \lambda\}$.

THEOREM 19. *Let (X, Y) be a semi-Pfaffian couple defined by degree 1 functions in the chain* (20). *Then, the following bounds can be established for the number of connected components of $(X, Y)_0$.*

Case 1. *If for all $\lambda > 0$, X_λ and Y_λ are effectively non-singular of dimension respectively d and k, we obtain from* (16) *the bound*

$$\sum_{p=0}^{d} 2^{q^2(n+r)^2/2}(4n+6)^{q(3n+2r)}q^{q(n+r)}. \tag{22}$$

Case 2. *If X and Y are the union of respectively M and N basic sets, the number of connected components of $(X, Y)_0$ is bounded by:*

$$MN\sum_{p=0}^{n-1} 2^{q^2(n+r)^2/2}(6n+6)^{q(3n+2r)}q^{q(n+r)}. \tag{23}$$

PROOF. These bounds are obtained using the results from Theorems 15 and 17, with $\alpha = 2$, $\beta = 1$ and $\ell = n + r$, and then bounding very bluntly the terms β_p and γ_p. □

3.3. Closure relative to the frontier of a fewnomial set. Let X be a semi-Pfaffian family such that for all $\lambda > 0$, the set X_λ is defined by $\mathcal{K}$-fewnomials. By definition of a family, the set ∂X_λ is contained in the domain of the Pfaffian chain, so by the results of [**5**], this set is a semi-Pfaffian set defined by functions in the chain, that is, polynomials of low additive complexity. Moreover, the format of ∂X_λ can be estimated from the format of X_λ. Applying those results together with those of Theorem 17, we can give estimates for the number of connected components of $(X, \partial X)_0$.

THEOREM 20. *Let X be a semi-Pfaffian family in a Pfaffian chain of type* (20). *If the format of X is of the form $(N, I, J, n, \ell = n + r, \alpha = 2, \beta = 1)$, the number of connected components of $(X, \partial X)$ is bounded by*

$$N^2(I+J)^{N+rO(n^2)}n^{(n+r)^{n^{O(n^2+nr)}}}. \tag{24}$$

PROOF. Following [**5**], the set ∂X_λ can be defined using the same Pfaffian chain as X_λ, using N' basic sets and functions of degree at most β', where, under the hypotheses above, the following bounds hold.

$$\beta' \leq n^{(n+r)^{O(n)}}, \quad N' \leq N(I+J)^{N+rO(n^2)}N^{(n+r)^{rO(n)}}.$$

The bound on the number of connected components follows readily. □

References

[1] R. Benedetti and J.-J. Risler. *Real Algebraic and Semi-algebraic Sets.* Hermann, 1990.

[2] L. van den Dries. *Tame Topology and O-minimal Structures.* LMS Lecture Note Series No. 248. Cambridge University Press, 1998.

[3] A. Gabrielov. Multiplicities of Pfaffian intersections, and the Łojasiewicz inequality. *Selecta Math. (N.S.),* 1(1):113–127, 1995.

[4] A. Gabrielov. Counter-examples to quantifier elimination for fewnomial and exponential expressions. *Preprint,* 1997.
Available at: `http://www.math.purdue.edu/~agabriel/preprint.html`

[5] A. Gabrielov. Frontier and closure of a semi-Pfaffian set. *Discrete Comput. Geom.,* 19(4):605–617, 1998.

[6] A. Gabrielov. Relative closure and the complexity of Pfaffian elimination. *Preprint,* 2000. Available at: `http://www.math.purdue.edu/~agabriel/preprint.html`

[7] A. Gabrielov and N. Vorobjov. Complexity of stratifications of semi-Pfaffian sets. *Discrete Comput. Geom.,* 14(1):71–91, 1995.

[8] A. Gabrielov and N. Vorobjov. Complexity of cylindrical decompositions of sub-Pfaffian sets. *J. Pure Appl. Algebra* 164:179–197, 2001.

[9] M. Karpinski and A. Macintyre, A generalization of Wilkie's theorem of the complement, and an application to Pfaffian closure. *Selecta Math. (N.S.)* 5(4):507–516, 1999.

[10] A. G. Khovanskiĭ. A class of systems of transcendental equations. *Dokl. Akad. Nauk SSSR,* 255(4):804–807, 1980.

[11] A. G. Khovanskiĭ. Fewnomials and Pfaff manifolds. In *Proceedings of the International Congress of Mathematicians, Vol. 1, 2 (Warsaw, 1983),* pp. 549–564, Warsaw, 1984. PWN.

[12] A. G. Khovanskiĭ. *Fewnomials.* American Mathematical Society, Providence, RI, 1991. Translated from the Russian by Smilka Zdravkovska.

[13] A. G. Khovanskiĭ. *Malochleny.* FAZIS, Moscow, 1997. (Russian.)

[14] J.-M. Lion and J.-P. Rolin. Homologie des ensembles semi-pfaffiens. *Ann. Inst. Fourier (Grenoble)* 46:723–741, 1996.

[15] J.-M. Lion and J.-P. Rolin. Volumes, Feuilles de Rolle de Feuilletages analytiques et Théorème de Wilkie, *Ann. Fac. Sci. Toulouse Math.* 7:93–112, 1998.

[16] J. Milnor, On the Betti numbers of real varieties. *Proc. Amer. Math. Soc.* 15:275–280, 1964.

[17] O. Oleĭnik and I. G. Petrovskiĭ. On the topology of real algebraic surfaces. *Izv. Akad. Nauk SSSR,* 13:389–402, 1949.

[18] S. Pericleous and N. Vorobjov. New complexity bounds for cylindrical decompositions of sub-Pfaffian sets. *Preprint,* 2001.

[19] P. Speissegger. The Pfaffian closure of an o-minimal structure. *J. Reine Angew. Math.,* 508:189–211, 1999.

[20] R. Thom. Sur l'homologie des variétés algébriques réelles. in *Differential and Combinatorial Topology,* S.S. Cairns Ed., pp. 255–265, Princeton Univ. Press, 1965.

[21] A. J. Wilkie. A theorem of the complement and some new o-minimal structures. *Selecta Math. (N.S.),* 5(4):397–421, 1999.

[22] T. Zell. Betti numbers of semi-Pfaffian sets. *J. Pure Appl. Algebra,* 139(1-3):323–338, 1999. Effective methods in algebraic geometry (Saint-Malo, 1998).

1395 Math Sciences Bldg West Lafayette IN 47907-1395 USA
E-mail address: `agabriel@math.purdue.edu`

E-mail address: `tzell@math.purdue.edu`

DIMACS Series in Discrete Mathematics
and Theoretical Computer Science
Volume **60**, 2003

How to show a set is not algebraic

Clint McCrory

ABSTRACT. We revisit Akbulut and King's first example of a compact semialgebraic set which satisfies Sullivan's local Euler characteristic condition, but which is not homeomorphic to an algebraic set. A nontrivial obstruction is computed using the link operator on the ring of constructible functions.

There are many local topological conditions satisfied by real algebraic sets. The simplest of these was discovered by Sullivan [**5**] more than thirty years ago: The link of every point has even Euler characteristic. In low dimensions this is the only obstruction for a set to be algebraic. More precisely, if a compact semialgebraic set of dimension less than three satisfies Sullivan's condition, then it is homeomorphic to an algebraic set. Akbulut and King [**1**] found an example of a compact 3-dimensional semialgebraic set satisfying Sullivan's condition which is not homeomorphic to an algebraic set, and this is the example we will discuss.

The method we use to compute the obstruction is due to Parusiński and the author [**2**]. We have also found an enormous list of independent obstructions in dimension four [**3**]. The yoga of algebraically constructible functions used to prove our obstructions vanish for algebraic sets is presented in the paper of Isabelle Bonnard in this volume.

Let X be a semialgebraic set in $\mathbb{R}^n$, and let $p \in X$. The *link* L_p of X at p is the intersection of X with a sphere of small radius $\epsilon > 0$ centered at p. For sufficiently small ϵ the topological type of L_p is independent of ϵ, and a neighborhood of p in X is homeomorphic to the cone on L_p. (The link L_p is homeomorphic to the boundary of the simplicial star of p in X for a semialgebraic triangulation of X.) Sullivan's theorem is that if X is algebraic then for all $p \in X$ the Euler characteristic $\chi(L_p)$ is even.

For example, consider the *Cartan umbrella* $X = \{(x, y, z) \mid x^2 = zy^2\}$ in $\mathbb{R}^3$ (Figure 1). For each real number t, let H_t be the horizontal plane $z = t$. For $t > 0$, $X \cap H_t$ is the two lines $x = \pm(\sqrt{t})y$, $z = t$. For $t = 0$, $X \cap H_t$ is the line $x = 0$, $z = 0$. For $t < 0$, $X \cap H_t$ is the point $x = 0$, $y = 0$, $z = t$.

The space X has five strata:

- two 2-strata $f = \{x^2 = zy^2, y < 0\}$, $f' = \{x^2 = zy^2, y > 0\}$,

1991 *Mathematics Subject Classification.* Primary 14P25, Secondary 14B05, 14P10 .

Key words and phrases. real algebraic set, semialgebraic set, Euler characteristic, link, constructible function.

- two 1-strata $d = \{x = 0, y = 0, z < 0\}$, $e = \{x = 0, y = 0, z > 0\}$,
- one 0-stratum $a = \{(0, 0, 0)\}$.

The local topology of X is constant along each of these strata. If $p \in X$, then

- if $p \in f \cup f'$, L_p is topologically a circle, so $\chi(L_p) = 0$;
- if $p \in e$, L_p is a graph with two vertices and four edges joining one vertex to the other, so $\chi(L_p) = 2 - 4 = -2$;
- if $p \in d$, L_p is two points, so $\chi(L_p) = 2$;
- if $p = (0, 0, 0)$, L_p is a graph with two vertices and two edges which are both loops at the same vertex (L_p is the union of a figure eight and a point), so $\chi(L_p) = 2 - 2 = 0$.

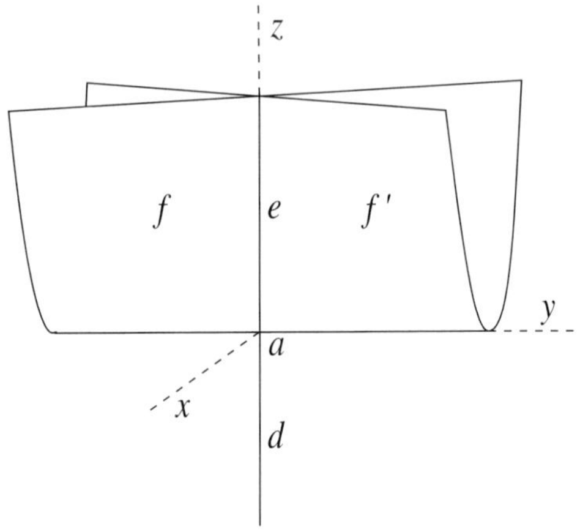

FIGURE 1. The Cartan umbrella

On the other hand, the *Whitney umbrella* $X = \{(x, y, z) \mid x^2 = zy^2,\ z \geq 0\}$ is not algebraic, because the link of the origin is a figure eight, which has Euler characteristic -1. (The Whitney umbrella is the image of the algebraic map $(u, v) \mapsto (uv, u, v^2)$.)

Now $\chi(L_p)$ is a topological invariant, since it can be computed from the local homology of X at p:

$$\chi(L_p) = 1 - \sum_k (-1)^k \operatorname{rank} H_k(X, X - \{p\}).$$

Thus the Whitney umbrella is not homeomorphic to a real algebraic set.

The known higher topological obstructions for a semialgebraic set to be algebraic are similar to Sullivan's obstruction. They are local—in fact they are mod 2 invariants of the links L_p—and they are combinatorial in the sense that they depend only on the dimensions and incidence of the cells of a semialgebraic cell structure on X.

Parusiński and the author [**2**] have found a way to describe such obstructions using the ring of constructible functions on X and an operator on this ring generalizing the Euler characteristic of the link. Here is a brief description of our method.

Let X be a semialgebraic set. An integer-valued function $\varphi : X \to \mathbb{Z}$ is *constructible* if it is a finite combination of characteristic functions of semialgebraic subsets X_i of X with integer coefficients: $\varphi = \sum_i n_i \mathbf{1}_{X_i}$. It follows that there exists a semialgebraic cell complex $\mathcal{K}$ on X such that φ is constant on each cell of $\mathcal{K}$. (A semialgebraic cell complex is a locally topologically trivial semialgebraic stratification such that each stratum is homeomorphic to a Euclidean space and has compact closure.)

The set of constructible functions on X is a ring, with $(\varphi+\psi)(p) = \varphi(p)+\psi(p)$ and $(\varphi\times\psi)(p) = \varphi(p)\times\psi(p)$. If $\varphi = \sum_i n_i \mathbf{1}_{X_i}$ with the sets X_i compact, the *Euler integral* of φ is defined by $\int_X \varphi = \sum_i n_i \chi(X_i)$, where χ is the Euler characteristic. If φ is constant on each cell of $\mathcal{K}$ and φ has compact support, then

$$\int_X \varphi = \sum_{C\in\mathcal{K}} (-1)^{\dim C} \varphi(C),$$

where $\varphi(C)$ is the value of φ on the cell C.

The *link operator* Λ on the ring of constructible functions on X is the Euler integral of the restriction of φ to the link of X at p,

$$(\Lambda\,\varphi)(p) = \int_{L_p} \varphi.$$

Sullivan's theorem says that if X is an algebraic set with characteristic function $\mathbf{1}_X$, the values of the function $\Lambda\,\mathbf{1}_X$ are even. In other words, if $\tilde{\Lambda} = \frac{1}{2}\,\Lambda$ then $\tilde{\Lambda}\,\mathbf{1}_X$ is integer-valued (and hence $\tilde{\Lambda}\,\mathbf{1}_X$ is a new constructible function on X).

The main result of [**2**] (Theorem 2.5, p. 536) implies that if X is an algebraic set, then *all* the functions obtained from $\mathbf{1}_X$ using the arithmetic operations $+$, $-$, $\times$, and the operator $\tilde{\Lambda}$ are integer-valued. Furthermore this property is topological ([**2**] A.7, p. 550).

The following statement is a corollary: If Y is an algebraic link (the link of a point in an algebraic set), then all the functions obtained from $\mathbf{1}_Y$ using the arithmetic operations $+$, $-$, $\times$, and the operator $\tilde{\Lambda}$ are integer-valued and have even Euler integral.

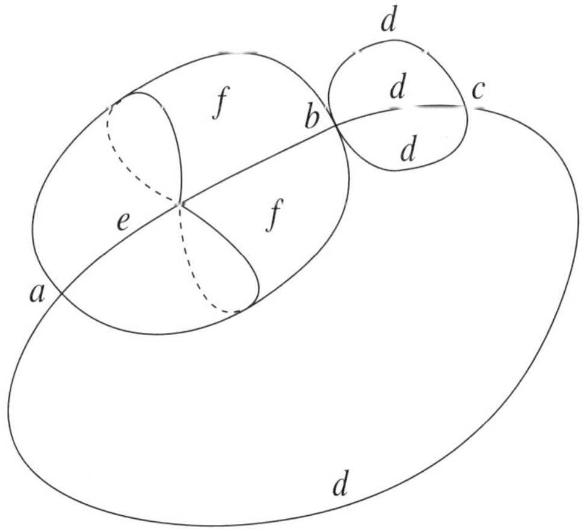

FIGURE 2. The space Y

Now we apply our method to Akbulut and King's example. Let Y be the topological space shown in Figure 2. The space Y is homeomorphic to an algebraic set in projective 3-space, the union of the projective Cartan umbrella $wx^2 = zy^2$ and the plane conic $(w+z)^2 + x^2 = z^2$, $y = 0$. Realized in this way, Y has a semialgebraic cell structure with three 0-cells, five 1-cells, and two 2-cells. The three 0-cells are labelled a, b, c in the figure. Four of the 1-cells are labelled d, and one of the 1-cells (incident to the 2-cells) is labelled e. The 2-cells are labelled f.

The table below summarizes the computation of a constructible function ψ obtained from $\mathbf{1}_Y$ using the arithmetic operations $+, -, \times$, and the operator $\tilde{\Lambda}$, such that $\int_Y \psi$ is odd.

The strategy for finding a suitable function ψ is based on the following observations. We define an operator $\tilde{\Omega}$ by $\tilde{\Omega}\,\xi = \xi - \tilde{\Lambda}\,\xi$. The interesting fact $\tilde{\Lambda}\,\tilde{\Lambda} = \tilde{\Lambda}$ implies that $\tilde{\Omega}\,\tilde{\Omega} = \tilde{\Omega}$, $\tilde{\Lambda}\,\tilde{\Omega} = 0$, and $\tilde{\Omega}\,\tilde{\Lambda} = 0$. If ξ has even-dimensional support then the support of $\tilde{\Lambda}\,\xi$ is of lower dimension. If If ξ has odd-dimensional support then the support of $\tilde{\Omega}\,\xi$ is of lower dimension.

Let d be the dimension of Y. For decreasing dimensions $k = d, d-1, d-2, \ldots$, we search through the functions with support dimension k which are obtained from $\mathbf{1}_Y$ using $+, -, \times, \tilde{\Lambda}$, until we find one with odd Euler integral.

In the example at hand, since the dimension of Y is 2 we start with $\varphi = \tilde{\Lambda}\,\mathbf{1}_Y$, which has support dimension 1, and we apply $\tilde{\Omega}$ to the powers of φ to get functions with support dimension zero.

	a	b	c	d	e	f	$\int$
$\mathbf{1}_Y$	1	1	1	1	1	1	0
$\varphi = \tilde{\Lambda}\,\mathbf{1}_Y$	0	1	2	1	-1	0	0
φ^2	0	1	4	1	1	0	0
$\tilde{\Lambda}\,\varphi^2$	1	2	2	1	1	0	0
$\tilde{\Omega}\,\varphi^2 = \varphi^2 - \tilde{\Lambda}\,\varphi^2$	-1	-1	2	0	0	0	0
$\psi = \varphi\,\tilde{\Omega}\,\varphi^2$	0	-1	4	0	0	0	3

In the table the columns a, b, c, d, e, f correspond to the types of cells of Y. Two cells have the same label if they have the same local topology at an interior point. This implies that any function obtained from $\mathbf{1}_Y$ using $+, -, \times, \tilde{\Lambda}$ will be constant on all cells with the same label. The rows of the table are such functions, and the entries of the table give the values of the function on each type of cell.

The last column is the Euler integral of the function. To compute the Euler integral we take into account that there are four cells of type d and two cells of type f. Thus, for example, for the first row we have $\int \mathbf{1}_Y = (1{+}1{+}1) - (4{\cdot}1{+}1) + (2{\cdot}1) = 0$, and for the second row we have $\int \varphi = (0+1+2) - (4\cdot 1 + (-1)) + (2\cdot 0) = 0$.

The computation of $\varphi = \tilde{\Lambda}\,\mathbf{1}_Y$ is similar to our first example, the Cartan umbrella. The link of the point a is the union of a figure eight and a point, which has Euler characteristic 0, so $\varphi(a) = 0$. The link of b is the union of a figure eight and three points, which has Euler characteristic 2, so $\varphi(b) = 1$. The link of c is four points so $\varphi(c) = 2$. The link of a point of type d is two points so $\varphi(d) = 1$. The link of a point of type e is a graph with two vertices and four edges from one vertex to the other, which has Euler characteristic -2, so $\varphi(e) = -1$. And the link of a point of type f is a circle, which has Euler characteristic 0, so $\varphi(f) = 0$.

The computation of $\tilde{\Lambda}\,\varphi^2$ is similar. For example, the link of a is the union of a figure eight and a point. The point is of type d, at which φ^2 takes the value 1. The vertex of the figure eight is of type e, at which φ^2 also takes the value 1. The

1-cells of the figure eight are of type f, at which φ^2 takes the value 0. Thus the Euler integral of φ^2 on the link of a is $(1+1)-0=2$, so $\tilde{\Lambda}\,\varphi^2(a)=1$.

The result of our computation is

$$\int_Y \psi = 3,\ \psi = (\tilde{\Lambda}\,\mathbf{1}_Y)\big((\tilde{\Lambda}\,\mathbf{1}_Y)^2 - \tilde{\Lambda}(\tilde{\Lambda}\,\mathbf{1}_Y)^2\big).$$

Thus Y is not homeomorphic to an algebraic link.

Akbulut and King's example X is the *suspension* of Y, the union of two cones with base Y. Now X can be realized as a semialgebraic set in a Euclidean space, with Y the link of the vertex of either cone. Since the Euler characteristic of Y is 0, X satisfies Sullivan's condition. Since Y is not homeomorphic to an algebraic link, it follows that X is not homeomorphic to an algebraic set.

A more direct way to prove that X is not homeomorphic to an algebraic set is to find a function η obtained from $\mathbf{1}_X$ using $+, -, \times, \tilde{\Lambda}$ such that η is not integer-valued. It is not hard to see that the operators $\tilde{\Lambda}$ and $\tilde{\Omega}$ are interchanged by restriction to a link (see [**3**] 1.3(d), p. 499). More precisely, if Y is the link of p in X, and ξ is a constructible function on X, then

$$(\tilde{\Lambda}\,\xi)|_Y = \tilde{\Omega}(\xi|_Y),\ (\tilde{\Omega}\,\xi)|_Y = \tilde{\Lambda}(\xi|_Y).$$

Now we apply this to the Akbulut-King example. Since $(\mathbf{1}_X)|_Y = \mathbf{1}_Y$, we have $(\tilde{\Omega}\,\mathbf{1}_X)|_Y = \tilde{\Lambda}\,\mathbf{1}_Y$. In other words, if $\varphi' = \tilde{\Omega}\,\mathbf{1}_X$ then $\varphi'|_Y = \varphi$. Thus $\varphi'\,\tilde{\Lambda}(\varphi')^2|_Y = \varphi\,\tilde{\Omega}\,\varphi^2$. So if p is one of the two suspension vertices of X then $\tilde{\Lambda}(\varphi'\,\tilde{\Lambda}(\varphi')^2)(p) = \frac{1}{2}\int_Y \varphi\,\tilde{\Omega}\,\varphi^2 = \frac{3}{2}$. The function $\eta = \tilde{\Omega}(\varphi'\,\tilde{\Lambda}(\varphi')^2)$ is supported at the two vertices of X, and at each of these vertices $\eta(p) = -\frac{3}{2}$.

Akbulut and King [**1**] analyze five local topological obstructions for a three-dimensional semialgebraic set to be algebraic—the local mod 2 Euler characteristic and four new invariants. They prove that these five obstructions are independent, and that the vanishing of these obstructions is a sufficient condition for a compact 3-dimensional semialgebraic set to be homeomorphic to an algebraic set.

The situation in dimension four is not so simple. There are at least $2^{43} - 43$ independent local topological obstructions [**3**], but it is not known whether the vanishing of all these obstructions is a sufficient condition for a compact 4-dimensional semialgebraic set to be homeomorphic to an algebraic set.

Significant progress toward the description of a set of sufficient conditions in dimension four has been made by Michelle Previte [**4**] using Akbulut and King's *resolution towers*. However, the relation between resolution tower obstructions and our constructible function invariants remains a mystery.

References

[1] S. Akbulut, H. King, *Topology of Real Algebraic Sets*, MSRI Publ. vol. 25, Springer, New York, 1992.

[2] C. McCrory, A. Parusiński, *Algebraically constructible functions*, Ann. Sci. École Norm. Sup. **30** (4) (1997), 527–552.

[3] C. McCrory, A. Parusiński, *Topology of real algebraic sets of dimension 4: necessary conditions*, Topology **39** (2000), 495–523.

[4] M. Previte, *The topology of four-dimensional real algebraic sets*, Ph.D. dissertation, University of Maryland, 2000.

[5] D. Sullivan, *Combinatorial invariants of analytic spaces*, Proc. Liverpool Singularities Symposium I, Lecture Notes in Math. vol. 192, Springer, Berlin, 1971, 165–169.

Mathematics Department, University of Georgia, Athens, Georgia 30602-7403, USA
E-mail address: clint@math.uga.edu

DIMACS Series in Discrete Mathematics
and Theoretical Computer Science
Volume **60**, 2003

Minimizing Polynomial Functions

Pablo A. Parrilo and Bernd Sturmfels

ABSTRACT. We compare algorithms for global optimization of polynomial functions in many variables. It is demonstrated that existing algebraic methods (Gröbner bases, resultants, homotopy methods) are dramatically outperformed by a relaxation technique, due to N.Z. Shor and the first author, which involves sums of squares and semidefinite programming. This opens up the possibility of using semidefinite programming relaxations arising from the Positivstellensatz for a wide range of computational problems in real algebraic geometry.

1. Introduction

This is an expository and experimental paper concerned with the following basic problem. Given a multivariate polynomial function $f \in \mathbb{R}[x_1, \ldots, x_n]$ which is bounded below on $\mathbb{R}^n$, find the global minimum f^* and a point p^* attaining it:

$$f^* \quad = \quad f(p^*) \quad = \quad \min\{\, f(p) \,:\, p \in \mathbb{R}^n \}. \tag{1.1}$$

Exact algebraic algorithms for this task find all the critical points and then identify the smallest value of f at any critical point. Such methods will be discussed in Section 2. The techniques include Gröbner bases, resultants, eigenvalues of companion matrices [**CLO98**], and numerical homotopy methods [**Li97**], [**EV99**].

An entirely different approach was introduced by N.Z. Shor ([**Sho87**], [**SS97**]) and further developed in the dissertation of the first author [**Par00, Par01**]. The idea is to compute the largest real number λ such that $f(x) - \lambda$ is a sum of squares in $\mathbb{R}[x_1, \ldots, x_n]$. Clearly, λ is a lower bound for the optimal value f^*. We show in Section 3 that, when the degree of f is fixed, the lower bound λ can be computed in polynomial time using semidefinite programming [**VB96**]. If $\lambda = f^*$ holds then this is certified by semidefinite programming duality, and the certificate yields the optimal point p^*. A similar scheme, focusing on the connection with the theory of moments, was presented recently by Lasserre in [**Las01**]. In our computational experiments, to be presented in Section 5, we found that $\lambda = f^*$ almost always holds, and we solved problems up to $n = 15$.

2000 *Mathematics Subject Classification.* Primary 13J30, 90C22 ; Secondary 13P10, 65H10.

Key words and phrases. Polynomials, real algebra, Gröbner bases, sums of squares, semidefinite programming, global optimization, Positivstellensatz.

The collaboration between the authors was made possible through a grant from the Center for Pure and Applied Mathematics at UC Berkeley. The second author was supported in part by NSF Grant DMS-9970254 and the Miller Institute at UC Berkeley.

The objective of this article is to provide a bridge between mathematical programming and algebraic geometry, demonstrating that algorithms from the former have the potential to play a major role in future algorithms in the latter. This will be underlined in Section 6, where we present open problems, and in Section 7 where we show that semidefinite programming in conjunction with the Positivstellensatz is applicable to a wide range of computational problems in real algebraic geometry.

2. Computational Algebra

In this section we discuss the following approach to our problem (1.1). Form the partial derivatives of the given polynomial f and consider the ideal they generate:

$$I \quad = \quad \langle \frac{\partial f}{\partial x_1}, \frac{\partial f}{\partial x_2}, \ldots, \frac{\partial f}{\partial x_n} \rangle \quad \subseteq \quad \mathbb{R}[x_1, \ldots, x_n] =: \mathbb{R}[\mathbf{x}]$$

The zeros of the ideal I in complex n-space $\mathbb{C}^n$ are the *critical points* of f. Their number (counting multiplicity) is the dimension over $\mathbb{R}$ of the residue ring:

$$\mu \quad = \quad \dim_{\mathbb{R}} \mathbb{R}[\mathbf{x}]/I \quad = \quad \# \mathcal{V}_{\mathbb{C}}(I).$$

We shall assume that μ is finite. (If $\mu = +\infty$ then one can apply perturbation techniques to reduce to the case $\mu < +\infty$). For instance, if f is a dense polynomial of even degree $2d$ then it follows from *Bézout's Theorem* that $\mu = (2d-1)^n$.

Consider the subset of *real critical points*:

$$\mathcal{V}_{\mathbb{R}}(I) \quad = \quad \{p^{(1)}, p^{(2)}, \ldots, p^{(\nu)}\} \quad \subset \quad \mathbb{R}^n.$$

This set is usually much smaller than the set of all complex critical points, i.e., typically we have $\nu \ll \mu$. If we know the set $\mathcal{V}_{\mathbb{R}}(I)$, then our problem is solved.

When $n \geq 2$, it may happen that the optimal value f^* is achieved only asymptotically at infinity (e.g., $f = x^2 + (1-xy)^2$). This can be avoided, for instance, by requiring that $f(\mathbf{x}) \geq \alpha ||\mathbf{x}||^2$, for some $\alpha > 0$ and all $\mathbf{x}$ sufficiently large.

LEMMA 2.1. *Under the assumptions above, the optimal value f^* is attained at a critical point:*

$$f^* \quad = \quad \min\{ f(p^{(1)}), f(p^{(2)}), \ldots, f(p^{(\nu)}) \} \tag{2.1}$$

The three techniques to be described in this section all compute the set $\mathcal{V}_{\mathbb{R}}(I)$ of real critical points. We will illustrate them for the following example:

$$\text{Minimize} \quad f(x,y,z) \quad = \quad x^4 + y^4 + z^4 - 4xyz + x + y + z \tag{2.2}$$

The optimal value for this problem is $f^* = -2.112913882$. By symmetry, there are three optimal points (x^*, y^*, z^*) attaining this value:

$$(0.988, -1.102, -1.102), \ (-1.102, 0.988, -1.102), \ (-1.102, -1.102, 0.988).$$

2.1. Gröbner bases and eigenvalues. We review the method of solving polynomial equations by means of Gröbner bases and eigenvalues [**CLO98**, §2.4]. We are free to choose an arbitrary term order $\prec$ on the polynomial ring $\mathbb{R}[x_1, \ldots, x_n]$. Let $\mathcal{G}$ be a Gröbner basis for the critical ideal I with respect to $\prec$. While computing Gröbner bases is a time-consuming task in general, this is not an issue in this paper, since in all our examples the n given generators $\partial f / \partial x_i$ already form a Gröbner basis in the total degree order. In our example the Gröbner basis is

$$\mathcal{G} \quad = \quad \{ \underline{x^3} - yz + 1/4, \ \underline{y^3} - xz + 1/4, \ \underline{z^3} - xy + 1/4 \} \tag{2.3}$$

A monomial $x_1^{u_1} \cdots x_n^{u_n}$ is *standard* if it is not divisible by the leading term of any element in the Gröbner basis $\mathcal{G}$. The set $\mathcal{B}$ of standard monomials is an $\mathbb{R}$-basis for the residue ring $\mathbb{R}[\mathbf{x}]/I$. The standard monomials for (2.3) are:

$$\mathcal{B} \quad = \quad \{ x^i y^j z^k \, : \, i,j,k \in \mathbb{N}, \, i,j,k \leq 2 \}, \qquad \mu = \#(\mathcal{B}) = 27. \tag{2.4}$$

For any polynomial $g \in \mathbb{R}[\mathbf{x}]$ consider the $\mathbb{R}$-linear endomorphism:

$$\mathrm{Times}_g \, : \, \mathbb{R}[\mathbf{x}]/I \rightarrow \mathbb{R}[\mathbf{x}]/I \, , \; h \mapsto g \cdot h.$$

This endomorphism is represented in the basis $\mathcal{B}$ by a real $\mu \times \mu$-matrix T_g. The entry of T_g with row index $\mathbf{x}^u \in \mathcal{B}$ and column index $\mathbf{x}^v \in \mathcal{B}$ is the coefficient of $\mathbf{x}^u$ in the normal form of $\mathbf{x}^v \cdot g(\mathbf{x})$ with respect to $\mathcal{G}$.

PROPOSITION 2.2. *The optimal value f^* is an eigenvalue of the matrix T_f. Any eigenvector of T_f with eigenvalue f^* defines an optimal point $p^* = (p_1^*, \ldots, p_n^*)$ by the eigenvector identities $\mathrm{T}_{x_i} \cdot v = p_i \cdot v$ for $i = 1, \ldots, n$.*

PROOF. This follows from Lemma 2.1 and Theorem (4.5), Chapter 2 in the book of Cox-Little-O'Shea [**CLO98**]; see also [**CLO98**, Exercise 17, page 62]. □

The resulting algorithm is to compute symbolically the matrices T_f and T_{x_i} for $i = 1, \ldots, n$, then compute numerically its eigenvalues (and matching eigenvectors) of T_f, and finally determine f^* and p^* as in the proposition.

In our example the matrix T_f has format 27×27 with rows and columns indexed by (2.4). Of its 729 entries only 178 are nonzero. For instance, the column indexed by xyz has four nonzero entries, namely, the coefficients of

$$\mathrm{normalform}_{\mathcal{G}}(xyz \cdot f) \quad = \quad \frac{3}{4}x^2yz + \frac{3}{4}xy^2z + \frac{3}{4}xyz^2 \; - \; x^2y^2z^2.$$

The matrix T_f has maximal rank 27. Of its eigenvalues only three are real:

$$-0.8692394998, \; -0.8702981639, \; -2.112913879$$

The three real eigenvalues have multiplicity 3,1, and 3 respectively.

2.2. Resultants and discriminants. One algebraic method for solving polynomial equations is to use resultants. Closely related to resultants are discriminants. They express the condition on a hypersurface to have a singularity, by means of a polynomial in the coefficients of its defining equation. Let t be a new indeterminate and form the discriminant of the polynomial $f(x) - t$ with respect to $x_1, \ldots, x_n$:

$$\delta(t) \quad := \quad \Delta_x\big(f(x_1, \ldots, x_n) - t \big)$$

Here Δ_x refers to the *A-discriminant*, defined in [**GKZ94**, Chapter 9], where A is the support of f together with the origin. From [**GKZ94**, §10.1.H] we conclude that the discriminant $\delta(t)$ equals the characteristic polynomial of the matrix T_f.

COROLLARY 2.3. *The optimal value f^* is a real root of $\delta(t)$.*

In our example, $\delta(t)$ is $256t^3 - 512t^2 - 96t + 473$ times the third power of

$$65536t^6 + 393216t^5 + 1056768t^4 + 1011712t^3 - 421376t^2 - 437152t + 166419.$$

The optimal value $f^* = -2.11...$ is a root of this sextic. This sextic has Galois group S_6, so f^* cannot be expressed in radicals over the rationals.

The suggested algorithm is to compute $\delta(t)$, and minimal polynomials for the coordinates x_i^* of the optimal point, by elimination of variables using matrix formulas for resultants and discriminants [**GKZ94**, Chapter 13]. The subsequent numerical computation is to find the roots of a univariate polynomial.

2.3. Homotopy methods. The critical equations form a square system: n equations in n variables having finitely many roots. Such a system is well-suited for *numerical homotopy continuation methods.* For an introduction to this subject see the papers of Li [**Li97**] and Verschelde [**EV99**]. The basic idea is to introduce a deformation parameter τ into the given system. For instance, we might replace (2.3) by the following system which depends on a complex parameter τ:

$$(2.5)\qquad H_\tau \;:\; x^3 - \tau \cdot yz + 1/4 \;=\; y^3 - \tau \cdot xz + 1/4 \;=\; z^3 - \tau \cdot xy + 1/4 \;=\; 0.$$

The solutions $\big(x(\tau), y(\tau), z(\tau)\big)$ are algebraic functions of τ. Our goal is to find them for $\tau = 1$. It is easy to find the solutions for $\tau = 0$:

$$(x(0), y(0), z(0)) \quad = \quad \big(4^{-1/3} \cdot \eta_1,\, 4^{-1/3} \cdot \eta_2,\, 4^{-1/3} \cdot \eta_3 \big), \quad \eta_i^3 = 1.$$

Homotopy methods trace the full set of solutions from $\tau = 0$ to $\tau = 1$ along a suitable path in the complex τ-plane. We determine f^* by evaluating the objective function $f(x, y, z)$ at all the real solutions for $\tau = 1$.

Homotopy methods are frequently set up so that the system at $\tau = 0$ breaks up into several systems, each of which consists of binomials. If the input polynomials are sparse, then these are the *polyhedral homotopies* which take the Newton polytopes of the given equations into consideration. In the sparse case, the number μ will be the *mixed volume* of the Newton polytopes. For an introduction to these polyhedral methods see [**CLO98**, Chapter 7] and the references given there.

2.4. How large is the Bézout number ? A common feature of all three algebraic algorithms in this section is that their running time is controlled by the number μ of complex critical points. In the eigenvalue method we must perform linear algebra on matrices of size $\mu \times \mu$, in the discriminant method we must find and solve a univariate polynomial of degree μ, and in the homotopy method, we are forced to trace μ paths from $\tau = 0$ to $\tau = 1$. Each of these three methods becomes infeasible if the number μ is too big; for instance, $\mu \geq 10{,}000$ might be too big.

Suppose that the given polynomial f in $\mathbb{R}[x_1, \ldots, x_n]$ has even degree $2d$ and is dense. This will be the case in the family of examples studied in Section 5. Then μ coincides with the Bézout number $(2d-1)^n$. Some small values for the Bézout number are listed in Table 1. Most entries in this table are bigger than $10,000$. We are led to believe that the algebraic methods will be infeasible for quartics if $n \geq 8$.

Each entry in the first row of Table 1 is a one. This means we can minimize quadratic polynomial functions by solving a system of linear equations (in polynomial time). The punchline of this paper is to reduce or approximate our problem by a semidefinite programming problem, which can also be solved in polynomial time for fixed d.

3. Sums of Squares and Semidefinite Programming

We present the method introduced by N.Z. Shor ([**Sho87**], [**SS97**]), and further extended in [**Par00, Par01**] and [**Las01**], for minimizing polynomial functions. This method is a relaxation: it always produces a lower bound for the value of f^*. However, as we shall see in Section 5, this bound very frequently agrees with f^*.

$2d \setminus n$	3	5	7	9	11	13	15
2	1	1	1	1	1	1	1
4	27	243	2187	19683	177147	1594323	...
6	125	3125	78125	1953125	48828125	...	...
8	343	16807	823543	40353607	1977326743	...	...
10	729	59049	4782969	387420489	...	...	...
12	1331	161051	19487171	...	...	...	...

TABLE 1. The Bézout number $\mu = (2d-1)^n$ for the critical equations.

We may assume that the given polynomial $f(x_1, \ldots, x_n)$ has even degree $2d$. Let X denote the column vector whose entries are all the monomials in $x_1, \ldots, x_n$ of degree at most d. The length of the vector X equals the binomial coefficient

$$N = \binom{n+d}{d}.$$

Let $\mathcal{L}_f$ denote the set of all real symmetric $N \times N$-matrices A such that $f(\mathbf{x}) = X^T \cdot A \cdot X$. This is an affine subspace in the space of real symmetric $N \times N$-matrices. Assume that the constant monomial 1 is the first entry of X. Let E_{11} denote the matrix unit whose only nonzero entry is a one in the upper left corner.

LEMMA 3.1. *For any real number λ, the following two are equivalent:*

- *The polynomial $f(\mathbf{x}) - \lambda$ is a sum of squares in $\mathbb{R}[\mathbf{x}]$.*
- *There is a matrix $A \in \mathcal{L}_f$ such that $A - \lambda \cdot E_{11}$ is positive semidefinite, that is, all eigenvalues of this symmetric matrix are non-negative reals.*

PROOF. The matrix $A - \lambda \cdot E_{11}$ is positive semidefinite if and only if there exists a real Cholesky factorization $A - \lambda \cdot E_{11} = B^T \cdot B$. If this holds then

$$f(\mathbf{x}) - \lambda = X^T \cdot A \cdot X - \lambda = X^T \cdot (A - \lambda \cdot E_{11}) \cdot X = X^T \cdot B^T B \cdot X = (BX)^T \cdot (BX)$$

is a sum of squares, and every sum of squares representation arises in this way. □

We write f^{sos} for the largest real number λ for which the two equivalent conditions are satisfied. Unless $f^{sos} = -\infty$ (which may happen, see below), f^{sos} is achieved, since the cone of sum of squares is closed. We always have $f^* \geq f^{sos}$, and this inequality may be strict. An example where $f^{sos} = -\infty$ is Motzkin's polynomial:

$$m(x,y) \quad = \quad x^4y^2 + x^2y^4 - 3x^2y^2. \tag{3.1}$$

It satisfies $m(x,y) \geq -1$, but $m(x,y) - \lambda$ is not a sum of squares for *any* $\lambda \in \mathbb{R}$. We refer to [**Rez00**] for an excellent survey of the problem of representing a polynomial as a sum of squares, and the important role played by Motzkin's example.

Sums of squares are crucial for us because of the following result.

THEOREM 3.2. *Fix* $\deg(f) = 2d$ *and let the number of variables n vary. Then f^{sos} can be obtained from f by solving a semidefinite program with a polynomial (in n) number of variables and contraints. The analogous statement holds if n is fixed and d varies.*

Semidefinite programming (SDP) is the study of optimization problems over the cone of all positive semidefinite matrices. This branch of optimization has received a lot of attention in recent years, both for its theoretical elegance and its practical

$2d \setminus n$	3	5	7	9	11	13	15
2	4	6	8	10	12	14	16
4	10	21	36	55	78	105	136
6	20	56	120	220	364	560	816
8	35	126	330	715	1365	2380	3876
10	56	252	792	2002	4368	8568	15504
12	84	462	1716	5005	12376	27132	54264

TABLE 2. The matrix size $N = \binom{n+d}{d}$ for the semidefinite programs.

applications. Semidefinite programs can be solved to an arbitrary tolerance ϵ in *polynomial time* (in the problem size and $1/\epsilon$), using interior point methods; see [**NN94**], [**WSV00**], [**VB96**], though we note that the exact complexity status of SDP remains open, for both the Turing and the real model of computation.

This complexity result (together with Lemma 3.1) implies Theorem 3.2 because the quantity $N = \binom{n+d}{d} = \binom{n+d}{n}$ grows polynomially if either n or d is fixed. This result appears in [**Par00**].

Available implementations of interior-point methods for semidefinite programming perform extremely well in practice, say, for problems involving matrices up to 500 rows and columns (provided there are not too many variables). This allows for the efficient computation of f^{sos}, and as we shall see in Section 4, SDP duality furnishes an efficient test to check whether $f^* = f^{sos}$ and for computing the optimal point p^* in the affirmative case. A comparison of Tables 1 and 2 suggests that SDP has the potential to compute much larger instances than algebraic methods. Section 5 will show that this is indeed the case.

Our example (2.2) has parameters $d = 2, n = 3$. The affine space $\mathcal{L}_f$ consists of all 10×10-matrices $A(\lambda, \mathbf{c})$ with $\lambda = 0$ and $c_i \in \mathbb{R}$ arbitrary in the family

$$\begin{bmatrix} -\lambda & 1/2 & 1/2 & 1/2 & c_1 & -c_2 & c_3 & -c_4 & -c_5 & c_6 \\ 1/2 & -2\,c_1 & c_2 & c_4 & 0 & c_7 & -c_8 & -c_9 & c_{10} & c_{12} \\ 1/2 & c_2 & -2\,c_3 & c_5 & -c_7 & c_8 & 0 & c_{13} & c_{14} & c_{15} \\ 1/2 & c_4 & c_5 & -2\,c_6 & c_9 & -c_{10}-c_{13}-2 & -c_{14} & -c_{12} & -c_{15} & 0 \\ c_1 & 0 & -c_7 & c_9 & 1 & 0 & c_{16} & 0 & c_{17} & c_{18} \\ -c_2 & c_7 & c_8 & -c_{10}-c_{13}-2 & 0 & -2\,c_{16} & 0 & -c_{17} & -c_{19} & -c_{11} \\ c_3 & -c_8 & 0 & -c_{14} & c_{16} & 0 & 1 & c_{19} & 0 & c_{20} \\ -c_4 & -c_9 & c_{13} & -c_{12} & 0 & -c_{17} & c_{19} & -2\,c_{18} & c_{11} & 0 \\ -c_5 & c_{10} & c_{14} & -c_{15} & c_{17} & -c_{19} & 0 & c_{11} & -2\,c_{20} & 0 \\ c_6 & c_{12} & c_{15} & 0 & c_{18} & -c_{11} & c_{20} & 0 & 0 & 1 \end{bmatrix}$$

The rows and columns of this matrix are indexed by the entries of the vector

$$X \quad = \quad \begin{bmatrix} 1 & x & y & z & x^2 & xy & y^2 & xz & yz & z^2 \end{bmatrix}^T.$$

We invite the reader to check the identity

$$X^T \cdot A(\lambda, \mathbf{c}) \cdot X \quad = \quad f(x,y,z) - \lambda \qquad \text{for all } c_1, \ldots, c_{20} \in \mathbb{R}.$$

The lower bound f^{sos} is the largest real number λ such that, for some choice of $c_1, \ldots, c_{20} \in \mathbb{R}$, the matrix $A(\mathbf{c}, \lambda)$ has all eigenvalues nonnegative. We find that

$$f^{sos} \quad = \quad f^* \quad = \quad -2.112913882,$$

and the optimal matrix (to five digits) is given by:

$$\begin{bmatrix}
2.1129 & 0.5000 & 0.5000 & 0.5000 & -0.4678 & -0.0922 & -0.4678 & -0.0922 & -0.0922 & -0.4678 \\
0.5000 & 0.9356 & 0.0922 & 0.0922 & -0.0000 & 0.0892 & -0.0892 & 0.0892 & -0.6666 & -0.0892 \\
0.5000 & 0.0922 & 0.9357 & 0.0922 & -0.0892 & 0.0892 & 0.0000 & -0.6667 & 0.0892 & -0.0892 \\
0.5000 & 0.0922 & 0.0922 & 0.9356 & -0.0892 & -0.6666 & -0.0892 & 0.0892 & 0.0892 & 0.0000 \\
-0.4678 & -0.0000 & -0.0892 & -0.0892 & 1.0000 & -0.0000 & -0.3180 & 0.0000 & 0.0554 & -0.3181 \\
-0.0922 & 0.0892 & 0.0892 & -0.6666 & -0.0000 & 0.6360 & 0.0000 & -0.0554 & -0.0554 & 0.0554 \\
-0.4678 & -0.0892 & 0.0000 & -0.0892 & -0.3180 & 0.0000 & 1.0000 & 0.0554 & -0.0000 & -0.3180 \\
-0.0922 & 0.0892 & -0.6667 & 0.0892 & 0.0000 & -0.0554 & 0.0554 & 0.6361 & -0.0554 & 0.0000 \\
-0.0922 & -0.6666 & 0.0892 & 0.0892 & 0.0554 & -0.0554 & -0.0000 & -0.0554 & 0.6360 & 0.0000 \\
-0.4678 & -0.0892 & -0.0892 & 0.0000 & -0.3181 & 0.0554 & -0.3180 & 0.0000 & 0.0000 & 1.0000
\end{bmatrix}.$$

This matrix is positive semidefinite. By computing a factorization $B^T \cdot B$ as in the proof of Lemma 3.1, we can express $f - f^{sos}$ as a sum of squares. In the next section we show how to recover the points at which the optimal value is achieved.

Note that the number 20 of free parameters is the case "$n = 3, d = 2$" of:

REMARK 3.3. *The dimension of $\mathcal{L}_f$ equals the number of linearly independent quadratic relations among the monomials of degree $\leq d$ in n variables. It equals*

$$\textit{dim } \mathcal{L}_f \quad = \quad \frac{1}{2}\left[\binom{n+d}{d}^2 + \binom{n+d}{d}\right] - \binom{n+2d}{2d}.$$

The codimension (with respect to the space of symmetric matrices) is equal to

$$\textit{codim } \mathcal{L}_f \quad = \binom{n+2d}{2d}.$$

4. Semidefinite Programming Duality

In Section 3 we demonstrated that computing f^{sos} is equivalent to minimizing a linear functional over the intersection of the affine space $\mathcal{L}_f$ with the cone of positive semidefinite $N \times N$-matrices. In our discussion we have represented the space $\mathcal{L}_f$ by a spanning set of matrices. For numerical efficiency reasons it is usually preferable to represent $\mathcal{L}_f$ by its defining equations (unless n and d are very small).

Duality is a crucial feature of semidefinite programming. It plays an important role in designing the most efficient interior-point algorithms. In what follows we review the textbook formulation of SDP duality, in terms of matrices. Thereafter we present a reformulation in algebraic geometry language, and we then explain how to test the condition $f^{sos} = f^*$ and how to recover the optimal point p^*.

4.1. Matrix Formulation. Let $\mathcal{S}^N$ denote the real vector space of symmetric $N \times N$-matrices, with the inner product $A \bullet B := \text{trace}\,(A\,B)$, and the Löwner partial order given by $A \preceq B$ if $B - A$ is *positive semidefinite.* Recall that $A \in \mathcal{S}^N$ is positive semidefinite if $x^T A x \geq 0$, for all $x \in \mathbb{R}^N$. This condition is equivalent to nonnegativity of all eigenvalues of A, to the nonnegativity of all principal minors, and to the existence of a Cholesky factorization $A = C^T \cdot C$.

The general SDP problem ([**VB96**], [**WSV00**]) can be expressed in the form:

$$(4.1) \qquad \begin{array}{lrcl} \text{minimize} & F \bullet X & & \\ \text{subject to} & \mathcal{G}X & = & b \\ & X & \succeq & 0 \end{array}$$

where $X, F \in \mathcal{S}^N$, $b \in \mathbb{R}^M$, and $\mathcal{G} : \mathcal{S}^N \longrightarrow \mathbb{R}^M$ is a linear operator. This is usually called the *primal* form, in analogy with the linear programming (LP) case.

Notice that (4.1) is a *convex* optimization problem, since the objective function is linear and the feasible set is convex. There is an associated *dual* problem:

$$\text{(4.2)} \qquad \begin{array}{ll} \text{maximize} & b^T y \\ \text{subject to} & F - \mathcal{G}^* y \succeq 0 \end{array}$$

where $y \in \mathbb{R}^M$ and $\mathcal{G}^* : \mathbb{R}^M \longrightarrow \mathcal{S}^N$ is the operator adjoint to $\mathcal{G}$. Any feasible solution of the dual problem is a lower bound of the optimal value of the primal:

$$F \bullet X - b^T y \;=\; F \bullet X - y^T \mathcal{G} X \;=\; (F - \mathcal{G}^* y) \bullet X \geq 0.$$

The last inequality holds since the inner product of two positive semidefinite matrices is nonnegative. The converse statement (primal feasible solutions give upper bounds on the optimal dual value) is obviously also true. The inequality above is called *weak duality*. Under certain conditions (notably, the existence of strictly feasible solutions), *strong duality* also holds: the optimal values of the primal and the dual problems coincide. If strong duality holds, then at optimality the matrix $X \cdot (F - \mathcal{G}^* y)$ is zero, since $A, B \succeq 0, \operatorname{trace}(AB) = 0$ implies $AB = 0$. This can be interpreted as a generalization of the usual *complementary slackness* LP conditions.

Practical implementations of SDP (we will use SeDuMi [**Stu99**]) simultaneously compute both the optimal matrix X for (4.1) and the optimal vector y for (4.2).

4.2. Polynomial Formulation. We set $m = N - 1$ and we identify $\mathcal{S}^N$ with the real vector space $\mathbb{R}[x]_2 = \mathbb{R}[x_0, x_1, \ldots, x_m]_2$ of quadratic forms in $m+1$ variables. The vector space dual to $\mathcal{S}^N$ is now denoted $\mathbb{R}[\partial]_2 = \mathbb{R}[\partial_0, \partial_1, \ldots, \partial_m]_2$. The dual pairing is given by differentiation and is denoted $\bullet$. For any $f \in \mathbb{R}[\partial]_2$ and any real vector $p = (p_0, \ldots, p_m) \in \mathbb{R}^{m+1}$, the following familiar identity holds:

$$\text{(4.3)} \qquad f(\partial_0, \ldots, \partial_m) \bullet \frac{1}{2}\Big(\sum_{i=0}^{m} p_i x_i\Big)^2 \;=\; f(p_0, \ldots, p_m).$$

We consider the general quadratic programming problem:

$$\text{(4.4)} \qquad \text{Minimize} \quad f(p) \quad \text{subject to} \quad g_0(p) = 1 \text{ and } g_1(p) = \cdots = g_r(p) = 0,$$

where $f, g_0, \ldots, g_r \in \mathbb{R}[\partial]_2$ are given and we are looking for an optimal point $p \in \mathbb{R}^{m+1}$. This problem can be relaxed to the following **primal SDP:**

$$\begin{array}{c} \text{Minimize} \quad f(\partial) \bullet q(x) \quad \text{subject to} \quad q(x) \succeq 0 \\ \text{and} \quad g_0(\partial) \bullet q(x) = 1 \text{ and } g_1(\partial) \bullet q(x) = \cdots = g_r(\partial) \bullet q(x) = 0. \end{array}$$

The inequality $q(x) \succeq 0$ means that q is non-negative on $\mathbb{R}^{m+1}$, i.e., q is in the positive semidefinite cone in $\mathbb{R}[x]_2$. In view of (4.3), the optimal value of (4.4) is greater than or equal to the optimal value of the primal SDP, and equality holds if and only if there is an optimal solution of the form $q(x) = \frac{1}{2}(\sum_{i=0}^m p_i x_i)^2$.

Every semidefinite programming problem comes with a dual problem, as in the previous subsection; see also [**VB96**]. In our case the **dual SDP** takes the form:

$$\begin{array}{c} \text{Maximize the first coordinate } \; \lambda \; \text{ of the vector } \; (\lambda, \mu_1, \ldots, \mu_r) \in \mathbb{R}^{r+1} \\ \text{subject to the conditions} \quad f(\partial) + \sum_{i=1}^r \mu_i \cdot g_i(\partial) - \lambda \cdot g_0(\partial) \succeq 0 \end{array}$$

Assuming the existence of a strictly feasible solution, the maximum value in the **dual SDP** is always equal to the minimum value in the **primal SDP**.

PROPOSITION 4.1. *If the primal SDP has an optimal solution of the form* $q(x) = \frac{1}{2}(\sum_{i=0}^{m} p_i x_i)^2$ *then the vector* $(p_0, \ldots, p_m)$ *is an optimal solution for (4.4).*

4.3. Minimizing Quadratic Functions over Toric Varieties. A *toric variety* is an algebraic variety, in affine space or projective space, which has a parametric representation by monomials. Equivalently, a toric variety is an irreducible variety which is cut out by *binomial equations*, that is, differences of monomials. Here we will be interested in those projective toric varieties which are defined by quadratic binomials. This class includes many examples from classical algebraic geometry, such as Veronese and Segre varieties. See [**Stu95**] for an introduction.

Let X be a toric variety in projective m-space whose defining prime ideal is generated by quadratic binomials $g_1, \ldots, g_r$ in $\mathbb{R}[\partial_0, \ldots, \partial_m]$. Each generator has the form $\partial_i \partial_j - \partial_k \partial_l$ for some $i, j, k, l \in \{0, 1, \ldots, m\}$. We set $g_0(\partial) = \partial_0^2$. Then the equation $g_0(\partial) = 1$ on X defines an affine toric variety $\tilde{X}$, such that X is the projective closure of $\tilde{X}$. Every quadratic polynomial function on the affine variety $\tilde{X}$ is represented by a quadratic form $f \in \mathbb{R}[\partial]_2$ as above. This representation is unique modulo the $\mathbb{R}$-linear span of $g_1, \ldots, g_r$. Our problem (4.4) is hence equivalent to minimizing a quadratic function over an affine toric variety defined by quadrics:

$$\text{Minimize} \quad f(p) \quad \text{subject to} \quad p \in \tilde{X} \tag{4.5}$$

The optimal value of the **dual SDP** relaxation in Subsection 4.2 is the largest real number λ such that $f - \lambda$ is a sum of squares in the coordinate ring of $\tilde{X}$.

Let us now return to our original problem (1.1) where the given polynomial is dense of degree $2d$ in n variables. Here X is the *Veronese variety* in projective N-dimensional space which is parameterized by all monomials of degree at most d. (If the polynomial in (1.1) is sparse then another toric variety can be used.) Writing our given polynomial as a quadratic form in homogeneous coordinates on X, our minimization problem (1.1) is precisely the quadratic toric problem (4.5).

Our approach to (4.5) consists of simultaneously solving the primal and dual SDP relaxations in Subsection 4.2. If the optimal value λ of the dual SDP agrees with the true minimum of f over $\tilde{X}$ then the primal SDP will have an optimal solution $q(x) = \frac{1}{2}(\sum_{i=0}^{m} p_i x_i)^2$ which exhibits an optimal point $(p_0, \ldots, p_m) \in X$ at which f is minimized.

In our running example, we have $m = 9$ and $r = 20$, and X is the quadratic Veronese three-fold in projective 9-space which is given parametrically as

$$(x_0 : x_1 : \cdots : x_9) \quad = \quad \big(1 : r : s : t : r^2 : rs : s^2 : rt : st : t^2\big).$$

It is cut out by twenty quadratic binomials such as $x_0 x_5 - x_1 x_2$. These binomials correspond to the parameters c_i in the 10×10-matrix $A(\lambda, \mathbf{c})$ in Section 2.

5. Experimental Results

We now present our computational experience with Shor's relaxation for global minimization of polynomial functions. As mentioned earlier, the computational advantages of our method are based on the following three independent facts:

- The dimension N of the matrix required in the sum of squares formulation is much smaller than the Bézout number μ, since it only scales polynomially with the number of variables. See Tables 1 and 2 above.

- Semidefinite programming provides an efficient algorithm for deciding whether a polynomial is a sum of squares, and to find such representations for polynomials whose coefficients may depend linearly on parameters.
- The lower bound f^{sos} very often coincides with the exact solution f^* of our problem (1.1), at least for the class of problems analyzed here.

The experimental results in this section strongly support the validity of these facts.

5.1. The test problems. For our computations, we fix a positive integer K, and we sample from the following family of polynomials of degree $2d$ in n variables:

$$f(x_1,\ldots,x_n) \quad = \quad x_1^{2d}+x_2^{2d}+\cdots+x_n^{2d} \;+\; g(x_1,\ldots,x_n) \tag{5.1}$$

where $g \in \mathbb{Z}[x_1,\ldots,x_n]$ is a random polynomial of total degree $\leq 2d-1$ whose $\binom{n+2d-1}{n}$ coefficients are independently and uniformly distributed among integers between $-K$ and K. Thus our family depends on three parameters: n, d and K.

This family has been selected to ensure three important properties:

Boundedness: The highest order terms x_i^{2d} ensure that f is bounded below, and that the minimum value f^* is achieved at some point $p^* \in \mathbb{R}^n$.

Efficient basis computation: When solving polynomial systems, the calculation of a Gröbner basis is a time-consuming task. The structure of the polynomial (5.1) allows us to bypass this expensive step, since the set of n scaled partial derivatives $x_i^{2d-1}+\frac{1}{2d}\cdot\partial g/\partial x_i$ is already a Gröbner basis with respect to total degree; cf. [**CLO97**, §2.9, Proposition 4].

Simplicity: A main reason for this choice of model is its simplicity. While more sophisticated choices have other desirable mathematical properties (such as invariance under certain transformations), we preferred to analyze here, as a first step, a relatively easy to describe set of instances.

An important question is whether the structure of the ensemble of polynomials (5.1) is somehow "biased" towards the application of sum of squares methods. This is a relevant issue, since the performance of algorithms on "random instances" sometimes provides more information on the problem family, rather than on the algorithm itself. Concerning this question, we limit ourselves to notice that, for K sufficiently large, the family (5.1) does indeed include polynomials f for which $f^{sos} < f^*$. A simple example is $f(x,y) = x^8+y^8+2700\,m(x,y)$, where $m(x,y)$ is the Motzkin polynomial (3.1).

The polynomials in our family have global minima that generally have large negative values, of the order of $-K^{2d}$. This leads to ill-conditioning of the symmetric matrices described in Lemma 3.1, and hence to numerical problems for the interior-point algorithm. Our remedy is a simple homogeneous scaling of the form

$$f_s(x_1,\ldots,x_n) \quad = \quad \alpha^{-2d}\cdot f(\alpha x_1,\ldots,\alpha x_n), \qquad \text{for some } \alpha > 0.$$

Obviously, this does not affect the properties of being a sum of squares, or whether $f^* = f^{sos}$. However, as is generally the rule in numerical optimization, this scaling step greatly affects both the speed and the accuracy of the SDP solution.

5.2. Algorithms and software. Most of the test examples were run on a Pentium III 733Mhz with 256 MB, running Linux version 2.2.16-3, and using MATLAB version 5.3. Because of physical memory limitations, our largest examples (quartics in fifteen variables), were run on a Pentium III 650Mhz with 320 MB, under Windows 2000. The semidefinite programs were solved using the SDP solver

$2d \setminus n$	3	5	7	9	11	13	15
4	0.67	28.9	526	-	-	-	-
6	12.3	2643	-	-	-	-	-
8	70.6	-	-	-	-	-	-
10	508	-	-	-	-	-	-

TABLE 3. Running time (in seconds) for the homotopy method.

SeDuMi [**Stu99**], written by Jos Sturm. It is currently one of the most efficient codes available, at least for the restricted class of problems relevant here. SeDuMi can be run from within MATLAB, and implements a self-dual embedding technique. The default parameters are used, and the solutions computed are typically exact to machine precision (SeDuMi provides an estimate of the quality of the solution).

The MATLAB Optimization toolbox was used for the implementation of a local search approach, to be described in Section 5.3. For the numerical homotopy method, we used the software PHCpack [**Ver99**], written by Jan Verschelde. The computation of the sparse matrix T_f was done using Macaulay 2 [**GS**], and its eigenvalues were numerically computed using MATLAB.

We do not make strong claims about the efficiency of our implementations: while reasonably fast, for large scale problems considerable speedups are possible at the expense of customized algorithms. Nevertheless, we believe that the issues raised regarding the applicability of algebra-based techniques to problems with large Bézout number remain valid, independently of the particular software employed.

5.3. Standard local optimization. An alternative approach to the problem is given by traditional (nonconvex) numerical optimization. There exist many variations, but arguably the most successful methods for relatively small problems such as the present ones are based on local gradient and Hessian information. Typical algorithms in this class employ an iterative scheme, combining the Newton search direction in combination with a line search [**NW99**]. These methods are reasonably fast in converging to a *local* minimum. For the larger problems in our family, they usually converge to a stationary point within 10 seconds. However, they often end up in the wrong solution, unless a very accurate starting point is given.

The drawbacks of local optimization methods are well-known: lacking convexity, there are no guarantees of global (or even local) optimality. Worse, even if in the course of the optimization we actually reach the global minimum, there is usually no computationally feasible way of verifying optimality.

Nevertheless, local optimization is an important tool for polynomial problems, as is the use of homotopy methods to trace the optimal value under small changes in the input data. It would interesting to investigate how these local numerical techniques can be best combined with the computations to be described next.

5.4. Experimental results using computational algebra. In Table 3 we present typical running times for the homotopy based approach, described in Section 2.3. These were obtained running PHCpack in "black-box" mode (`phc -b`), that requires no user-specified parameters. The software traces all solutions (not necessarily real), its number being equal to the Bézout number. Comparing with Table 1, we can notice the adverse effect of large Bézout numbers in the practical performance of the algorithm, in spite of Verschelde's impressive implementation.

$2d \setminus n$	3	5	7	9	11	13	15
4	2000	2000	2000	200	20	20	2
6	2000	200	20	-	-	-	-
8	2000	20	-	-	-	-	-
10	2000	-	-	-	-	-	-

TABLE 4. Number of random instances in each category ($K = 100, 1000, 10000$).

For the eigenvalue approach outlined in Section 2.1, we compute the matrix T_f using a straightforward implementation in Macaulay 2: the endomorphism Times_f is constructed, and applied to the elements of the monomial basis $\mathcal{B}$. The resulting matrix, in a sparse floating point representation, is sent to a file for further processing. We found that the construction of the matrix T_f takes a surprisingly long time. For instance, it took Macaulay 2 over 10 minutes to produce the 125×125-matrix for $2d = 6$, $n = 3$. The eigenvalue problem itself is solved using MATLAB; it exploits the sparsity of the matrix, and runs reasonably fast. However, it appears that even a more efficient implementation of this method will not be able to compete with the timings in Table 3, let alone the timings in Table 5.

After several discouraging attempts for small examples, we did not pursue a full implementation for the resultant-based methods sketched in Section 2.2.

5.5. Experimental results using semidefinite programming. We ran several instances of polynomials in the family described above, for values of K equal to 100, 1000, and 10000. In Table 5 the typical running times for the semidefinite programming based approach on a single instance are presented. These are fairly constant across instances, and no special structure is exploited (besides what SeDuMi does internally).

The number of random instances for each combination of the parameters is shown in Table 4. These values were chosen in order to keep the total computation time for a given category in the order of a few hours.

Regarding the accuracy of the relaxation, in *all cases tested* the condition $f^{sos} = f^*$ was satisfied. As explained in the previous section, this can be numerically verified by checking whether the solution of the corresponding SDP has rank one, from which a candidate global minimizer is obtained. Evaluating the polynomial at this point provides an upper bound on the optimal value, that can be compared with the lower bound f^{sos}. In all our instances, the difference between these two quantities was extremely small, and within the range of numerical error.

As an additional check, when we used different methods for solving the same instance, we have verified the solutions against each other. As expected, the solutions were numerically close, in many cases up to machine precision.

In particular, it is noted that the approach can handle in a reasonable time (less than 35 minutes) the case of a quartic polynomial in thirteen variables. Our largest examples have the same degree ($2d = 4$) and fifteen variables, correspond to an SDP with a matrix of dimensions 136×136 with 3876 auxiliary variables, and can be solved in a few hours. A quick glance at the corresponding Bézout number in Table 1 makes clear the advantages of the presented approach.

$2d \setminus n$	3	5	7	9	11	13	15
4	0.2	0.5	4.4	52	361	1994	27400*
6	0.3	21.2	1046	-	-	-	-
8	1.2	669	-	-	-	-	-
10	6.6	-	-	-	-	-	-

TABLE 5. Running time (in seconds) for the semidefinite programs. The instance with an asterisk was solved on different hardware than the rest.

6. What Next ?

We have demonstrated that the sums of squares relaxation is a powerful and practical technique in polynomial optimization. There are many open questions, both algorithmic and mathematical, which are raised by our experimental results. One obvious question is how often does it occur that $f^{sos} = f^*$? This can be studied for our simple model (5.1), or, perhaps better, for various natural probability measures on the space of polynomials of bounded degree. This question is closely related to understanding the inclusion of the convex cone of forms that are sums of squares inside the cone of positive semidefinite forms. For the three-dimensional family of symmetric sextics, this problem was studied in detail by Choi, Lam and Reznick [**CLR87**]. Their work is an inspiration, but it also provides a warning as to how difficult the general case will be, even for ternary sextics without symmetry.

We hope to pursue some of the following directions of inquiry in the near future.

6.1. Sparseness and Symmetry. Most polynomial systems one encounters are sparse in the sense that there are only few monomials with nonzero coefficients. Methods involving Newton polytopes, such as sparse resultants [**GKZ94**] and polyhedral homotopies [**EV99**], are designed to deal with such problems. Symmetry with respect to finite matrix groups is another feature of many polynomial problems arising in practise. The book of Gatermann [**Gat00**] is an excellent first reference.

We wish to adapt our semidefinite programming approach to input polynomials f which are sparse or symmetric or both. For instance, our polynomial example (2.2) is both sparse and invariant under permutation of the variables x, y, z. Both Newton polytope techniques and representation theory can be used to reduce the size of the matrices and the number of free parameters in the semi-definite programs.

6.2. Higher degree relaxations. If we are unlucky, then the output produced by SeDuMi will not satisfy the hypothesis of Proposition 4.1, and we conclude that the bound f^{sos} is probably strictly smaller than the optimal solution f^*. In that event we redo our computation in higher degree, now with a larger SDP. The key idea is that even though $f(x) - \lambda$ may not be a sum of squares, if there exists a positive polynomial $g(x)$ such that $g(x) \cdot (f(x) - \lambda)$ is a sum of squares, then $\lambda \leq f^*$. The choice of g can be either made *a priori* (for instance, $g = \sum_{i=1}^{n} x_i^{2k}$), or as a result of an optimization step (see [**Par00**] for details). The Positivstellensatz (see Section 7) ensures that f^* will be found if the degree of g is large enough.

6.3. Solving polynomial equations. A natural application of Shor's relaxation, hinted at in [**Sho87**], is solving polynomial systems $g_1(x) = \cdots = g_r(x) = 0$. The polynomial $f(x) := \sum_{i=1}^{r} g_i^2(x)$ satisfies $f^* \geq f^{sos} \geq 0$, and $f^* = 0$ holds if and only if the system has a real root. Clearly, $f^{sos} > 0$ is a sufficient condition for the nonexistence of real roots. An important open problem, essentially raised in [**Sho87**], is to characterize inconsistent systems $\{g_1, \ldots, g_r\}$ with $f^{sos} = 0$. On the other hand, if $f^* = f^{sos} = 0$ holds then it is possible, at least in principle, to obtain a numerical approximation of real roots using SDP. However, for a robust implementation, perturbation arguments are required and some important numerical issues arise, so the perspectives for practical applications are still unclear.

6.4. Minimizing polynomials over polytopes. Consider a compact set

$$P \quad = \quad \{ x \in \mathbb{R}^n \;:\; \ell_1(x) \geq 0, \ldots, \ell_s(x) \geq 0 \},$$

where ℓ_i is a linear form plus a constant, say, P is a polytope with s facets. Handelman's Theorem [**Han88**] states that every polynomial which is strictly positive on P can be expressed as a positive linear combination of products $\ell_1(x)^{i_1} \cdots \ell_s(x)^{i_s}$. Suppose we wish to minimize a given polynomial function $f(x)$ over P. For $D \in \mathbb{N}$ we define the *D-th Handelman bound* $f^{(D)}$ to be the largest $\lambda \in \mathbb{R}$ such that

$$f(x) - \lambda \quad = \sum_{i_1+\cdots+i_s \leq D} c_{i_1\cdots i_s} \cdot \ell_1(x)^{i_1} \cdots \ell_s(x)^{i_s} \qquad \text{for some } c_{i_1\cdots i_s} \geq 0.$$

Handelman's Theorem states that the increasing sequence $f^{(D)}, f^{(D+1)}, f^{(D+2)}, \ldots$ converges to the minimum of f over P. Each bound $f^{(D)}$ can be computed using **linear programming only**. It would be interesting to study the quality of these bounds, and the running time of these linear programs, and to see how things improve as we augment the approach with semidefinite programming techniques.

6.5. Which semialgebraic sets are semidefinite ? The feasible set of an SDP can be expressed by a linear matrix inequality, as in the dual formulation in Section 4.2. It would be interesting to study these feasible sets using techniques from real algebraic geometry, and to identify characteristic features of these sets. Here is a very concrete problem whose solution, to the best of our knowledge, is not known. Fix three real symmetric matrices A, B and C of size $N \times N$. Then

$$S \quad = \quad \{ \, (x, y) \in \mathbb{R}^2 \;:\; xA + yB + C \text{ is positive semidefinite} \, \}$$

is a closed, convex, semialgebraic subset of the plane $\mathbb{R}^2$. The problem is to find a good characterization of those subsets S. Given a semialgebraic subset $S \subset \mathbb{R}^2$ which is closed and convex, how to decide whether a "semidefinite representation" exists, and, in the affirmative case, how to find matrices A, B, C of minimum size.

7. Numerical Real Algebraic Geometry and the Positivstellensatz

The first part of the above title refers to a paper by Sommese and Wampler [**SW96**]. This paper and other more recent ones suggest that numerical algorithms will play an increasingly important role in computational (complex) algebraic geometry, and that polynomial systems will become much more visible in the context of *Scientific Computation*. Along the same lines, the fastest software for computing Gröbner bases, due to Faugère [**Fau99**], no longer uses the Buchberger algorithm but replaces it by sophisticated numerical linear algebra. Faugère's scheme

Gröbner Bases + *Numerical Linear Algebra* $\longrightarrow$ *Polynomial Problems over* $\mathbb{C}$

has the potential of entering the standard repertoire of Scientific Computation.

Following [**Par00**] we propose an analogous scheme for the field of real numbers:

Positivstellensatz + *Semidefinite Programming* $\rightarrow$ *Polynomial Problems over* $\mathbb{R}$.

In what follows, we shall explain this relationship and why we see the Positivstellensatz as the main catalyst for a future role of real algebra in scientific computation.

The Positivstellensatz [**BCR98**] is a common generalization of Linear Programming Duality (for linear inequalities) and Hilbert's Nullstellensatz (for an algebraically closed field). It states that, for a system of polynomial equations and inequalities, either there exists a solution in $\mathbb{R}^n$, or there exists a certain polynomial identity which bears *witness* to the fact that no solution exists. For instance, a single polynomial inequality $f(x) < 0$ either has a solution $x \in \mathbb{R}^n$, or there exists an identity $g(x)f(x) = h(x)$ where g and h are sums of squares. See [**BS89**] for an exposition of the Positivstellensatz from the perspective of computational geometry. Finding a witness by linear programming is proposed in [**BS89**, §7.3].

Here is our punchline, first stated in the dissertation of the first author [**Par00, Par01**]: *A Positivstellensatz witness of bounded degree can be computed by semidefinite programming. Here we can also optimize linear parameters in the coefficients.* This suggests the following algorithm for deciding a system of polynomial equations and inequalities: decide whether there exists a witness for infeasibility of degree $\leq D$, for some $D \gg 0$. If our system is feasible, then we might like to minimize a polynomial $f(x)$ over the solution set. The *D-th SDP relaxation* would be to ask for the largest real number λ such that the given system together with the inequality $f(x) - \lambda < 0$ has an infeasibility witness of degree D. This generalizes what was proposed in Sections 6.2, 6.3 and 6.4.

It is possible, at least in principle, to use an a priori bound for the degree D in the Positivstellensatz, however, the currently known bounds are still very large. Lombardi and Roy recently announced a bound which is triply-exponential in the number n of variables. We hope that such bounds can be further improved, at least for some natural families of polynomial problems arising in optimization.

Here is a very simple example in the plane to illustrate our method:

$$f := x - y^2 + 3 \geq 0, \qquad g := y + x^2 + 2 = 0. \tag{7.1}$$

By the Positivstellensatz, the system $\{f \geq 0,\, g = 0\}$ has no solution $(x, y) \in \mathbb{R}^2$ if and only if there exist polynomials $s_1, s_2, s_3 \in \mathbb{R}[x, y]$ that satisfy the following:

$$s_1 + s_2 \cdot f + 1 + s_3 \cdot g \equiv 0, \quad \text{where } s_1 \text{ and } s_2 \text{ are sums of squares.} \tag{7.2}$$

The D-th SDP relaxation of the polynomial problem $\{f \geq 0,\, g = 0\}$ asks whether there exists a solution (s_1, s_2, s_3) to (7.2) where the polynomial s_1 has degree $\leq D$ and the polynomials s_2, s_3 have degree $\leq D - 2$. For each fixed integer $D > 0$ this can be tested by semidefinite programming. For $D = 2$ we find the solution

$$s_1 = \tfrac{1}{3} + 2\left(y + \tfrac{3}{2}\right)^2 + 6\left(x - \tfrac{1}{6}\right)^2, \qquad s_2 = 2, \qquad s_3 = -6.$$

The resulting identity (7.2) proves the inconsistency of the system $\{f \geq 0,\, g = 0\}$.

Acknowledgements. We thank the feedback and suggestions of Eberhard Becker, Alicia Dickenstein and Jean-Charles Faugère, as well as the two anonymous reviewers for their useful and constructive comments.

References

[BCR98] J. Bochnak, M. Coste, and M-F. Roy, *Real Algebraic Geometry*, Springer, 1998.

[BS89] J. Bokowski and B. Sturmfels, *Computational Synthetic Geometry*, Lecture Notes in Mathematics, vol. 1355, Springer-Verlag, 1989.

[CLO97] D. A. Cox, J. B. Little, and D. O'Shea, *Ideals, Varieties and Algorithms*, Undergraduate Texts in Mathematics, Springer-Verlag, 1997.

[CLO98] ______, *Using Algebraic Geometry*, Graduate Texts in Mathematics, vol. 185, Springer-Verlag, 1998.

[CLR87] M.D. Choi, T.Y. Lam, and B. Reznick, *Even symmetric sextics*, Mathematische Zeitschrift **195** (1987), no. 4, 559–580.

[EV99] I. Z. Emiris and J. Verschelde, *How to count efficiently all affine roots of a polynomial system*, Discrete Appl. Math. **93** (1999), no. 1, 21–32, 13th European Workshop on Computational Geometry CG'97 (Würzburg, 1997). MR **2000a:**65064

[Fau99] J.C. Faugère, *A new efficient algorithm for computing Gröbner bases* (F_4), Journal of Pure and Applied Algebra **139** (1999), 61–88.

[Gat00] K. Gatermann, *Computer Algebra Methods for Equivariant Dynamical Systems*, Lecture Notes in Mathematics, vol. 1728, Springer-Verlag, 2000.

[GKZ94] I. M. Gel'fand, M. Kapranov, and A. Zelevinsky, *Discriminants, Resultants, and Multidimensional Determinants*, Birkhäuser, 1994.

[GS] D. R. Grayson and M. E. Stillman, *Macaulay 2, a software system for research in algebraic geometry*, Available at `http://www.math.uiuc.edu/Macaulay2`.

[Han88] D. Handelman, *Representing polynomials by positive linear functions on compact convex polyhedra*, Pacific Journal of Mathematics **132** (1988), 35–62.

[Las01] J. B. Lasserre, *Global optimization with polynomials and the problem of moments*, SIAM J. Optim. **11** (2001), no. 3, 796–817. MR 1 814 045

[Li97] T. Y. Li, *Numerical solution of multivariate polynomial systems by homotopy continuation methods*, Acta Numerica **6** (1997), 399–436.

[NN94] Y. E. Nesterov and A. Nemirovski, *Interior Point Polynomial Methods in Convex Programming*, Studies in Applied Mathematics, vol. 13, SIAM, Philadelphia, PA, 1994.

[NW99] J. Nocedal and S. J. Wright, *Numerical Optimization*, Springer, 1999.

[Par00] P. A. Parrilo, *Structured Semidefinite Programs and Semialgebraic Geometry Methods in Robustness and Optimization*, Ph.D. thesis, California Institute of Technology, May 2000, Available at `http://www.cds.caltech.edu/~pablo/`.

[Par01] ______, *Semidefinite programming relaxations for semialgebraic problems*, *Mathematical Programming*, to appear. Preprint available at `http://www.aut.ee.ethz.ch/~parrilo/`, 2001.

[Rez00] B. Reznick, *Some concrete aspects of Hilbert's 17th problem*, Contemporary Mathematics, vol. 253, American Mathematical Society, 2000, pp. 251–272.

[Sho87] N. Z. Shor, *Class of global minimum bounds of polynomial functions*, Cybernetics **23** (1987), no. 6, 731–734, (Russian orig.: Kibernetika, No. 6, (1987), 9–11).

[SS97] N. Z. Shor and P. I. Stetsyuk, *The use of a modification of the r-algorithm for finding the global minimum of polynomial functions*, Cybernetics and Systems Analysis **33** (1997), 482–497.

[Stu95] B. Sturmfels, *Gröbner Bases and Convex Polytopes*, University Lectures, vol. 8, American Mathematical Society, 1995.

[Stu99] J.F. Sturm, *Using SeDuMi 1.02, a MATLAB toolbox for optimization over symmetric cones*, Optimization Methods and Software **11-12** (1999), 625–653, Version 1.03 is available at `http://fewcal.kub.nl/sturm/software/sedumi.html`.

[SW96] A.J. Sommese and C.W. Wampler, *Numerical algebraic geometry*, The Mathematics of Numerical Analysis (Park City, UT, 1995), Lectures in Applied Mathematics, vol. 32, American Mathematical Society, 1996, pp. 749–763.

[VB96] L. Vandenberghe and S. Boyd, *Semidefinite programming*, SIAM Review **38** (1996), no. 1, 49–95.

[Ver99] J. Verschelde, *Algorithm 795: PHCpack: A general-purpose solver for polynomial systems by homotopy continuation*, ACM Transactions on Mathematical Software **25** (1999), no. 2, 251–276, Software available at `http://www.math.uic.edu/~jan/`.

[WSV00] H. Wolkowicz, R. Saigal, and L. Vandenberghe (eds.), *Handbook of Semidefinite Programming*, Kluwer, 2000.

Department of Control and Dynamical Systems, California Institute of Technology, Pasadena, CA 91125-8100, USA.

Current address: Automatic Control Laboratory, Swiss Federal Institute of Technology (ETH), CH-8044 Zürich, Switzerland.

E-mail address: parrilo@aut.ee.ethz.ch

Department of Mathematics, University of California, Berkeley, CA 94720, USA.

E-mail address: bernd@math.berkeley.edu

DIMACS Series in Discrete Mathematics
and Theoretical Computer Science
Volume **60**, 2003

Patterns of Dependence among Powers of Polynomials

Bruce Reznick

ABSTRACT. Let $F = \{f_1, \dots, f_r\}$ be a family of polynomials and let the ticket of F, $T(F)$, denote the set of integers m so that $\{f_j^m\}$ is linearly dependent. We show that $|T(F)| \le \binom{r-1}{2}$ and present many concrete tickets. For example, let ζ denote a primitive fifth root of unity, and let $f_j(x,y) = \zeta^j x^2 + ixy + \zeta^{-j} y^2$, $0 \le j \le 4$. Then $T(\{f_0, \dots, f_4, xy\}) = \{1, 2, 3, 4, 8, 14\}$.

1. Introduction and definitions

The motivation for this paper is a remarkable example due to A. H. Desboves in 1880 (see [**3**, p.684]) and, independently, N. Elkies (see [**2**, p.542]) in 1995.

EXAMPLE 1. Let

$$f_1(x,y) = x^2 + \sqrt{2}xy - y^2, \quad f_2(x,y) = ix^2 - \sqrt{2}xy + iy^2,$$
$$f_3(x,y) = -x^2 + \sqrt{2}xy + y^2, \quad f_4(x,y) = -ix^2 - \sqrt{2}xy - iy^2. \tag{1.1}$$

Then $\sum_j f_j^5 = 0$. It is not hard to show that $\sum_j f_j = \sum_j f_j^2 = 0$ as well; however, $\{f_j^m\}$ is linearly independent for other integers m. This is surprising, and not only because of the gap in the exponents. Let $M_m(t_1, t_2, t_3, t_4) = \sum_j t_j^m$. Then (f_1, f_2, f_3, f_4) lies in the intersection $\{M_1 = 0\} \cap \{M_2 = 0\} \cap \{M_5 = 0\} \subset \mathbb{C}^4$. This is not what one would expect geometrically. However, Newton's Theorem for symmetric forms implies that M_5 is in the ideal (M_1, M_2), so $\{M_5 = 0\}$ is contained in $\{M_1 = 0\} \cap \{M_2 = 0\}$. We shall return to this example throughout the paper from varying points of view, and prove the assertions of this paragraph.

Let $F = \{f_j\}$ be a finite set of polynomials. The *ticket* of F, $T(F)$, is defined by

$$T(F) = \{m \in \mathbb{N} : \{f_j^m\} \text{ is linearly dependent}\}.$$

EXAMPLE 2. Here are some families with tickets that are easy to verify by hand.

$$T(\{x, y, z\}) = \emptyset,$$
$$T(\{x, y, x+y\}) = \{1\}, \tag{1.2}$$
$$T(\{x^2 - y^2, 2xy, x^2 + y^2\}) = \{2\}.$$

1991 *Mathematics Subject Classification.* Primary 11D41, 11E76, 14Q15, 32H25; Secondary 11P05, 15A99, 30D35.

We let $\overline{m}(F)$ denote the maximum element of $T(F)$, with the understanding that $\overline{m}(F) = 0$ if $T(F) = \emptyset$. More specifically, for integers $r \geq 2$, $n \geq 2$, $d \geq 1$, let $\mathcal{F}(r, n, d)$ denote the set of families $F = \{f_1, \ldots, f_r\}$, where each f_j is a homogeneous polynomial over $\mathbb{C}$ in n variables of degree d, and no two f_j's are proportional. Let

$$\mathcal{T}(r, n, d) = \{T(F) : F \in \mathcal{F}(r, n, d)\}.$$

Let $\overline{m}_r$ denote the maximum of $\overline{m}(F)$, given $|F| = r$. It is known that $\overline{m}_r \leq r^2 - 2r$. We shall say that $F \in \mathcal{F}(r, n, d)$ is a *dysfunctional* family if $|T(F)| > r - 2$. (The term "dysfunctional" refers to excessive dependence, as in Example 1.)

One of the principal goals of this research is to determine $\mathcal{T}(r, n, d)$. We are able to do so in the cases $r \leq 3$, $(n, d) = (2, 1)$, and $(r, n, d) = (4, 2, 2)$.

It is not difficult to show that any finite set of integers can appear as a ticket. Indeed, let ζ denote a primitive d-th root of unity, and let

$$F_d^* := \{x^d + y^d, x^d + \zeta y^d, x^{d-1}y, \ldots, xy^{d-1}\} \in \mathcal{F}(d+1, 2, d).$$

We show in Example 13 that $T(F_d^*) = \{d\}$. Now suppose that $A = \{m_j\}$ is given and consider $F = \cup F_{m_j}^*$, where each form in $F_{m_j}^*$ is binary in the variables (x_j, y_j). It is then easy to see that $T(F) = A$. More generally, if $T(F_1) = T_1$ and $T(F_2) = T_2$, and if F_1 and F_2 involve disjoint sets of variables, then $T(F_1 \cup F_2) = T_1 \cup T_2$. Since this seems like cheating, we shall say that a family F is *indecomposable* if there does not exist a nontrivial partition $F = F_1 \cup F_2$ so that $T(F) = T(F_1) \cup T(F_2)$.

CONJECTURE 1.1. Every finite subset of $\mathbb{N}$ appears as the ticket of some indecomposable family.

There does not seem to be much in the literature on tickets, *per se*, except for some discussion of $\overline{m}_r$.

If $r = 3$, then after scaling, $m \in T(F)$ implies that $f_1^m + f_2^m - f_3^m = 0$. In 1879, J. Liouville proved (see [**14**, p.263]) that Fermat's Last Theorem is true for non-constant polynomials; considering (1), we see that $\overline{m}_3 = 2$.

J. Molluzzo, in his 1972 thesis [**10**] (cited in [**11**]) shows that $\overline{m}_r \geq \lfloor r^2/4 \rfloor - 1$. We shall discuss Molluzzo's construction in Example 12. The bound for $\overline{m}_r$ is slightly improved in Example 9 for $r \leq 6$.

M. Green proved in 1975 [**4**, p.71] that if $\{\phi_j\}$, $1 \leq j \leq r$, are holomorphic functions in n complex variables, no two of which are proportional, and $\sum_{j=1}^{r} \phi_j^m = 0$, then $m \leq (r-1)^2 - 1$. This implies that $\overline{m}_r \leq r(r-2)$.

D. J. Newman and M. Slater [**11**] discussed Waring's Problem for polynomials; upon homogenizing their polynomials, one obtains information about $T(F)$ if one of the f_j's is the d-th power of a linear form. Their work was generalized by J. F. Voloch [**15**] to equations over function fields.

There is also a connection to Nevanlinna theory – see [**8**], [**9**]. If $m \in T(F)$ for $F \in \mathcal{F}(r, 2, d)$, then by dehomogenizing and transposing the linear relation among f_j^m, we obtain an equation $\sum_{j=1}^{r-1} \phi_j^m(z) = 1$ in rational functions ϕ_j of a single complex variable. A very recent survey of this work is [**6**]. W. Hayman [**8**] proved in 1985 that if $\sum_{j=1}^{3} \phi_j^m(z) = 1$ for rational functions ϕ_j, then $r \leq 7$; thus, $\overline{m}_4 \leq 7$. Example 1 shows that $\overline{m}_4 \geq 5$. G. Gundersen [**5**] has found meromorphic (not rational) functions g_j for which $g_1^6(z) + g_2^6(z) + g_3^6(z) = 1$.

We now collect some trivial remarks, which will be used without citation throughout the rest of the paper. If $f_j = \alpha f_i$, then f_i^m and f_j^m are dependent for every m, so $T(F)$ is uninteresting. This explains the non-proportionality condition in the definition of $\mathcal{F}(r,n,d)$.

The dimension of the vector space of forms of degree D in n variables is $\binom{n+D-1}{n-1}$. Thus, if $F \in \mathcal{F}(r,n,d)$ and if $r > \binom{n+md-1}{n-1}$, then $\{f_j^m\}$ must be dependent. In this case we say that m is *forced* to be in $T(F)$.

There is no *a priori* reason to assume that the polynomials in a given family F are all homogeneous and of the same degree, but this restriction is harmless. Suppose $F = \{f_j\}$, $1 \le j \le r$, where each f_j is a polynomial in n variables of degree $\le d$. Homogenize F by defining

$$f_j'(x_1, \dots, x_{n+1}) = x_{n+1}^d f_j\left(\frac{x_1}{x_{n+1}}, \dots, \frac{x_n}{x_{n+1}}\right), \qquad 1 \le j \le r,$$

and let $F' = \{f_j'\} \in \mathcal{F}(r, n+1, d)$. It is easy to see that $T(F) = T(F')$. Similarly, if we dehomogenize $G = \{g_j\} \in \mathcal{F}(r,n,d)$, via

$$g_j'(x_1, \dots, x_{n-1}) = g_j(x_1, \dots, x_{n-1}, 1), \qquad 1 \le j \le r,$$

and let $G' = \{g_j'\}$, then $T(G) = T(G')$.

If $F = \{f_j\}$ and

$$f_j'(x_1, \dots, x_n) = f_j\left(\sum_{k=1}^n \alpha_{1k}x_k + \beta_1, \cdots \sum_{k=1}^n \alpha_{nk}x_k + \beta_n\right), \qquad 1 \le j \le r,$$

where $\det([\alpha_{jk}]) \ne 0$, and $F' = \{f_j'\}$, then clearly $T(F) = T(F')$: the ticket of a family is not changed by a simultaneous affine invertible linear change of variables.

As a special case of this, if $F = \{f_j\}$, $f_j' = \lambda_j f_j$, where $\lambda_j \ne 0$ and $F' = \{f_j'\}$, then $T(F) = T(F')$. That is, we may normalize the members of a family without affecting its ticket. In particular, if $\sum_j c_j f_j^m = 0$, $c_j \ne 0$, then we may assume without loss of generality that $c_j = \pm 1$ at our convenience. (Of course, we can only do this for *one* such exponent $m \in T(F)$.)

Given $\{f_j\}$ in n variables and n polynomials $\psi_k(y_1, \dots, y_m)$, we can define

$$f_j'(y_1, \dots, y_m) := f_j(\psi_1(y), \dots, \psi_n(y)).$$

Then $\sum_{j=1}^r \lambda_j f_j = 0$ implies $\sum_{j=1}^r \lambda_j f_j' = 0$. It is possible that the f_j''s might not be pairwise non-proportional, but it might also be possible that a judicious choice of ϕ_k will allow for the ticket to be extended. For example, $T(\{x+y, x-y\}) = \emptyset$ and $4xy = (x+y)^2 - (x-y)^2$ is not a square, but if we replace x and y by x^2 and y^2 and add $2xy$ to the family, we put 2 into the ticket; c.f. (1).

If $0 \ne g$ is a form, then $T(\{f_j\}) = T(\{gf_j\})$. Thus, $d < d'$ implies trivially that $\mathcal{T}(r,n,d) \subseteq \mathcal{T}(r,n,d')$.

If $n < n'$, then by viewing $f_j(x_1, \dots, x_n)$ as a form in $(x_1, \dots, x_n, x_{n+1}, \dots, x_{n'})$, we see that $\mathcal{T}(r,n,d) \subseteq \mathcal{T}(r,n',d)$.

Regrettably, it is not true that $r < r'$ implies $\mathcal{T}(r,n,d) \subseteq \mathcal{T}(r',n,d)$. For example, $T(\{x,y\}) = \emptyset$, so $\emptyset \in \mathcal{T}(2,2,1)$. However, $m = 1$ is forced to be in $T(F)$ for every $F \in \mathcal{F}(3,2,1)$. We suspect that the only exceptions to the inclusion $\mathcal{T}(r,n,d) \subseteq \mathcal{T}(r',n,d)$ are consequences of such forcing.

Here is an outline of the rest of this paper. In section two, we prove our main theoretical result: if $F \in \mathcal{F}(r,n,d)$, then $|T(F)| \le \binom{r-1}{2}$. The proof gives an explicit non-zero polynomial in $(m, x_1, \dots, x_n)$ which vanishes identically when $m \in T(F)$. This is a modest improvement on the bound $|T(F)| \le r(r-2)$, which follows immediately from Green's bound on $\overline{m}_r$. In section three, we analyze $\mathcal{T}(r,n,d)$ in the "easy" cases $r \le 3$, and $(n,d) = (2,1)$ and $(r,n,d) = (4,2,2)$, and discuss $(n,d) = (\ge 3, 1)$. We also show that $\{1,2,3\} \notin \mathcal{T}(4,n,d)$. In section four we work out some concrete tickets. The most dysfunctional family we have found is $F \in \mathcal{F}(2k,2,2)$ for which $|T(F)| = 3k-3$ (and $\overline{m}(F) = 4k-4$); see Example 8. For each integer a, we give $F_a \in \mathcal{F}(a+2,2,a)$ so that $T(F_a)$ consists precisely of the divisors of a. We also give explicit $F, F' \in \mathcal{F}(6,2,2)$ with $T(F) = \{1,2,8\}$ and $T(F') = \{1,2,3,4,8,14\}$. These generalize (1.1). In fact, for every $v \ge 2$, there is a non-trivial identity equating two sums of the $(3v-1)$-st powers of v binary quadratic forms. We close the paper with generalizations, speculations, open questions and acknowledgments.

2. How long can a ticket be?

In this section we prove that $|T(F)| \le \binom{r-1}{2}$. The exposition is simplified by assuming that the f_j's are not homogeneous. We assume that $f_j = f_j(x)$, where $x = (x_1, \dots, x_n)$, although n does not play an explicit role in the proof. The proof is constructive in the sense that, for each F, we define a non-zero polynomial of degree $\binom{r-1}{2}$ in (m,x) which vanishes identically in x whenever $m \in T(F)$.

THEOREM 2.1. *Suppose $F = \{f_1, \dots, f_r\}$ is a set of polynomials over $\mathbb{C}$, no two of which are proportional, with* $\deg(f_j) \le d$. *Then $|T(F)| \le \binom{r-1}{2}$. If $d \le r-3$, then $|T(F)| < \binom{r-1}{2}$.*

PROOF. For a polynomial g and $k \ge 0$, let $(g)_k$ denote the degree k homogeneous part of g, so that $g = \sum_{k \ge 0} (g)_k$. If $\{g_1, \dots, g_r\}$ are linearly dependent polynomials, then $\sum_{j=1}^r \lambda_j g_j = 0$ implies that $\sum_{j=1}^r \lambda_j (g_j)_k = 0$ for all k. That is,

$$(2.1) \qquad \begin{pmatrix} (g_1)_0 & (g_2)_0 & \cdots & (g_r)_0 \\ (g_1)_1 & (g_2)_1 & \cdots & (g_r)_1 \\ \cdots & \cdots & \ddots & \cdots \\ (g_1)_{r-1} & (g_2)_{r-1} & \cdots & (g_r)_{r-1} \end{pmatrix} \begin{pmatrix} \lambda_1 \\ \lambda_2 \\ \vdots \\ \lambda_r \end{pmatrix} = \begin{pmatrix} 0 \\ 0 \\ \vdots \\ 0 \end{pmatrix}.$$

Therefore, as a polynomial in x,

$$(2.2) \qquad W(g_1, \dots, g_r; x) := \begin{vmatrix} (g_1)_0 & (g_2)_0 & \cdots & (g_r)_0 \\ (g_1)_1 & (g_2)_1 & \cdots & (g_r)_1 \\ \cdots & \cdots & \ddots & \cdots \\ (g_1)_{r-1} & (g_2)_{r-1} & \cdots & (g_r)_{r-1} \end{vmatrix} = 0.$$

(This familiar and elementary Wronskian argument is assuredly not a *sufficient* condition for dependence. Given (2.2), the only null-vectors satisfying (2.1) might well have non-constant components. It might be the case that $\deg(g_j) < r-1$ for all j, rendering (2.2) trivial, and it might happen that $\sum_{j=1}^r \lambda_j g_j$ has terms of degree $\ge r$. Nevertheless, the weak necessary condition (2.2) will suffice for our needs. Wronskian arguments can also be found in [**11**] and [**15**].)

We turn to F. Since f_i/f_j is not constant for $i \ne j$, there exists $k = k(i,j)$ so that

$$(2.3) \qquad \frac{\partial}{\partial x_k}\left(\frac{f_i}{f_j}\right) = \frac{1}{f_j^2}\left(f_j \frac{\partial f_i}{\partial x_k} - f_i \frac{\partial f_j}{\partial x_k}\right)$$

is not identically zero. For precision, let $k(i,j)$ be the smallest such index. Further, there exists a point $P \in \mathbb{C}^n$ which is *not* contained in the finite union

$$\left(\bigcup_{j=1}^{r}\{f_j = 0\}\right) \bigcup \left(\bigcup_{1\le i<j\le r}\left\{f_j\frac{\partial f_i}{\partial x_{k(i,j)}} - f_i\frac{\partial f_j}{\partial x_{k(i,j)}} = 0\right\}\right). \tag{2.4}$$

By translation, we may assume that $P = 0$. If we now let $\gamma_j = (f_j(0))^{-1}$, and replace f_j by $\gamma_j f_j$, then we have changed neither the definition of the varieties in (2.4) nor $T(F)$. Accordingly, we may assume without loss of generality that $f_j(0) = 1$ for all j. By evaluating (2.3) at $P = 0$, we see that $\frac{\partial f_i}{\partial x_{k(i,j)}}(0) \neq \frac{\partial f_j}{\partial x_{k(i,j)}}(0)$, hence the first-order terms in the Taylor series of the f_j's at 0 are different. To emphasize this point, we write $(f_j)_1 = L_j$.

We now compute $(f_j^m)_k$ explicitly:

$$\begin{aligned} f_j^m &= (1 + L_j + (f_j)_2 + \cdots + (f_j)_d)^m \\ &= \sum_{\substack{\ell_0+\cdots+\ell_d=m\\ \ell_0\ge 0,\ldots,\ell_d\ge 0}} \frac{m!}{\ell_0!\ldots\ell_d!}1^{\ell_0}L_j^{\ell_1}((f_j)_2)^{\ell_2}\ldots((f_j)_d)^{\ell_d}, \end{aligned} \tag{2.5}$$

and so $(f_j^m)_k$ will consist of those terms from (2.5) in which $\sum_{i=1}^d i\ell_i = k$. Write $(N)_s := \frac{N!}{(N-s)!} = N(N-1)\cdots(N-(s-1))$. Since $\ell_0 = m - \sum_{i=1}^d \ell_i$, we have

$$(f_j^m)_k = \sum_{\substack{\ell_1+\cdots+\ell_d\le m\\ \ell_1+\cdots+d\ell_d=k}} \frac{(m)_{\ell_1+\cdots+\ell_d}}{\ell_1!\ldots\ell_d!}L_j^{\ell_1}((f_j)_2)^{\ell_2}\ldots((f_j)_d)^{\ell_d}. \tag{2.6}$$

Consider (2.6) as presenting $(f_j^m)_k$ as a polynomial in m, whose coefficients are polynomials in x. ¿From this perspective, each summand has m-degree $\ell_1 + \cdots + \ell_d$. Since $\ell_1 + 2\ell_2 + \cdots + d\ell_d = k$ is fixed, the maximum value of this m-degree is k, achieved uniquely (when $\ell_1 = k$ and $\ell_i = 0$, $i \ge 2$) in the term $\frac{(m)_k}{k!}L_j^k$. Thus,

$$(f_j^m)_k - \tfrac{L_j^k}{k!}\cdot m^k + \mathcal{O}_x(m^{k-1}), \qquad 0 \le j \le r-1. \tag{2.7}$$

It is also easy to see that the term in $(f_j^m)_k$ with minimum m-degree occurs when $\ell_d = \lfloor k/d \rfloor$, and all other ℓ_i's vanish, except for $\ell_{k-d\lfloor \frac{k}{d}\rfloor} = 1$, (unless $d \mid k$). Let $u(k,d) := \lfloor \frac{k-1}{d} \rfloor$. This smallest degree is $1 + u(k,d)$ and implies that

$$m(m-1)\cdots(m-u(k,d)) \mid (f_j^m)_k, \qquad 1 \le k \le r-1.$$

(More directly, since f_j^m has degree $\le dm$, $(f_j^m)_k = 0$ if $k > md$; that is, if $m \le \frac{k-1}{d}$.) Note that, in particular, $m \mid (f_j^m)_k$ if $k \ge 1$.

We apply (2.2) with $g_j = f_j^m$. Let $W(m;x) := W(f_1^m, \ldots, f_r^m; x)$. If $m \in T(F)$, then $W(m;x) = 0$ as a polynomial in x. In view of (2.7), $W(m;x)$ equals

$$\begin{vmatrix} 1 & \cdots & 1 \\ L_1\cdot m & \cdots & L_r \cdot m \\ \frac{L_1^2}{2!}\cdot m^2 + \mathcal{O}_x(m) & \cdots & \frac{L_r^2}{2!}\cdot m^2 + \mathcal{O}_x(m) \\ \vdots & \ddots & \vdots \\ \frac{L_1^{r-1}}{(r-1)!}\cdot m^{r-1} + \mathcal{O}_x(m^{r-2}) & \cdots & \frac{L_r^{r-1}}{(r-1)!}\cdot m^{r-1} + \mathcal{O}_x(m^{r-2}) \end{vmatrix}. \tag{2.8}$$

Then $W(m;x)$ has m-degree $\le \binom{r}{2}$, and the coefficient of $m^{\binom{r}{2}}$ is the Vandermonde determinant

$$\frac{1}{2!\cdots(r-1)!}\begin{vmatrix} 1 & \cdots & 1 \\ L_1(x) & \cdots & L_r(x) \\ . & \ddots & \cdots \\ L_1^{r-1}(x) & \cdots & L_r^{r-1}(x) \end{vmatrix} = \frac{1}{2!\cdots(r-1)!}\prod_{1\le i<j\le r}(L_j(x)-L_i(x)).$$

Since the L_j's are distinct linear forms, there exists $y \in \mathbb{C}^n$ so that $L_i(y) \ne L_j(y)$ for $i \ne j$. Thus, $W(m;y)$ is a non-zero polynomial in m of exact degree $\binom{r}{2}$. Since m^{r-1} is a factor, $W(m;y)$ (and so $W(m;x)$) can have at most $\binom{r}{2}-(r-1)=\binom{r-1}{2}$ distinct factors of the shape $m-m_k$. (Recall that $T(F) \subseteq \{m_k\}$.) Moreover, the k-th row of (2.8) is divisible by $(m)_{1+u(k-1,d)}$, hence $W(m;y)$ is divisible by

$$\prod_{k=0}^{r-1}(m)_{1+u(k,d)} = \prod_{k=0}^{r-1} m(m-1)\dots(m-u(k,d)) = \prod_{i=0}^{u(r-1,d)}(m-i)^{r-id-1}.$$

In particular, if $d \le r-3$, then $m^{r-1}(m-1)^2$ divides $W(m;y)$, so $|T(F)| < \binom{r-1}{2}$. Indeed, if $F \in \mathcal{F}(r,n,d)$, then

$$|T(F)| \le \binom{r}{2} - (r-1) - \sum_{i=1}^{u(r-1,d)}(r-2-id).$$

If $d=2$, this bound works out to be $\lfloor r^2/4 \rfloor - 1$. For larger d it is $\approx \left(\frac{d-1}{2d}\right)r^2$. □

If $r=3$, then Theorem 2.1 implies that $|T(F)| \le 1$; see Theorem 3.4 below.

If $r=4$, then Theorem 2.1 implies that $|T(F)| \le 3$, and the bound is achieved in Example 1, and elsewhere. Example 5 will discuss $F \in \mathcal{T}(4,2,2)$ with $T(F) = \{1,2,4\}$. Theorem 3.7 will show that there is no family in any $\mathcal{F}(4,n,d)$ for which $T(F) = \{1,2,3\}$. (However, by Theorem 3.3, this is the ticket for any $F \in \mathcal{F}(5,2,1)$.)

Theorem 2.1 simplifies the task of finding the ticket: one may compute $W(m,x)$ and looks for factors in m. Once a finite set of candidate exponents is found, the dependence of the f_j^m's can be checked directly. We illustrate this in a special case, which will lead back to Example 1.

If $(n,d)=(1,2)$ and $r=4$, then (2.5) becomes explicitly

$$\begin{aligned}(1+at+bt^2)^m &= 1+mat+\left(\tfrac{m(m-1)}{2}a^2+mb\right)t^2 \\ &\quad + \left(\tfrac{m(m-1)(m-2)}{6}a^3+m(m-1)ab\right)t^3+\dots,\end{aligned} \tag{2.9}$$

(Note the factor of $(m)_0$ in $(f^m)_1$ and $(f^m)_2$, and the factor of $(m)_1$ in $(f^m)_3$.) After substituting (2.9) into (2.2) for $f_j(t) = 1+a_jt+b_jt^2$, $1 \le j \le 4$, and taking common factors out of the rows, we find that

$$W(m;t) = \frac{m^3(m-1)}{12}t^6\begin{vmatrix} 1 & \dots & 1 \\ a_1 & \dots & a_4 \\ (m-1)a_1^2+2b_1 & \dots & (m-1)a_4^2+2b_4 \\ (m-2)a_1^3+6a_1b_1 & \dots & (m-2)a_4^3+6a_4b_4 \end{vmatrix}. \tag{2.10}$$

Let $\Delta(m)$ denote the determinant in (2.10); if $1 \ne m_0 \in T(F)$, then $\Delta(m_0)=0$.

EXAMPLE 1 (PART TWO). We dehomogenize (1.1) by setting $(x,y) \mapsto (1,t)$ and normalizing so that $f_j(0) = 1$. This gives

$$(2.11)\qquad \begin{aligned} f_1(t) &= 1 + \sqrt{2}\,t - t^2, & f_2(t) &= 1 + i\sqrt{2}\,t + t^2,\\ f_3(t) &= 1 - \sqrt{2}\,t - t^2, & f_4(t) &= 1 - i\sqrt{2}\,t + t^2. \end{aligned}$$

Taking $a_j = i^{j-1}\sqrt{2}$ and $b_j = (-1)^j$ in (2.10), we find that $\Delta(m)$ is the determinant

$$\begin{vmatrix} 1 & 1 & 1 & 1 \\ \sqrt{2} & i\sqrt{2} & -\sqrt{2} & -i\sqrt{2} \\ 2(m-2) & -2(m-2) & 2(m-2) & -2(m-2) \\ 2\sqrt{2}(m-5) & -2i\sqrt{2}(m-5) & -2\sqrt{2}(m-5) & 2i\sqrt{2}(m-5) \end{vmatrix}.$$

A computation shows that $\Delta(m) = -128i(m-2)(m-5)$, which vanishes only for $m = 2, 5$. Remarkably, each root corresponds to a linear dependence among the $\{f_j^m\}$. If we replace $\sqrt{2}$ in (2.11) by a parameter $\mu > 0$, then the roots of $\Delta(m)$ work out to be $m = 1 + 2\mu^{-2}$, and $m = 2 + 6\mu^{-2}$. We shall show in Example 1 (Part Three) that, apart from $\mu = \sqrt{2}$, these roots correspond to a linear dependence only when $(\mu, m) = (\sqrt{6}, 3)$ and $(\sqrt{2/3}, 4)$. (See also (3.9) and (4.5).)

3. $\mathcal{T}(r, n, d)$ for small r, n, d

In this section, we describe $\mathcal{T}(r, n, d)$ in some simple cases.

3.1. Families of binary linear forms. It is very easy to see that $\mathcal{T}(r, 2, 1) = \{\{1, 2, \dots, r-2\}\}$. If $m \le r - 2$, then $r > \binom{2+m\cdot 1-1}{2-1} = m + 1$, hence $m \in T(F)$ is forced. Suppose $m = r - 1$ and write $f_j(x,y) = \alpha_j x + \beta_j y$ for $1 \le j \le r$. Then the matrix giving $\{f_j^{r-1}\}$ with respect to the basis $\{\binom{r-1}{\ell} x^{r-1-\ell} y^\ell\}$ has Vandermonde determinant

$$\begin{vmatrix} \alpha_1^{r-1} & \alpha_1^{r-2}\beta_1 & \dots & \beta_1^{r-1} \\ \alpha_2^{r-1} & \alpha_2^{r-2}\beta_2 & \dots & \beta_2^{r-1} \\ \vdots & \vdots & \ddots & \vdots \\ \alpha_r^{r-1} & \alpha_r^{r-2}\beta_r & \dots & \beta_r^{r-1} \end{vmatrix} = \prod_{i<j} (\alpha_i\beta_j - \alpha_j\beta_i) \ne 0,$$

since the f_j's are pairwise non-proportional. It follows that $r - 1 \notin T(F)$. If $m \ge r$, then take $f_j(x,y) = \alpha_j x + \beta_j y$ for $r + 1 \le j \le m + 1$ so that $F' := F \cup \{f_{r+1}, \dots, f_{m+1}\}$ consists of $m + 1$ pairwise non-proportional forms. Since $m \notin T(F')$ it follows that $m \notin T(F)$.

3.2. Families of linear forms in $n \ge 3$ variables. It follows from the foregoing that $\{1, \dots, r-2\} \in \mathcal{T}(r, n, 1)$ for $n \ge 3$. Can $\mathcal{T}(r, n, 1)$ contain anything else? We need a simple lemma.

LEMMA 3.1. *Suppose*

$$(3.1)\qquad \sum_{j=1}^{r} \lambda_j \left(\alpha_{j1}x_1 + \dots + \alpha_{jn}x_n\right)^m = 0.$$

Then for every $\gamma = (\gamma_1, \dots, \gamma_n) \in \mathbb{C}^n$,

$$(3.2)\qquad \sum_{j=1}^{r} \lambda_j \left(\sum_{k=1}^{n} \alpha_{jk}\gamma_k\right) (\alpha_{j1}x_1 + \dots + \alpha_{jn}x_n)^{m-1} = 0.$$

PROOF. Apply the differential operator $\frac{1}{m}\sum_{k=1}^{n} \gamma_k \frac{\partial}{\partial x_k}$ to (3.1). □

LEMMA 3.2. *Suppose* $F = \{f_j\} \in \mathcal{F}(r, n, 1)$ *and* $f_j(x) = \alpha_{j1}x_1 + \cdots + \alpha_{jn}x_n$ *for* $1 \le j \le r$.

(1) *If* $m \in T(F)$ *for* $F \in \mathcal{F}(r, n, 1)$, *then* $m - 1 \in T(F)$.
(2) *If* $F \in \mathcal{F}(r, n, 1)$, *then* $\overline{m}(F) \le r - 2$.

PROOF. (1) If $m \in T(F)$, then some non-trivial equation (3.1) holds, with, say, $\lambda_{j_0} \ne 0$. Since $f_{j_0} \ne 0$, there exists ℓ so that $\lambda_{j_0}\alpha_{j_0\ell} \ne 0$. By differentiating (3.1) with respect to x_ℓ, we see that $m - 1 \in T(F)$.

(2) Suppose (3.1) holds. We claim by induction on $m \in T(F)$ that $m \le r - 2$. If $m = 1$, then the assumption that the forms are pairwise non-proportional implies that $r \ge 3$. Suppose now that the claim is true for $m - 1$. There exists γ so that $\sum_{k=1}^n \alpha_{rk}\gamma_k = 0$, but $\sum_{k=1}^n \alpha_{jk}\gamma_k \ne 0$ for $j < r$ (as no α_j is proportional to α_r) and so the summand in (3.2) for $j = r$ disappears. By the inductive hypothesis, $m - 1 \le r - 3$, hence $m \le r - 2$. □

If $r > \binom{n+m-1}{n-1}$, then $\{1, \dots, m\}$ is forced to be in $T(F)$. Let $m(r, n)$ denote the largest integer m so that $r > \binom{n+m-1}{n-1}$, and note that $m(r, 2) = r - 2$. We have thus established the following theorem:

THEOREM 3.3. *If* $F \in \mathcal{F}(r, n, 1)$, *then* $T(F) = \{1, \dots, k\}$, *where* $m(r, n) \le k \le r - 2$.

EXAMPLE 3. Fix (r, n) and let $m = m(r, n)$, so that $\binom{n+m}{n-1} \ge r > \binom{n+m-1}{n-1}$. By Biermann's Theorem (see [**12**, p.31]),

$$\{(i_1x_1 + \cdots + i_nx_n)^{m+1} : 0 \le i_k \in \mathbb{Z},\ i_1 + \cdots + i_n = m + 1\} \tag{3.3}$$

is a linearly independent set with $\binom{n+m}{n-1} \ge r$ elements. Let F be any subset of r of the linear forms in (3.3). Then $m + 1 \notin T(F)$, but $\{1, \dots, m\}$ is forced to be in $T(F)$. Thus $T(F)$ is precisely $\{1, \dots, m(r, n)\}$ in this case.

We strongly suspect that every ticket allowed by Theorem 3.3 is achieved, since the extremes appear. We also strongly suspect that a proof of this result should not be difficult.

An alternative presentation of the foregoing can be made using "Serret's Theorem" (see [**1**] or [**12**, p.26]): given $\alpha_j \in \mathbb{C}^n$, $1 \le j \le r$, $\alpha_j = (\alpha_{j1}, \dots, \alpha_{jn})$, the forms $\{(\sum_k \alpha_{jk}x_k)^m\}$ are linearly independent if and only if there exist "dual" forms $h_i(x_1, \dots, x_n)$ of degree m so that $h_i(\alpha_j) = 0$ if $i \ne j$, but $h_i(\alpha_i) \ne 0$. Theorem 2 in [**1**] implies Lemma 3.2(2).

3.3. Families of $r \le 3$ forms. This case is easily analyzed.

THEOREM 3.4. *For* $r \le 3$, $\mathcal{T}\{r, n, d\}$ *is given as follows:*

(1) $\mathcal{T}\{2, n, d\} = \{\emptyset\}$;
(2) $\mathcal{T}\{3, 2, 1\} = \{\{1\}\}$;
(3) $\mathcal{T}\{3, n, 1\} = \{\emptyset, \{1\}\}$, *if* $n \ge 3$;
(4) $\mathcal{T}\{3, n, d\} = \{\emptyset, \{1\}, \{2\}\}$, *if* $d \ge 2$.

PROOF. If $r = 2$, $0 = \lambda_1 f_1^m + \lambda_2 f_2^m$ and $\lambda \ne (0, 0)$, then f_1 and f_2 are proportional. If $r = 3$ and $d = 1$, then Theorem 3.3 and Example 3 imply (2) and (3). In any event, Theorem 2.1 implies that $|T(F)| \le 1$, "Fermat's Last Theorem" implies that $\overline{m}(F) \le 2$, and since $T(\{x^2 - y^2, 2xy, x^2 + y^2\}) = \{2\}$, it follows that $\{2\} \in \mathcal{T}\{3, n, d\}$ for $n, d \ge 2$. □

3.4. Families of four binary quadratic forms. The simplest remaining case is $(r, n, d) = (4, 2, 2)$. The heavy lifting for this is done in the forthcoming [**13**], which identifies all families $F \in \mathcal{F}(4, 2, 2)$ which have $m \in T(F)$ with $m \ge 3$. Note also that $4 > \binom{2+2-1}{2-1}$, so $1 \in T(F)$ is forced, and $|T(F)| \le 3$ by Theorem 2.1.

If $T(F) \subseteq \{1, 2\}$, then the possibilities are $\{1\}$ or $\{1, 2\}$. These are achieved, respectively by

$$\{x^2, y^2, (x+y)^2, (x-y)^2\}, \qquad \{x^2, y^2, x^2 + y^2, x^2 - y^2\}.$$

Suppose $m \in T(F)$ with $m \ge 3$. Then we may scale $\{f_j\}$ so that

$$f_1^m + f_2^m = f_3^m + f_4^m, \qquad m \ge 3. \tag{3.4}$$

is an equation in binary quadratic forms. Write the common sum in (3.4) as p, then $p \ne 0$, $\{f_1^m, f_2^m\} \ne \{f_3^m, f_4^m\}$, and there do not exist $\alpha_j \in \mathbb{C}$ and $g \in \mathbb{C}[x, y]$ so that $\alpha_1^m + \alpha_2^m = \alpha_3^m + \alpha_4^m$ and $f_j = \alpha_j g$. This is the equation studied in [**13**], where the uniqueness assertions in the rest of this subsection are established. (Uniqueness is taken up to invertible linear changes of variable.)

EXAMPLE 4. For $m = 3$, the solutions to (3.4) satisfying the non-triviality conditions come from a single one-parameter family, first given, in a different form, by J. Young in 1832 (see [**3**, p.554]). For $\alpha \notin \{0, 1, -1\}$ and $\omega = \zeta_3 = e^{2\pi i/3}$, let

$$\begin{aligned} f_1(x,y) &= \alpha x^2 - xy + \alpha y^2, & f_2(x,y) &= -x^2 + \alpha xy - y^2, \\ f_3(x,y) &= \omega\alpha x^2 - xy + \omega^2\alpha y^2, & f_4(x,y) &= -\omega x^2 + \alpha xy - \omega^2 y^2. \end{aligned} \tag{3.5}$$

Then

$$f_1^3(x,y) + \alpha f_2^3(x,y) = f_3^3(x,y) + \alpha f_4^3(x,y), \tag{3.6}$$

and it is not hard to show that $T(\{f_1, f_2, f_3, f_4\}) = \{1, 3\}$ for every α. (Further, ω and ω^2 can be permuted in (3.5), giving a third pair of cubes in (3.6) with the same sum.)

EXAMPLE 5. For $m = 4$, there are two non-trivial solutions to (3.4). Let

$$\begin{aligned} f_1(x,y) &= x^2 + y^2, & f_2(x,y) &= \omega x^2 + \omega^2 y^2, \\ f_3(x,y) &= \omega^2 x^2 + \omega y^2, & f_4(x,y) &= xy. \end{aligned} \tag{3.7}$$

Then it is easy to verify that

$$f_1 + f_2 + f_3 = 0, \quad f_1^2 + f_2^2 + f_3^2 = 6f_4^2, \quad f_1^4 + f_2^4 + f_3^4 = 18f_4^4.$$

It follows from Theorem 2.1 that $T(\{f_1, f_2, f_3, f_4\}) = \{1, 2, 4\}$. The "obvious" generalization of (3.7) is presented in Example 8.

After taking the linear transformation $(x, y) \mapsto (i(x - \omega y), x - \omega^2 y)$ in (3.7) and scaling, we obtain an equivalent version of F in $\mathbb{Z}[x, y]$:

$$\begin{aligned} f_1(x,y) &= x^2 + 2xy, & f_2(x,y) &= x^2 - y^2, \\ f_3(x,y) &= 2xy + y^2, & f_4(x,y) &= x^2 + xy + y^2. \end{aligned} \tag{3.8}$$

EXAMPLE 6. The other solution to (3.4) for $m = 4$ gives a family with ticket $\{1, 4\}$:

$$\sum_{\pm} (\sqrt{3}\, x^2 \pm \sqrt{2}\, xy - \sqrt{3}\, y^2)^4 = \sum_{\pm} (\sqrt{3}\, x^2 \pm i\sqrt{2}\, xy + \sqrt{3}\, y^2)^4. \tag{3.9}$$

This example corresponds to $(\mu, m) = (\sqrt{2/3}, 4)$ in Example 1 (Part Two).

The only solution to (3.4) for $m = 5$ is the family discussed in Example 1.

We summarize this discussion with a result whose proof relies on [**13**].

THEOREM 3.5.

$$\mathcal{T}(4, 2, 2) = \{\{1\}, \{1, 2\}, \{1, 3\}, \{1, 4\}, \{1, 2, 4\}, \{1, 2, 5\}\}.$$

Regarding $\mathcal{T}(4, 2, d)$ for larger d, Example 13 shows that $\{3\} \in \mathcal{T}(4, 2, 3)$ and Example 14 shows that $\{4\} \in \mathcal{T}(4, 2, 7)$.

3.5. If $r = 4$, then $\{1, 2, 3\}$ is not a ticket. We begin with a lemma.

LEMMA 3.6. *Suppose $\{f_j\} \in \mathcal{F}(r, n, d)$ is two-dimensional; that is, there exist f and g, not proportional, so that $f_j = \alpha_j f + \beta_j g$, $1 \le j \le r$. If $f'_j = \alpha_j x + \beta_j y$, then $T(\{f_j\}) = T(\{f'_j\})$. In particular, $T(\{f_j\}) \subseteq \{1, \dots, r-2\}$.*

PROOF. One inclusion is obvious. For the other, since no two f_j's are proportional, the same can be said about the (α_j, β_j)'s. Suppose $\sum_j \lambda_j f_j^m = 0$ and let $P(x, y) = \sum_j \lambda_j (\alpha_j x + \beta_j y)^m$, so that $0 = P(f, g)$. If $P = 0$, then $m \in T(\{f'_j\})$. Otherwise, $P(x, y) = \prod_i (\gamma_i x - \delta_i y)$ splits into linear factors, and $P(f, g) = 0$ implies that f and g are proportional, a contradiction. □

THEOREM 3.7. *For every (n, d), $\{1, 2, 3\} \notin \mathcal{T}(4, n, d)$.*

PROOF. Let $F = \{f_1, f_2, f_3, f_4\}$ and suppose there are three non-trivial equations

$$0 = \lambda_{11} f_1 + \lambda_{12} f_2 + \lambda_{13} f_3 + \lambda_{14} f_4, \tag{3.10}$$

$$0 = \lambda_{21} f_1^2 + \lambda_{22} f_2^2 + \lambda_{23} f_3^2 + \lambda_{24} f_4^2, \tag{3.11}$$

$$0 = \lambda_{31} f_1^3 + \lambda_{32} f_2^3 + \lambda_{33} f_3^3 + \lambda_{34} f_4^3. \tag{3.12}$$

If the f_j's were to satisfy a linear relation different from (3.10), then F would be two-dimensional and Lemma 3.6 would imply $3 \notin T(F)$. Note also that Theorem 3.4 implies that $\lambda_{3j} \neq 0$ for all j.

Suppose first that one of λ_{1k}'s equals zero, say $\lambda_{14} = 0$. Then after scaling the f_j's, we can assume that $f_1 + f_2 = f_3$, and by Theorem 2.1 with $r = 3$, $\lambda_{2j} \neq 0$. Thus, after dividing by $\lambda_{\ell 4}$, and relabeling, (3.11) and (3.12) imply that

$$\beta_{21} f_1^2 + \beta_{22} f_2^2 + \beta_{23} (f_1 + f_2)^2 = f_4^2, \tag{3.13}$$

$$\beta_{31} f_1^3 + \beta_{32} f_2^3 + \beta_{33} (f_1 + f_2)^3 = f_4^3. \tag{3.14}$$

If $\beta_{21} u^2 + \beta_{22} v^2 + \beta_{23} (u + v)^2$ were a square, then (3.13) would imply a linear relation among f_1, f_2 and f_4, different from (3.10) and as noted above, this would be a contradiction. Otherwise, (3.13) and (3.14) together imply that

$$\left(\beta_{21} f_1^2 + \beta_{22} f_2^2 + \beta_{23} (f_1 + f_2)^2\right)^3 = \left(\beta_{31} f_1^3 + \beta_{32} f_2^3 + \beta_{33} (f_1 + f_2)^3\right)^2. \tag{3.15}$$

Since f_1 and f_2 are not proportional, (3.15) and the argument of Lemma 3.6 imply that

$$(\beta_{21}u^2 + \beta_{22}v^2 + \beta_{23}(u+v)^2)^3 = (\beta_{31}u^3 + \beta_{32}v^3 + \beta_{33}(u+v)^3)^2. \tag{3.16}$$

But if $\Phi = G^3 = H^2$, then any irreducible factor of Φ must occur to the sixth power. Thus, (3.16) implies that $\beta_{21}u^2 + \beta_{22}v^2 + \beta_{23}(u+v)^2$ is a square, a contradiction.

Therefore, we may assume that $\lambda_{1k} \neq 0$, and so after scaling, $f_4 = f_1 + f_2 + f_3$ and

$$\begin{aligned} \beta_{21}f_1^2 + \beta_{22}f_2^2 + \beta_{23}f_3^2 + \beta_{24}(f_1+f_2+f_3)^2 &= 0, \\ \beta_{31}f_1^3 + \beta_{32}f_2^3 + \beta_{33}f_3^3 + \beta_{34}(f_1+f_2+f_3)^3 &= 0. \end{aligned} \tag{3.17}$$

By Bezout's Theorem, the intersection of the two curves

$$\begin{aligned} G_2(u,v,w) &:= \beta_{21}u^2 + \beta_{22}v^2 + \beta_{23}w^2 + \beta_{24}(u+v+w)^2 = 0, \\ G_3(u,v,w) &:= \beta_{31}u^3 + \beta_{32}v^3 + \beta_{33}w^3 + \beta_{34}(u+v+w)^3 = 0 \end{aligned}$$

is (projectively) at most six lines, unless the curves share a component. Since (f_1, f_2, f_3) parameterizes the intersection by (3.17) and no two f_j are proportional, either $G_2 \mid G_3$ or both share a linear factor ℓ. However, if $G_2 = \ell_1\ell_2$ for linear ℓ_j, then $G_2(f_1, f_2, f_3) = 0$ implies that $\ell_j(f_1, f_2, f_3) = 0$ for some j; this is different from $f_4 = f_1 + f_2 + f_3$ and gives a contradiction as before.

The only remaining possibility is that $G_2 \mid G_3$ and G_2 is irreducible. After relabeling once again, we see that this case is

$$\begin{aligned} &\big(\sigma_1u^2 + \sigma_2v^2 + \sigma_3w^2 + 2\sigma_4(uv + uw + vw)\big)\big(\gamma_1u + \gamma_2v + \gamma_3w\big) \\ &\qquad = \tau_1u^3 + \tau_2v^3 + \tau_3w^3 + \\ &\tau_4(3u^2v + 3uv^2 + 3uw^2 + 6uvw + 3uw^2 + 3v^2w + 3vw^2). \end{aligned} \tag{3.18}$$

Since $\beta_{3j} \neq 0$, we may assume in (3.18) that $\tau_4 \neq 0$ (hence $(\gamma_1, \gamma_2, \gamma_3) \neq (0,0,0)$) and that $\tau_j \neq \tau_4$ for $j = 1,2,3$. By considering the coefficients of $u^2v, u^2w, uv^2, v^2w, uw^2, vw^2$ and uvw in (3.18), we find that

$$\begin{aligned} 3\tau_4 &= 2\gamma_1\sigma_4 + \gamma_2\sigma_1 = 2\gamma_1\sigma_4 + \gamma_3\sigma_1 = 2\gamma_2\sigma_4 + \gamma_1\sigma_2 = 2\gamma_2\sigma_4 + \gamma_3\sigma_2 \\ &= 2\gamma_3\sigma_4 + \gamma_1\sigma_3 = 2\gamma_3\sigma_4 + \gamma_2\sigma_3 = (\gamma_1 + \gamma_2 + \gamma_3)\sigma_4. \end{aligned} \tag{3.19}$$

Since $\tau_4 \neq 0$, it follows that $\sigma_4 \neq 0$. It follows immediately from (3.19) that

$$0 = (\gamma_2 - \gamma_3)\sigma_1 = (\gamma_3 - \gamma_1)\sigma_2 = (\gamma_1 - \gamma_2)\sigma_3. \tag{3.20}$$

If $\gamma_1 = \gamma_2 = \gamma_3 := \gamma \neq 0$, then (3.19) reduces to

$$3\tau_4 = \gamma(\sigma_1 + 2\sigma_4) = \gamma(\sigma_2 + 2\sigma_4) = \gamma(\sigma_3 + 2\sigma_4) = 3\gamma\sigma_4.$$

This implies that $\sigma_1 = \sigma_2 = \sigma_3 = \sigma_4 := \sigma$, so $G_2(u,v,w) = (u+v+w)^2$, a contradiction.

If the γ_j's are not all equal, then after permutation of indices if necessary, we may assume that $\gamma_1 \neq \gamma_2$ and $\gamma_1 \neq \gamma_3$, hence $\sigma_2 = \sigma_3 = 0$ by (3.20). Then (3.19) becomes

$$3\tau_4 = 2\gamma_1\sigma_4 + \gamma_2\sigma_1 = 2\gamma_1\sigma_4 + \gamma_3\sigma_1 = 2\gamma_2\sigma_4 = 2\gamma_3\sigma_4 = (\gamma_1 + \gamma_2 + \gamma_3)\sigma_4.$$

Thus $\gamma_2 = \gamma_3$, and so $2\gamma_2\sigma_4 = (\gamma_1 + \gamma_2 + \gamma_3)\sigma_4$ implies $\gamma_1 = 0$. It then follows that $\sigma_1 = 2\sigma_4$, hence $G_2(u,v,w) = 2\sigma_4(u+v)(u+w)$, a final contradiction. □

4. A few interesting tickets

The examples in this section indicate some of the diversity that is possible in tickets of indecomposable families. They are inspired, for the most part, by the examples from [**13**]. The families consist of binomials or highly symmetric trinomial binary forms, and rely heavily on the orthogonality properties of roots of unity.

Example 8 presents highly dysfunctional families $F \in \mathcal{F}(2v+2,2,2)$ with $|T(F)| = 3v$. Example 9 gives families $F \in \mathcal{F}(r,2,2)$ so that $T(F)$ contains $\{1,\dots,r-2\}$ as well as $r+s$, for any single integer s, $2 \le s \le r-1$. This is less dysfunctional than the previous example, but with a larger jump. The families in Example 10 are not dysfunctional, but have an even larger jump; for example, $\{1,2,8\} \in \mathcal{T}(6,2,2)$. One special case combines Examples 9 and 10 to show that $\{1,2,3,4,8,14\} \in \mathcal{T}(6,2,2)$. Example 11 presents $F \in \mathcal{F}(a+2,2,a)$ so that $T(F)$ consists precisely of the divisors of a. Example 12 combines Examples 7 and 11 and includes Molluzzo's example showing that $\overline{m}_r \ge \lfloor r^2/4 \rfloor - 1$. Example 13 verifies the claim from the Introduction that $\{m\} \in \mathcal{T}(m+1,2,m)$, while Example 14, due to Euler, shows that $\{4\} \in \mathcal{T}(4,2,7)$.

For integral $q \ge 2$, write $\zeta_q = e^{\frac{2\pi i}{q}}$. We shall say that a set of polynomials $\{f_0,\dots,f_{q-1}\}$ is *q-cyclotomic* on $\{g_0,\dots,g_{q-1}\}$ if

$$f_j = \sum_{k=0}^{q-1} \zeta_q^{jk} g_k, \qquad 0 \le j \le q-1, \tag{4.1}$$

where the g_k's involve disjoint sets of monomials. Note that, if $\{f_j\}$ is q-cyclotomic on $\{g_k\}$, then $\{f_j^m\}$ is q-cyclotomic on $\{g_{m,k}\}$, where

$$g_{m,k} = \sum_{\sum i_j \equiv k \mod q} g_{i_1}\cdots g_{i_m}, \tag{4.2}$$

provided the $g_{m,k}$'s involve disjoint monomials. This will occur if, for example, $f_j = f_j(x,y)$ and all terms in g_i have y-degree congruent to i (mod q).

LEMMA 4.1. *If $\{f_0,\dots,f_{q-1}\}$ is q-cyclotomic on $\{g_0,\dots,g_{q-1}\}$, then*

$$\mathit{Span}(f_0,\dots,f_{q-1}) = \mathit{Span}(g_0,\dots,g_{q-1}).$$

In particular, $\{f_j\}$ is linearly dependent if and only if some $g_k = 0$.

PROOF. By the orthogonality properties of roots of unity, (4.1) inverts to

$$g_\ell = \frac{1}{q}\sum_{j=0}^{q-1} \zeta_q^{-j\ell} f_j, \qquad 0 \le \ell \le q-1. \tag{4.3}$$

Thus the two spans are equal. Since the g_k's involve different monomials, they can only be trivially dependent. □

EXAMPLE 1 (PART THREE). Continuing the discussion in Example 1 (Part Two), suppose

$$\begin{aligned} f_1(t) &= 1 + \mu t - t^2, \qquad & f_2(t) &= 1 + i\mu t + t^2, \\ f_3(t) &= 1 - \mu t - t^2, \qquad & f_4(t) &= 1 - i\mu t + t^2. \end{aligned} \tag{4.4}$$

That is, $F = \{f_j\}$ is 4-cyclotomic on $\{1, \mu t, -t^2, 0\}$. To ensure pairwise non-proportionality in (4.4), we require $\mu \ne 0$. Suppose $1 < m \in T(F)$. Then it follows from Lemma 4.1 that $g_{m,k} = 0$ for some k. But for $m > 1$ and integral v,

$$g_{m,0} = 1 + \dots, \qquad g_{m,1} = m\mu t + \dots,$$
$$g_{2v+1,2} = \dots + t^{4v+2}, \qquad g_{2v,3} = \dots - 2v\mu t^{4v-1},$$

and since these $g_{m,k}$'s all have at least one non-zero term, the only $g_{m,k}$'s which might vanish are $g_{2v,2}$ and $g_{2v+1,3}$. We have

$$\begin{aligned} &g_{2,2}(t) = (\mu^2 - 2)t^2, \qquad g_{3,3}(t) = (\mu^3 - 6\mu)t^3, \\ &g_{4,2}(t) = (6\mu^2 - 4)(t^2 + t^6), \qquad g_{5,3}(t) = (10\mu^3 - 20\mu)(t^3 + t^7), \\ &g_{6,2}(t) = (15\mu^2 - 6)t^2 + (\mu^6 - 30\mu^4 + 90\mu^2 - 20)t^6 + (15\mu^2 - 6)t^{10}, \\ &g_{7,3}(t) = (35\mu^3 - 42\mu)t^3 + (\mu^7 - 42\mu^5 + 210\mu^3 - 140\mu)t^7 + (35\mu^3 - 42\mu)t^{11}. \end{aligned} \tag{4.5}$$

We see from (4.5) that $g_{m,k} = 0$ for $m \le 7$ only in the following cases; $\mu^2 = 2$ and $m = 2$ or 5; $\mu^2 = 6$ and $m = 3$; and $\mu^2 = 2/3$ and $m = 4$. (It is evident that $g_{6,2}$ and $g_{7,3}$ can never vanish identically. Green's bound implies that we need only check $m \le 8$; the computation of $g_{8,2}$ is left to the reader.)

For $\mu = \sqrt{2}$, we've already discussed F in Example 1 (Part Two). (This includes the seemingly accidental $g_{2,2} \mid g_{5,3}$.)

If $\mu = \sqrt{2/3}$, then after some simplification, we obtain (3.9).

If $\mu = \sqrt{6}$, then after homogenizing and inverting by (4.2), we obtain the identity

$$\begin{aligned} 6\sqrt{6}(x^5 y + xy^5) &= (x^2 + \sqrt{6}xy - y^2)^3 - (x^2 - \sqrt{6}xy - y^2)^3 \\ &= (ix^2 - \sqrt{6}xy + iy^2)^3 - (ix^2 + \sqrt{6}xy + iy^2)^3. \end{aligned} \tag{4.6}$$

It turns out that $6\sqrt{6}(x^5 y + xy^5)$ has four other representations as a sum of two cubes, involving 24-th roots of unity (see [**13**].) Noam Elkies notes that this follows from the many symmetries of the form $x^5 y + xy^5$, whose zeros are the vertices of a regular octahedron in the Riemann sphere. It is possible to derive (4.6) from (3.5) using an unpleasant change of variables and an ugly choice of α.

EXAMPLE 7. Let $f_j(x, y) = x + \zeta_q^j y$. Then $\{f_0^m, \dots, f_{q-1}^m\}$ is q-cyclotomic on $\{g_{m,0}, \dots, g_{m,q-1}\}$, where

$$g_{m,k}(x, y) = \sum_{\substack{i \equiv k \mod q \\ 0 \le i \le m}} \binom{m}{i} x^{m-i} y^i.$$

If $m < q - 1$, then $g_{m,q-1} = 0$, hence $\{f_j^m\}$ is linearly dependent and so $m \in T(F)$. If $m \ge q - 1$, then $g_{m,k} \ne 0$, so $m \notin T(F)$. This is a special case of Theorem 3.3.

EXAMPLE 8. Suppose q is odd and let $f_j(x, y) = \zeta_q^j x^2 + \zeta_q^{-j} y^2$, $0 \le j \le q - 1$. (Since q is odd, no two f_j's are proportional.) Then

$$f_j^m(x, y) = \sum_{i=0}^{m} \binom{m}{i} \zeta_q^{(m-2i)j} x^{2m-2i} y^{2i},$$

so

$$g_{m,k}(x, y) = \sum_{m - 2i \equiv k \mod q} \binom{m}{i} x^{2m-2i} y^{2i}. \tag{4.7}$$

Now let

$$F_q = \{f_0(x,y), \dots, f_{q-1}(x,y), xy\} \in \mathcal{F}(q+1,2,2).$$

(Without xy, this is essentially Example 7.) By Lemma 4.1,

$$\text{Span}(f_0^m, \dots, f_{q-1}^m, (xy)^m) = \text{Span}(g_{m,0}, \dots, g_{m,q-1}, (xy)^m). \tag{4.8}$$

It follows from (4.8) that $m \in T(F_q)$ if and only if for some i, either $g_{m,i} = 0$, or $g_{m,i}$ is a multiple of $(xy)^m$.

In the first case, if $m \ge q-1$, then there will exist i, $0 \le i \le m$, satisfying $m - 2i \equiv k \mod q$, and $g_{m,i} \ne 0$. If $m \le q-2$, then the equation $m - 2i \equiv m+2 \mod q$ implies that $i \equiv q-1 \mod q$, hence $g_{m,m+2} = 0$. Therefore, $T(F_q)$ contains $\{1, \dots, q-2\}$. (Alternatively, $f_j(x,y) = g_j(x^2,y^2)$ for linear g_j, and $m \le q-2$ is forced in $T(\{g_j\})$.)

In the second case, $(xy)^m$ can only occur in $g_{m,i}$ if m is even, $i = m/2$ and $k = 0$. Putting this into (4.7), we see that $(xy)^m$ is the only term in $g_{m,0}$ if and only if $i = m/2$ is the only solution to $2i \equiv m \mod q$ with $0 \le i \le m$, and this occurs if and only if $m/2 < q$; that is, $m < 2q$. Thus $T(F_q)$ contains the even integers $\{2, \dots, 2q-2\}$. Writing $q = 2v+1$, we see that $r = |F_q| = 2v+2$ and

$$T(F_{2v+1}) = \{1, 2, \dots, 2v-1, 2v, 2v+2, \dots, 4v-2, 4v\}.$$

so that $|T(F_{2v+1})| = 3v = \frac{3}{2}r - 3$. The family defined in (3.8) is, in fact, F_3. The explicit linear relations are not very interesting:

$$\sum_{j=0}^{q-1} \left(\zeta_q^j x^2 + \zeta_q^{-j} y^2\right)^m = 0, \qquad \text{if } m < q \text{ is odd};$$

$$\sum_{j=0}^{q-1} \left(\zeta_q^j x^2 + \zeta_q^{-j} y^2\right)^m = q\binom{m}{m/2}(xy)^m, \qquad \text{if } m < 2q \text{ is even.}$$

EXAMPLE 9. This next example is a variation on Example 8, giving a smaller ticket, but one with a potentially larger gap. We no longer assume that q is odd. Let $\alpha \in \mathbb{C}$, let

$$f_{j,q,\alpha}(x,y) = \zeta_q^j x^2 + \alpha xy + \zeta_q^{-j} y^2, \qquad 0 \le j \le q-1, \tag{4.9}$$

and let

$$F_{q,\alpha} = \{f_{0,q,\alpha}(x,y), \dots, f_{q-1,q,\alpha}(x,y), xy\} \in \mathcal{F}(q+1,2,2).$$

Since

$$\left(\zeta_q^j x^2 + \alpha xy + \zeta_q^{-j} y^2\right)^m = \sum_{a+b+c=m} \frac{m!}{a!b!c!} \zeta_q^{j(a-c)} \alpha^b x^{2a+b} y^{b+2c},$$

by setting $b = m - a - c$ above, we see that

$$g_{m,k}(x,y) = \sum_{a-c \equiv k \mod q} \frac{m!}{a!c!(m-a-c)!} \alpha^{m-a-c} x^{m+(a-c)} y^{m-(a-c)}. \tag{4.10}$$

Again, $m \in T(F_{q,\alpha})$ if and only if $g_{m,k} = 0$ or is a multiple of $(xy)^m$. We shall not attempt a complete analysis of $T(F_{q,\alpha})$. Note, however, that the coefficient of a monomial in $g_{m,k}$ is a polynomial in α, and may vanish when α is suitably chosen.

If $m \le q-1$, then $a \equiv c \mod q$ and $m \ge |a-c|$ imply that $a = c$ so that (4.10) implies that

$$g_{m,0}(x,y) = \left(\sum_{a=0}^{\lfloor m/2 \rfloor} \frac{m!}{(a!)^2(m-2a)!}\alpha^{m-2a} \right) x^m y^m,$$

so $m \in T(F_{q,\alpha})$ for any α. Suppose now that $q \ge 3$ and $m = q+s$, $2 \le s \le q-1$. We fix $k = 0$. Then $a - c \equiv 0 \mod q$ implies that $a - c \in \{-q, 0, q\}$ and by (4.10), $g_{q+s,0}(x,y)$ is equal to

$$\left(\sum_{a=0}^{\lfloor s/2 \rfloor} \frac{(q+s)!}{a!(a+q)!(s-2a)!}\alpha^{s-2a} \right) (x^{s+2q}y^s + x^s y^{s+2q}) + B(\alpha)(xy)^{s+q}. \tag{4.11}$$

for some irrelevant polynomial $B(\alpha)$.

Since $s \ge 2$, there exists $\alpha = \alpha_0$ which kills the coefficient of $x^{s+2q}y^s + x^s y^{s+2q}$ in (4.11), so that $g_{q+s,0}$ is a multiple of $(xy)^{s+q}$. In this case,

$$\{1, \dots, q-1, q+s\} \subseteq T(F_{q,\alpha_0}).$$

We cannot rule out the possibility that $T(F_{q,\alpha_0})$ has other elements if α_0 is a root of other polynomials appearing in different $g_{m,k}$'s. Unless that happens, we have $r = |F_{q,\alpha_0}| = q+1$; if $s = q-1$ is maximal, then $\overline{m}(F_{q,\alpha_0}) \ge q+s = 2q-1$. Thus, $\overline{m}_r \ge 2r-3$, and this is larger than $\lfloor r^2/4 \rfloor - 1$ for $r \le 6$. The stronger bound $\overline{m}_6 \ge 14$ will follow from (4.15).

The simplest concrete realization of this phenomenon occurs when $q = 3$, $s = 2$ and $m = 5$. We have $\zeta_3 = \omega$ and (4.11) becomes

$$g_{5,0} = (5 + 10\alpha^2)(x^8y^2 + x^2y^8) + (a^5 + 20a^3 + 30a)x^5y^5.$$

Taking $\alpha_0 = \sqrt{-1/2}$, we define

$$\begin{aligned} f_{0,3,\alpha_0}(x,y) &= x^2 + \alpha_0 xy + y^2, \quad & f_{1,3,\alpha_0}(x,y) &= \omega x^2 + \alpha_0 xy + \omega^2 y^2, \\ f_{2,3,\alpha_0}(x,y) &= \omega^2 x^2 + \alpha_0 xy + \omega y^2, \quad & f_3(x,y) &= xy. \end{aligned} \tag{4.12}$$

Then $T(F_{3,\alpha_0})$ contains $\{1,2,5\}$. After the linear change $(x,y) \mapsto (i(y-cx), x+cy)$ for $c = \frac{\sqrt{6}-\sqrt{2}}{2} = \sqrt{2-\sqrt{3}}$, (4.12) becomes a scaled and permuted version of (1.1). Example 1 strikes again!

EXAMPLE 10. As a variation on the previous example, we again use (4.9), but specify $q = 2v \ge 4$ to be even and let

$$F'_{q,\alpha} = \{f_{0,q,\alpha}(x,y), \dots, f_{q-1,q,\alpha}(x,y)\} \in \mathcal{F}(2v, 2, 2).$$

Since $xy \notin F'_{q,\alpha}$, $m \in T(F'_{q,\alpha})$ only if there exists k so that $g_{m,k} = 0$. If $m \le v-1$, then $2v > \binom{2m+1}{1}$ so m is forced in $T(F'_{q,\alpha})$. With a careful choice of α, another exponent can be added to the ticket. We are interested in the largest possible jump. If $m = 3v-1$, then by (4.10), $g_{3v-1,v}(x,y)$ is equal to

$$\left(\sum_{a=0}^{v-1} \frac{(3v-1)!}{a!(a+v)!(2v-1-2a)!}\alpha^{2v-1-2a} \right) (x^{4v-1}y^{2v-1} + x^{2v-1}y^{4v-1}). \tag{4.13}$$

Since $v \ge 2$, there exists $\alpha = \alpha_0$ so that $g_{3v-1,v} = 0$; hence $\{1, ..., v-1, 3v-1\}$ is contained in $T(F'_{2v,\alpha_0})$. (As (4.13) gives an equation of degree $v-1$ satisfied by

α^2, we cannot expect "nice" roots α_0 for $v \ge 4$.) By applying the inversion formula (4.3) to $g_{3v-1,v} = 0$, we find that

$$0 = \sum_{j=0}^{2v-1} \zeta_{2v}^{-jv} f_{j,v,\alpha_0}^m = \sum_{j=0}^{2v-1} (-1)^j f_{j,v,\alpha_0}^m. \tag{4.14}$$

By transposing the terms in (4.14) with odd j, we obtain two sets of v quadratic forms with the property that the sum of their $3v-1$-st powers are equal.

The simplest case is $v = 2$, where (4.13) becomes $(10\alpha^3 + 20\alpha)(x^7y^3 + x^3y^7)$, so $\alpha_0 = \sqrt{-2}$. If we set $y \mapsto iy$, we get yet another appearance of Example 1.

If $v = 3$, then

$$g_{8,3}(x,y) = 56\alpha(3 + 5\alpha^2 + \alpha^4)(x^{11}y^5 + x^5y^{11}).$$

By setting $\alpha_0 = i\sqrt{\frac{5+\sqrt{13}}{2}}$ and noting $\zeta_6 = -\omega^2$, we see that (4.14) becomes, explicitly,

$$\begin{aligned}(x^2 + \alpha_0 xy + y^2)^8 + (\omega x^2 + \alpha_0 xy + \omega^2 y^2)^8 + (\omega^2 x^2 + \alpha_0 xy + \omega y^2)^8 = \\ (x^2 - \alpha_0 xy + y^2)^8 + (\omega x^2 - \alpha_0 xy + \omega^2 y^2)^8 + (\omega^2 x^2 - \alpha_0 xy + \omega y^2)^8 = \\ -3\left(\tfrac{1+\sqrt{13}}{2}\right)^4 x^2y^2(4x^{12} - 13^{3/2}x^6y^6 + 4y^{12}).\end{aligned}$$

For $F = \{\omega^k x^2 \pm \alpha_0 xy + \omega^{-k} y^2\} \in \mathcal{F}(6,2,2)$, we have $T(F) = \{1,2,8\}$.

A computer check of $g_{3v-1,v}$ for $v \le 40$ shows that it is irreducible, except for $v = 5$ which has a factor of $\alpha^2 + 1$. Letting $\zeta = \zeta_5$, and transposing as in (4.14), we find:

$$\sum_{j=0}^{4} (\zeta^j x^2 + ixy + \zeta^{-j} y^2)^{14} = \sum_{j=0}^{4} (\zeta^j x^2 - ixy + \zeta^{-j} y^2)^{14} = 5^7 (xy)^{14} \tag{4.15}$$

It is miraculous, and wholly unexpected, that the common sum is also a 14-th power. In the language of Example 9, with $q = 5$, $g_{14,4}$ is a linear combination of $(x^{19}y^9 + x^9y^{19})$ and $(x^{24}y^4 + x^4y^{24})$, in which both coefficients have $\alpha^2 + 1$ as a factor.

Taking advantage of this opportunity, we let $f_j = \zeta^j x^2 + ixy + \zeta^{-j} y^2$, $0 \le j \le 4$ and $f_5 = \sqrt{-5}\, xy$, and define $\dot{F} = \{f_0, \dots, f_5\}$, then

$$0 = \sum_{j=0}^{4} f_j - \sqrt{5} f_5 = \sum_{j=0}^{5} f_j^2 = \sum_{j=0}^{4} f_j^3 + \sqrt{5} f_5^3 = \sum_{j=0}^{5} f_j^4 = \sum_{j=0}^{5} f_j^8 = \sum_{j=0}^{5} f_j^{14}. \tag{4.16}$$

A computer check, using $\overline{m}_6 \le 24$, shows no other relations. Thus, we have $T(\dot{F}) = \{1,2,3,4,8,14\}$ and $\overline{m}_6 \ge 14$.

Finally, if we put $(x,y) = (1,-1)$ into (4.15) and use the closed form of $\cos(\frac{2\pi k}{5})$, we derive a peculiar numerical identity:

$$2\left(\sqrt{5} + 1 + 2i\right)^{14} + 2\left(-\sqrt{5} + 1 + 2i\right)^{14} + (4 - 2i)^{14} = 20^7.$$

EXAMPLE 11. This example is pleasantly uncomputational. Let

$$\widehat{F}_a = \{x^a + y^a, x^a, x^{a-1}y, \dots, xy^{a-1}, y^a\} \in \mathcal{F}(a+2, 2, a).$$

If $m \in T(\widehat{F}_a)$, then there is a non-trivial equation

$$\sum_{k=0}^{a} \lambda_k (x^{a-k}y^k)^m + \lambda_{a+1}(x^a + y^a)^m = 0.$$

But this is impossible if $\lambda_{a+1} = 0$, so we may scale to $\lambda_{a+1} = -1$ and consider

$$\sum_{k=0}^{a} \lambda_k (x^{a-k} y^k)^m = (x^a + y^a)^m = \sum_{\ell=0}^{m} \binom{m}{\ell} x^{ma-\ell a} y^{\ell a}.$$

This could hold if and only if $m \mid \ell a$ for each ℓ, $0 \le \ell \le m$; that is, if and only if m is a divisor of a. Thus, $T(\widehat{F}_a)$ consists of the divisors of a and $|T(\widehat{F}_a)|$ is the arithmetic function $d(a) \ll a$. This family is far from dysfunctional, but by taking $a = p^e$ for prime p, we see that $T(\widehat{F}_{p^e}) = \{1, p, p^2, \dots, p^e\}$. Thus, there is no bound on the ratio of consecutive elements in the ticket of an indecomposable family.

EXAMPLE 12. We combine Examples 7 and 11. For parameters $a, q \ge 2$, let

$$\widetilde{F}_{a,q} = \{x^a + y^a, x^a + \zeta_q y^a, \dots, x^a + \zeta_q^{q-1} y^a, x^a, x^{a-1}y, \dots, xy^{a-1}, y^a\}.$$

Then $\widetilde{F}_{a,q} \in \mathcal{F}(q+a+1, 2, a)$. By Lemma 4.1, $m \in T(\widetilde{F}_{a,q})$ if and only if there exists k, $0 \le k \le q-1$, so that

$$\sum_{i \equiv k \mod q} \binom{m}{i} x^{(m-i)a} y^{ia}$$

is a linear combination of $\{x^{m(a-j)} y^{mj} : 0 \le j \le a\}$. For k, $0 \le k \le q-1$, let

$$S_k(m; a, q) = \{ia : i \equiv k \mod q,\ 0 \le i \le m\} = \{(k + bq)a : 0 \le b \le \lfloor \tfrac{m-k}{q} \rfloor\}.$$

and let $Z(m; a, q)$ denote the set of k for which $S_k(m; a, q)$ only contains multiples of m. Then $m \in T(\widetilde{F}_{a,q})$ if and only if some $Z(m; a, q)$ is non-empty.

If $m \le q-1$, then $S_0(m; a, q) = \{0\}$, so $m \in T(\widetilde{F}_{a,q})$. Since $S_0(m; a, q)$ consists of multiples of qa, if $m \mid qa$, then $0 \in Z(m; a, q)$, so $m \in T(\widetilde{F}_{a,q})$. If $|S_k(m; a, q)| \ge 2$ and $k \in Z(m; a, q)$, then m divides ka and $(k+q)a$, so m is a divisor of qa.

In the remaining case, $m \ge q$ is not a divisor of qa, but there exists k so that $m \mid ka$ and $S_k(m; a, q)$ is a singleton, so that $k \ge m - (q-1)$. The existence of such m depends on the arithmetic properties of a and q. If $a = q = p$ is prime, then no such m exists, and $T(\widetilde{F}_{p,p}) = \{1, \dots, p, p^2\}$. On the other hand, if $a = q = 6$, say, then it is not hard to show that $m = 8, 10$ satisfy the criteria and

$$T(\widetilde{F}_{6,6}) = \{1, 2, 3, 4, 5, 6, 8, 9, 10, 12, 18, 36\}.$$

Note that $F_{6,6}$ is (barely) dysfunctional.

Molluzzo ([**10**], see [**11**, p.485]) gives an equation, which when homogenized, is closely related to the family $\widetilde{F}_{a,q}$, with many monomials deleted. As we have seen,

$$\sum_{j=0}^{q-1} (x^a + \zeta_q^j y^a)^a = q \sum_{\ell=0}^{\lfloor a/q \rfloor} \binom{a}{\ell q} x^{a(a-q\ell)} y^{aq\ell}.$$

Accordingly, we let $t = \lfloor a/q \rfloor$ for short and define

$$\breve{F}_{a,q} = \{x^a + \zeta_q^j y^a\} \cup \{x^a, x^{a-q}y^q, \dots, x^{a-tq}y^{tq}\} \in \mathcal{F}(q+t+1, 2, a),$$

so that $a \in T(\breve{F}_{a,q})$. The maximum possible value of a for fixed q, t occurs when $a = q(t+1) - 1$, and so it follows that

$$\overline{m}_{q+t+1} \ge q(t+1) - 1. \tag{4.17}$$

Fix $r = q + t + 1$. If $r = 2v$ is even, then (4.17) with $q = t + 1 = v$ shows that $\overline{m}_{2v} \geq v^2 - 1$; if $r = 2v + 1$ is odd, then (4.17) with $\{q, t + 1\} = \{v, v + 1\}$ shows that $\overline{m}_{2v+1} \geq v^2 + v - 1$. These combine to give the bound $\overline{m}_r \geq \lfloor r^2/4 \rfloor - 1$.

EXAMPLE 13. This example is another variation on Example 11. Recall the following family from the Introduction:

$$F_d^* := \{x^d + y^d, x^d + \zeta y^d, x^{d-1}y, \dots, xy^{d-1}\} \in \mathcal{F}(d + 1, 2, d),$$

where $\zeta = \zeta_d$. If $m \in T(F_d^*)$, then we have a non-trivial equation

$$\alpha(x^d + y^d)^m + \beta(x^d + \zeta y^d)^m + \sum_{j=1}^{d-1} \gamma_j (x^{d-j} y^j)^m = 0. \tag{4.18}$$

If $\alpha = \beta = 0$ in (4.18), then the γ_j's all vanish, so it is trivial. By considering the coefficients of x^{dm} and y^{dm}, we obtain the equations $\alpha + \beta = \alpha + \zeta^m \beta = 0$, hence $\beta = -\alpha$ and $\zeta^m = 1$, and so $m = sd$ for some integer s. We now consider the sum

$$\alpha\left((x^d + y^d)^{sd} - (x^d + \zeta y^d)^{sd}\right) + \sum_{j=1}^{d-1} \gamma_j (x^{d-j} y^j)^{sd}. \tag{4.19}$$

If $s > 1$, then the coefficient of $x^{sd^2-d}y^d$ in (4.19) will be $sd\alpha(1 - \zeta) \neq 0$. If $s = 1$, simply set $\alpha = 1$ and $\gamma_j = \binom{d}{j}(\zeta^j - 1)$ in (4.19) to set it equal to 0. We conclude that $T(F_d^*) = \{d\}$.

EXAMPLE 14. There are two famous parameterizations due to Euler which give singleton tickets. The Euler-Binet (complete) solution to (3.4) for $m = 3$ (see [**7**, p.200] gives $F \in \mathcal{F}(4, 3, 4)$ with $T(F) = \{3\}$. (But we already know that $F \in \mathcal{F}(4, 2, 3)$ has this ticket by the last example.)

Euler's binary septic solution to (3.4) for $m = 4$ ([**7**, p.201]) is

$$\begin{aligned}
f_1(x, y) &= x^7 + x^5y^2 - 2x^3y^4 + 3x^2y^5 + xy^6, \\
f_2(x, y) &= x^6y - 3x^5y^2 - 2x^4y^3 + x^2y^5 + y^7, \\
f_3(x, y) &= x^7 + x^5y^2 - 2x^3y^4 - 3x^2y^5 + xy^6, \\
f_4(x, y) &= x^6y + 3x^5y^2 - 2x^4y^3 + x^2y^5 + y^7.
\end{aligned}$$

Here, $F = \{f_j\} \in \mathcal{F}(4, 2, 7)$, and it can be shown that $T(F) = \{4\}$. This can be seen by a shortcut, avoiding the machinery of Theorem 2.1. Suppose $\sum_{j=1}^4 \lambda_j f_j^m = 0$. Then consideration of the terms x^{7m} and y^{7m} shows that $\lambda_3 = -\lambda_1$ and $\lambda_4 = -\lambda_2$. Hence by transposition, $\lambda_1(f_1^m - f_3^m) = \lambda_4(f_2^m - f_4^m)$. Evaluation at $(x, y) = (1, 1)$ implies that $\lambda_1(4^m - (-2)^m) = \lambda_4((-2)^m - 4^m)$, hence $\lambda_4 = -\lambda_1$, and the only possible relation is $f_1^m + f_2^m - f_3^m - f_4^m = 0$. This equation can now easily be checked for $m \leq 8$. (The family F_4^* has five binary quartic forms; the ticket $\{4\}$ appears here for a family with fewer forms, but in higher degree.)

It is apparently unknown whether there is a non-trivial solution to the equation $f_1^4 + f_2^4 = f_3^4 + f_4^4$ in real binary forms of degree d for $3 \leq d \leq 6$. It seems clear to the author that the vast, centuries-old literature of Diophantine parameterizations can be data-mined for many more interesting examples of tickets.

5. Generalizations, speculations and open questions

This final section collects a number of disconnected miscellaneous remarks.

There is no reason to restrict our attention to forms over $\mathbb{C}$; for any field k, one might just as easily define $\mathcal{F}_k(r,n,d)$, where the forms have coefficients in k. It would be particularly interesting to study these for $k = \mathbb{Q}$ or $\mathbb{R}$. It is proved in [**13**] that there is no solution to (3.4) for real polynomials when $m = 5$. Thus, $(1,2,5) \notin \mathcal{T}_{\mathbb{R}}(4,2,2)$; (3.8) shows that $(1,2,4) \in \mathcal{T}_{\mathbb{Q}}(4,2,2)$. The forms in the highly dysfunctional family F_q in Example 8 are $\mathbb{R}$-combinations of x^2+y^2, $i(x^2-y^2)$ and xy, and hence can be made real by taking $(x,y) \mapsto (x+iy, x-iy)$. It seems difficult to determine when a given family of forms can be made real or rational by an appropriate linear change of variables. This is probably a deep question.

One can also sharpen the definition of $T(F)$. Let

$$\delta_m(F) = |F| - \dim(\operatorname{span}(\{f_j^m\}),$$

so that $T(F) = \{m : \delta_m(F) > 0\}$.

CONJECTURE 5.1.

$$\sum_{m=1}^{\infty} \delta_m(F) \le \binom{r-1}{2}. \tag{5.1}$$

Conjecture 5.1 implies Theorem 2.1 of course, and is valid for the examples given where $|T(F)| = \binom{r-1}{2}$ for $r = 3,4$. Further, if $(n,d) = (2,1)$, then $\delta_m(F) = r-(m+1)$ for $m \le r-2$, and (5.1) is sharp.

We can weaken Conjecture 1.1 enough to make it very plausible: suppose $A = \{m_k\}$ is finite and $m' \notin A$. Does there exist indecomposable F so that $A \subseteq T(F)$ and $m' \notin T(F)$? If not, then the linear dependence of $\{f_j^{m_k}\}$ forces the dependence of $\{f_j^{m'}\}$. This seems unlikely.

EXAMPLE 1 (PART FOUR). We return to our favorite example one more time. Consider the intersection of the plane $M_1(t) = 0$ and the conic $M_2(t) = 0$. Then $t_4 = -(t_1+t_2+t_3)$, and so

$$t_1^2+t_2^2+t_3^2+(-(t_1+t_2+t_3))^2 = 2(t_1^2+t_2^2+t_3^2+t_1t_2+t_1t_3+t_2t_3) = 0. \tag{5.2}$$

It is not difficult to diagonalize (5.2) into

$$(t_1-t_2)^2+2(t_1+t_2)^2+(t_1+t_2+2t_3)^2 = 0. \tag{5.3}$$

Let

$$t_1-t_2 = x^2-y^2, \quad \sqrt{2}(t_1+t_2) = 2xy, \quad i(t_1+t_2+2t_3) = x^2+y^2$$

be the usual Pythagorean parameterization of (5.3). After solving for the t_i's, we discover that $t_j = f_j$, which is yet another way of deriving (1.1)!

The fact that $M_5(t) = 0$ as well can be explained in several ways. First,

$$t_1^5+t_2^5+t_3^5+(-(t_1+t_2+t_3))^5 =$$
$$-5(t_1+t_2)(t_1+t_3)(t_2+t_3)(t_1^2+t_2^2+t_3^2+t_1t_2+t_1t_3+t_2t_3),$$

so that (5.2) implies that $M_5(t) = 0$.

More generally, let e_k, $1 \le k \le 4$, denote the k-th elementary symmetric function in four variables. By Newton's Theorem, if $p(t_1,t_2,t_3,t_4)$ is a symmetric polynomial, then there is a polynomial P so that $p = P(e_1, e_2, e_3, e_4)$. If p is *any* symmetric form of degree five in four variables (not just M_5), then

$$p = \alpha_1 e_1^5 + \alpha_2 e_1^3 e_2 + \alpha_3 e_1^2 e_3 + \alpha_4 e_1 e_2^2 + \alpha_5 e_1 e_4 + \alpha_6 e_2 e_3. \tag{5.4}$$

If $M_1(t) = M_2(t) = 0$, then $e_1(t) = e_2(t) = 0$, and so $p = 0$ as well; p can be written as an element of the ideal (e_1, e_2) by solving for α_k in (5.4).

This construction generalizes. Suppose $r \geq 4$ and let A_r denote the set of integers which *cannot* be written in the form $a(r-1) + br$ for non-negative integers a and b. It is well-known that $|A_r| = \binom{r-1}{2}$ and the largest element in A_r is $r(r-1) - r - (r-1)$. Suppose

$$t_1^k + \cdots + t_r^k = 0, \qquad 1 \leq k \leq r-2. \tag{5.5}$$

Then the same argument given above implies that any homogeneous symmetric polynomial in r variables whose degree lies in A_r will vanish. If we could find pairwise non-proportional polynomials f_j, $1 \leq j \leq r$, satisfying (5.5), then we would thereby construct a family F of r polynomials with $T(F) = A_r$. This is a pipe dream. As several algebraic geometers have kindly pointed out, no such parameterization of (5.5) exists for $r \geq 5$ because it is then a smooth curve of positive genus. Many of the same people have also observed that this paper is "really" the study of rational curves lying on the intersection of several Fermat varieties of different degrees.

There is a different geometric interpretation of (1.1). Define the four pairs of complex numbers (β_j, γ_j) by the factorization $f_j(x,y) = \sigma_j(x - \beta_j y)(x - \gamma_j y)$. Under the standard stereographic projection of $\mathbb{C}$ to the unit sphere S^2, the four pairs (β_j, γ_j) map to the antipodal pairs of vertices of an inscribed cube. This is approach is somewhat reminiscent of Klein's classical work on the icosahedron. It would be very satisfying to understand (4.16) in the same detail as (1.1).

Example 8 shows that $|T(F)|$ can be as large as $\frac{3}{2}r - 3$. This is probably not maximal. Is $|T(F)| = \mathcal{O}(r)$? If not, what is the true growth rate? Can $c \cdot r^2$ be achieved for some $c > 0$? Does $\lim_{r\to\infty} r^{-2}\overline{m}_r$ exist?

Is it true that $T(F)$ contains at most $r-2$ consecutive integers? This is true in all the dysfunctional families presented here, and would subsume Theorem 3.7.

Finally, we make the following definitions:

$$\mathcal{T}(r,n,\infty) = \bigcup_d \mathcal{T}(r,n,d) = \lim_{d\to\infty} \mathcal{T}(r,n,d);$$

$$\mathcal{T}(r,\infty,d) = \bigcup_n \mathcal{T}(r,n,d) = \lim_{n\to\infty} \mathcal{T}(r,n,d);$$

$$\mathcal{T}(r,\infty,\infty) = \bigcup_{n,d} \mathcal{T}(r,n,d) = \lim_{n,d\to\infty} \mathcal{T}(r,n,d).$$

Since $\mathcal{T}(r,\infty,\infty)$ is finite, it must actually be achieved by some $\mathcal{T}(r,n,d)$. Can we compute "minimal" $d(n,r)$ so that $\mathcal{T}(r,n,d(n,r)) = \mathcal{T}(r,n,\infty)$, "minimal" $n(d,r)$ so that $\mathcal{T}(r,n(d,r),d) = \mathcal{T}(r,\infty,d)$ or a family of "minimal" $(n(r),d(r))$'s so that $\mathcal{T}(r,n(r),d(r)) = \mathcal{T}(r,\infty,\infty)$?

6. Acknowledgments

I thank my colleagues Mike Bennett, Sean Sather-Wagstaff and Jack Wetzel, the organizers of the Analytic Number Theory, Commutative Ring Theory and

Geometric Potpourri seminars at UIUC, for their tolerance in letting me woodshed this material in their groups. I also thank Ricky Pollack, Marie-Francoise Roy and Micha Sharir for inviting me to speak at the DIMACS workshop on Algorithmic and Quantitative Aspects of Real Algebraic Geometry in Mathematics and Computer Science in March 2001.

I thank my colleagues Nigel Boston, Dan Grayson and Marcin Mazur for useful conversations. Marcin alerted me to the work of Mark Green. I would also like to thank Andrew Granville, Mark Green, Colleen Kilker, János Kollár and Gerry Myerson for helpful email correspondence. Finally, I am deeply grateful to Noam Elkies and Gary Gundersen for careful readings of the preprint and many useful suggestions.

More than is customary in such acknowledgments, the author wants to emphasize that he is solely responsible for any errors in content or aberrations in mathematical taste.

References

[1] A. Chlebowicz, A. Sładek, and M. Wołowiec-Musiał, *Automorphisms of certain forms of higher degree over ordered fields*, Linear Algebra Appl., **331** (2001), 145-153.

[2] H. Darmon and A. Granville, *On the equations* $z^m = F(x, y)$ *and* $Ax^p + By^q = Cz^r$, Bull. London Math. Soc., **27** (1995), 513-543.

[3] L. E. Dickson, *History of the Theory of Numbers, vol. II, Diophantine Analysis*, Chelsea, New York, 1966.

[4] M. L. Green, *Some Picard theorems for holomorphic maps to algebraic varieties*, Amer. J. Mathematics, **97** (1975), 43-75.

[5] G. G. Gundersen, *Meromorphic solutions of* $f^6 + g^6 + h^6 = 1$, Analysis (Munich), **18** (1998), 285-290.

[6] G. G. Gundersen, *Complex functional equations*, Univ. Joensuu Math. Dept. Report Series, in the proceedings of the summer school, August 2000.

[7] G. H. Hardy and E. M. Wright, *An Introduction to the Theory of Numbers*, 5th Ed., The Clarendon Press, Oxford Univeristy Press, New York, 1979.

[8] W. Hayman, *Warings Problem für analytische Funktionen*, Bayer. Akad. Wiss. Math.-Natur. Kl. Sitzungsber. 1984, (1985), 1–13.

[9] W. Hayman, *Nevanlinna Theory 1945–1995*, XVIth Rolf Nevanlinna Colloquium (Joensuu, 1995) (I. Laine, O. Martio, eds.), W. de Gruyter, Berlin, New York, 1996, 1-11.

[10] J. Molluzzo, Doctoral Thesis, Yeshiva Univ., 1972.

[11] D. J. Newman and M. Slater, *Waring's Problem for the Ring of Polynomials*, J. Number Theory, **11** (1979), 477-487.

[12] B. Reznick, *Sums of even powers of real linear forms*, Mem. Amer. Math. Soc. Number 463, **96** (1992).

[13] B. Reznick, *Equal sums of higher powers of quadratic forms* in preparation. Seminar notes can be found at http://www.math.uiuc.edu/~reznick/92100notes.pdf.

[14] P. Ribenboim, *13 Lectures on Fermat's Last Theorem*, Springer-Verlag, New York-Heidelberg, 1979.

[15] J. F. Voloch, *Diagonal equations over function fields*, Bol. Soc. Bras. Math., **16**(2), (1985), 29-39.

Department of Mathematics, University of Illinois, Urbana, IL 61801
E-mail address: reznick@math.uiuc.edu

DIMACS Series in Discrete Mathematics
and Theoretical Computer Science
Volume **60**, 2003

Efficient Algorithms Based on Critical Points Method

Fabrice Rouillier

ABSTRACT. We consider the problem of computing one point in each semi-algebraically connected component of a real algebraic set.

1. Introduction

Many important theoretical results have been obtained recently in computational real algebraic geometry but only few symbolic algorithms are currently usable in practice. Efficient tools exist for studying real roots of univariate polynomials (Sturm/Sturm-Habicht [**GVLRR89**] sequences, Descartes' rule of sign [**BCL82**], etc.), and some algorithms have been implemented for solving symbolically zero-dimensional systems (Gröbner bases, Rational Univariate Representations [**Rou99**], Geometrical Resolutions [**GLS01**], etc.), but the only available method for the study of real roots of positive dimensional systems is CAD (Cylindrical Algebraic Decomposition [**Col75**]).

The CAD algorithm computes at least one sample point in each semi-algebraically connected component for each non-empty sign condition defined by a set of polynomials. It is a great tool for dealing with small but very general problems like quantifier elimination However, the CAD algorithm computes in every case the same basic data (a cylindrical decomposition of the variety) which makes it inefficient in practice for solving large sub-problems, like deciding the emptiness of a real algebraic set for which theoretical algorithms with a better theoretical complexity exist.

In this article, we consider the problem of computing one point in each semi-algebraically connected component of a real algebraic set. It is important to have an efficient algorithm for this problem, because such an algorithm serves as a a black box in algorithms for solving more complex problems like deciding emptiness of semi-algebraic sets ([**BPR96**], [**GV96**]), and also in solving directly problems arising in industry [**FdSMR98**]. Like many authors ([**GV88, Ren92, Can94, JH89, BPR96, BGHM00**]), our method of choice is to compute the critical points of a well chosen function to produce at least one point in each semi-algebraically connected component of the given variety. This article summarizes recent practical progress made regarding this method ([**RRS00, ARS00, RSS01, Din01**]) and we attempt to justify the numerous

Key words and phrases. Computer Algebra, Real Algebraic Geometry, Algorithms.

choices made in order to get an efficient implementation using existing computational tools and software.

The first part describes briefly the CAD algorithm in its original version (many variants exist) and the principle of the critical points method. It ends with a short comparison of the two methods in their basic versions.

In the second part, we recall the main basic mathematical objects (ideals, prime ideals, equi-dimensional decompositions, real roots etc.) that can be computed in practice and give information about the practical behavior of the various computational strategies. We then recall the main items of the method proposed in [**RRS00**] which contains a practical implementation of the critical points method in the case of varieties defined by a single equation. We finally end with the methods proposed in [**ARS00**] (several variants can be founded in [**Din01**]), which is, up to our knowledge, the most efficient implementation of the critical points method.

1.1. Notations. We denote by K an ordered field, R its real closure and C its algebraic closure. If I is an ideal of $K[X_1, \ldots, X_n]$, $V(I)$ will denote the zero set of I.

1.2. Cylindrical Algebraic Decomposition (CAD). Basically, Collins' CAD algorithm ([**Col75**]) computes one sample point in each semi-algebraically connected component for each non empty sign condition defined by a set of polynomials F.

The basic principle of the CAD algorithm is simple to describe. Start with a set of polynomials $P_1 = \{p_1^{(1)}, \ldots, p_{s_1}^{(1)}\}$, $p_i \in K[X_1, \ldots, X_n]$.

At level 1, consider all the polynomials of P_1 as univariate polynomials w.r.t. X_1 and describe "sign changing events" (multiple roots, roots at infinity, common roots, etc.). This define a sequence of real algebraic sets in $\mathbb{R}^{n-1}$ or equivalently a set ofpolynomials in $K[X_2, \ldots, X_n]$. This is the projection step. The computational process can be recursively defined as follows:

Projection step

At level k, we have a set P_k of polynomial of $K[X_k, \ldots, X_n]$. We construct $P_{k+1} = Proj(P_k)$ as being the smallest set such that:

- If $p \in P_k$, $\deg_{X_k}(p) = d \geq 2$, $Proj(P_k)$ contains all the $sr_j(p, \frac{\partial p}{\partial X_k})$ (non-constant) for $j = 0, \ldots, d$.
- If $p \in P_k$, $q \in P_k$, $Proj(P_k)$ contains $sr_j(p, q)$ (non-constant) for $j = 0, \ldots, \min(\deg_{X_k}(p), \deg_{X_k}(q))$.
- If $p \in P_k$, $\deg_{X_k}(p) \geq 1$ and $lc_{X_k}(p)$ non-constant, $Proj(P_k)$ contains $lc_{X_k}(p)$ and $Proj(P_k \setminus \{p\} \cup \{p - lc_{X_k}(p)\})$.
- If $p \in P_k$, $\deg_{X_k}(p) = 0$ and p non-constant, $Proj(P_k)$, contains p.

where $sr_j(p, q)$ denotes j-th sub-resultant coefficient associated with p and q (For example $sr_0(p, q)$ is the resultant of p and q), while $lc_X(p)$ denotes the leading coefficient of p w.r.t. X. We do not give details about the sub-resultant computation ([**Col67**]). On have just to retain that it is based on the reduction of Sylvester's matrix. Depending on the versions of the used algorithm, such a computation costs between $O(d \log^2 d)$and $O(d^2)$ arithmetic operations if d is a bound on the degrees of polynomials $\in P_k$, the degrees of the polynomials $\in P_{k+1}$ are in $O(d^2)$. The next projection step will be done using $P_{k+1} = \{p_1^{(k+1)}, \ldots, p_{s_{k+1}}^{(k+1)}\}$, $p_i^{(k+1)} \in K[X_{k+1}, \ldots, X_n]$.

Lifting step

After the n-th projection step, we have a set of univariate polynomials $P_n \in K[X_n]$ whose roots define a partition of R. We then compute recursively the cylindrical algebraic decomposition adapted to $P_{n-1}, \ldots, P_1$ (at least one sample point in each semi-algebraically connected component for each non-empty sign condition defined by a set of polynomials P_i). The main operations are:

(1) compute real roots of all polynomials of P_k and sort them;
(2) take one point on each interval between roots of (1);
(3) specialize X_k to (1) and (2) in $P_{k-1} \ldots P_1$.

Except for $k = n$, step (1) leads to the problem of isolating the real roots of a polynomial with real algebraic numbers as coefficients which is, in practice, a difficult task.

Also, the basic CAD algorithm can be easily described and implemented using exclusively operations with univariate polynomials. The exponential behavior of the method is mainly due to the projection step and in particular the increase of polynomial degrees due to sub-resultant computations.

1.3. Critical points Method. In [**GV88**] an algorithm is presented for deciding the emptiness of a semi-algebraic set with a single exponential complexity in the number of variables. In this method as well as in most of its variants [**Ren92, Can94, JH89, BPR94, BPR96, Roy96**], the key idea is to apply deformations so that the critical points of the projection function on one coordinate define a finite set that meets every semi-algebraic connected component of the deformed variety. The final result is then obtained by taking the limits of the points when the infinitesimal tend to zero.

Lets briefly describe the critical points method to compute (at least) one point on each (semi-algebraically) connected component of a real algebraic set.

The main objective is to avoid cascading projections as in CAD algorithm. The basic principle consists in choosing a Morse function defined on the variety, with a finite number of critical points and which intersects each semi-algebraically connected component of the studied variety. The problem is then reduced to the study of zero-dimensional systems.

The two following functions are the most frequently used:

- f = **coordinate function.** If V is compact and smooth, the set of critical points of f is finite and intersects every semi-algebraically connected components.
- f = **distance function to one point** A. If V is smooth and A generic enough, the set of critical points of f is finite and intersects every semi-algebraically connected components.

1.4. Discussion. In [**Hon91**] the author concludes that for small inputs, the running time of the Critical Points Method would be greater than 1 million years while CAD runs in less than 2 sec: *For solving the general case, critical point methods use various tricks (sum of squares, infinitesimal deformations etc.) in order to fit basic theorem's hypothesis or to keep a good theoretical complexity. This induce, in practice, huge computations.*

In [**HRS93**], the authors give examples showing that Critical Point Methods should give better results in practice since many optimizations can be done in particular cases.

For computing one point in each semi-algebraically connected component of a real algebraic set, CAD's output is doubly exponential in the number of variables while critical points method is single exponential.

Lets apply the strategy described in [**BPR94**] on an example (using a coordinate function as Morse function) to understand what happens in practice.

Let $P_1, \ldots, P_s$ be a system with $P_i \in R[X_1, \ldots, X_n]$ and take $Q = P_1^2 + \ldots + P_s^2$.

The basic idea is based on the following well known result: if $V_R(Q)$ is smooth and compact, the set of critical points of the coordinate function x_1 is finite and intersects each semi-algebraically connected component of $V_R(Q)$.

In order to deal with a smooth and compact hypersurface, the authors propose a set of two infinitesimal deformations:

- $Q_\omega = Q^2 + (X_1^2 + X_2^2 + \ldots, X_n + X_{n+1}^2 - \omega^2)$ where X_{n+1} is a new variable and ϵ an infinitesimal;
- $Q_{\omega,\epsilon} = (1-\epsilon)Q_\omega + \epsilon(X_1^{2(d_1+1)} + X_2^{2(d_2+1)} + \ldots + X_n^{2(d_n+1)} + X_{n+1}^6 - 3(\omega + 1)^{2(d_1+1)})$ where d_i , $i = 1 \ldots n+1$ is the degree of Q_ω in X_i and $\frac{1}{\omega}$ an infinitesimal.

The algorithm then computes the set of critical points (in $\mathbb{R}\langle\frac{1}{\omega}, \epsilon\rangle^n$) of the coordinate function x_1 restricted to $V(Q_{\omega,\epsilon})$: the limits when ϵ and $\frac{1}{\omega}$ go to zero of these points define points in each semi-algebraically connected component of $V_R(Q)$.

Computing such points in practice reduces to computing the zeroes of the ideal:

$$I = \langle Q_{\omega,\epsilon}, \frac{\partial}{\partial X_2} Q_{\omega,\epsilon}, \ldots, \frac{\partial}{\partial X_n} Q_{\omega,\epsilon}\rangle \tag{1.1}$$

From a practical point of view, this set of polynomials already defines a Gröbner basis for the degree lexicographic ordering, but the symbolic resolution of such a system (computing a lexicographic Gröbner basis by change of ordering [**FGLM93**] or a Rational Univariate Representation [**Rou99**]) needs, at least, to solve a linear system of equations in $R\langle\epsilon, \frac{1}{\omega}\rangle[X_1, \ldots, X_n]$ whose dimension is the dimension of the $R\langle\epsilon, \frac{1}{\omega}\rangle$-vector space $\frac{R\langle\epsilon, \frac{1}{\omega}\rangle[X_1, \ldots, X_n]}{\langle Q_{\omega,\epsilon}, \frac{\partial}{\partial X_2} Q_{\omega,\epsilon}, \ldots, \frac{\partial}{\partial X_n} Q_{\omega,\epsilon}\rangle}$.

Take for example $n = 3$, $s = 3$ and P_1, P_2, P_3 of degree 2 in each variable X_1, X_2, X_3. The polynomial Q_ω will be of degree 4 in X_1, X_2, X_3, of degree 2 in X_4 and ω, so that the polynomial $Q_{\omega,\epsilon}$ will be of degree 10 in the variables X_1, X_2, X_3, of degree 6 in X_4, of degree 10 in ω and of degree 1 in ϵ. Since $\{Q_{\omega,\epsilon}, \frac{\partial}{\partial X_2} Q_{\omega,\epsilon}, \ldots, \frac{\partial}{\partial X_4} Q_{\omega,\epsilon}\}$ is a Gröbner basis, we can easily see that the dimension of the $R\langle\epsilon, \frac{1}{\omega}\rangle$-vector space $\frac{R\langle\epsilon, \frac{1}{\omega}\rangle[X_1, \ldots, X_4]}{\langle Q_{\omega,\epsilon}, \frac{\partial}{\partial X_2} Q_{\omega,\epsilon}, \ldots, \frac{\partial}{\partial X_4} Q_{\omega,\epsilon}\rangle}$ is $4050 = 10 * 9 * 9 * 5$.

Such a computation is not possible in practice since we would have, at least, to invert a square matrix of dimension 4050 with coefficients in $R[\omega, \epsilon]$ of degree 10 in ω.

Of course, many improvements could be proposed like replacing, for example, the degrees $2(d_i + 1)$ by odd integers greater than d_i in the expression of $Q_{\omega,\epsilon}$ but nevertheless, the computation of the solutions of the system (1.1) would remain impossible for small problems.

2. The Critical Points Method in practice

In this section, we propose practical variants of the Critical Points method for computing at least one point in each semi-algebraically connected component of a real algebraic set. We will explain the numerous technical and strategic choices made, depending on the available existing computational methods and describe some algorithmic optimizations. Our main objective are:

- to keep a good size for the output (single exponential in the number of variables),
- to use **existing computational tools** (and computable mathematical objects),
- to provide an **algorithm** (no *generic choices*, certified result, verifiable hypothesis, etc.)
- provide an **exact result** (in our case, the rational univariate representations of the roots of the zero-dimensional systems defining a finite set of points).

In order to be able to perform the computations in practice, one must as much as possible:

- avoid artificial growth of degrees (sum of squares);
- use *simple* arithmetics (avoid infinitesimal as most as possible).

2.1. A quick review of computable objects and related algorithms. The goal of this section is to list the main objects that are computable in practice using existing software. This will help to understand some of the technical choices and strategies.

2.1.1. *Univariate polynomials.* Several algorithms exist for the study of real or complex roots of univariate polynomials. Among the exact or certified methods on can enumerate

- Sturm/Sturm-Habicht methods and variants,
- algorithms based on Descartes rule of sign,
- numerical or semi-numerical methods.

Since we are exclusively interested in computing, in an exact manner, real roots, we focus on exact methods such as Sturm [**Stu35**] and/or Descartes' rule of sign based algorithms [**CA76**]. One should mention that the most general method is the Sturm-Habicht [**GVLRR89**] method which allows one to deal with polynomials with various kind of coefficients with good properties of specialization [**GVLRR94**]. On the contrary, Descartes' rule of sign based methods assume that the coefficients lie in an archimedian field.

In terms of efficiency, Descartes' rule of sign based methods are the most efficient ones when dealing with polynomial with rational coefficients: this method can *solve* polynomials of degrees greater than 1000 with coefficients having several thousands of digits [**RZ01**]. Until now our experiments concerning computations of the real roots of parametric polynomials or polynomial with Puiseux series as coefficients have been limited to polynomials of low degree (less than 100), whatever the strategy used.

2.1.2. *Zero-dimensional algebraic systems.* A lot of work has been done on the subject during the 1990's. Recent progress on Gröbner bases [**Fau99**] makes it possible now to solve zero-dimensional systems with rational coefficients and several thousands of complex roots. Let $F = \{f_1, \dots, f_s\}$ with $f_i \in K[x_1, \dots, x_n]$ a set of polynomials defining a zero-dimensional ideal $I = < F >$.

The main step in *solving* a zero-dimensional system using exact computations is that of eliminating variables.

For the general case, the best exact representation of the complex roots of a zero-dimensional system is surely the Rational Univariate Representation ([**Rou99**] - called Geometric resolution in [**GLS01**]). With the notations above and following [**Rou99**], the Rational Univariate Representation of the roots of F has (always) the following shape:

$$\begin{cases} f(T) = 0 \\ x_1 = g_1(T)/g(T) \\ \qquad \vdots \\ x_n = g_n(T)/g(T) \end{cases}$$

T is a new variable, independent from $x_1, \ldots, x_n$, the polynomials $f, g, g_1, \ldots, g_n$ have coefficients in K, f and g are coprime and there is a bijection between the roots of f and the zeroes of I that preserves the multiplicities and the real roots.

Taking a Gröbner basis for any monomial ordering, a general algorithm without restrictions on the initial system is proposed in [**Rou99**].

Note that computing a Rational Univariate Representation is equivalent to computing Lexicographic Gröbner bases when the system is in the so called *shape position* (when the quotient algebra $K[x_1, \ldots, x_n]/I$ is cyclic), a lexicographic Gröbner basis have, modulo a linear change of variables, the following shape:

$$\begin{cases} f(x_1) \\ x_2 - g_2(x_1) \\ \qquad \vdots \\ x_n - g_n(x_1) \end{cases}$$

Note that a probabilistic "Gröbner-Free" alternative for such a computation is proposed in [**GLS01**]. In practice, the computation of the RUR can be done efficiently for systems with a large number of complex solutions (say up to about 1000) in the case of polynomials with integer coefficients (using the algorithm proposed in [**Rou99**]) or, in the case of ideals in shape position, by computing a lexicographic Gröbner basis (like suggested in [**FGLM93**]).

Computing with Puiseux series or parameters as coefficients prevents one from using many optimizations, and hence only small systems can be solved in these cases.

2.1.3. *Positive dimensional algebraic systems.* The number of efficient tools for computing properties of positive dimensional algebraic systems is poor. There are mainly three kind of implemented algorithms:

- Gröbner bases;
- Triangular sets;
- Geometric Resolutions (Probabilistic).

We mostly focus on the use of Gröbner bases and Triangular sets. In contrast to the zero-dimensional situation, the term of resolution in the case of positive dimensional systems is not well defined. For example, the knowledge of a Gröbner basis allows one to compute the degree and dimension easily but the obtained information about the the localization of the roots and especially those of the real roots is poor.

The Triangular sets method produce perhaps the most friendly output. However, a triangular set does not represent in a straightforward manner the solutions of the system. More precisely, if $F = \{f_1, \ldots, f_s\}$ with $f_i \in K[x_1, \ldots, x_n]$ is a set of polynomials, one can compute a set of triangular sets T_i:

$$T_i = \begin{cases} t_{1,i}(x_1) \\ t_{2,i}(x_1, x_2) \\ \vdots \\ t_{n,i}(x_1, \ldots, x_{d+1}, x_{d+2}, \ldots, x_n) \end{cases}$$

(some $t_{j,i}$ may be identically null) that *represents* the complex roots of the system. The relation between the roots of T_i and the roots of F is not straightforward. We describe some basic facts about triangular sets; more details can be found in [**Aub99, ALM99, Mor97**]. For $p \in K[X_1, \ldots, X_n] \setminus K$, we denote by $\mathrm{mvar}(p)$ (and we call *main variable* of p) the greatest variable appearing in p w.r.t. a fixed lexicographic ordering.

DEFINITION 2.1. A set of polynomials $T = (t_{d+1}, \ldots, t_n) \subset K[X_1, \ldots, X_n] \setminus K$ is said to be a *triangular set* if

$$\forall (t_i, t_j) \in T \times T \mid t_i \neq t_j \ , \quad \mathrm{mvar}(t_i) \neq \mathrm{mvar}(t_j).$$

In the following, we denote by:

- h_i the leading coefficient of t_i when it is seen as a univariate polynomial in its main variable), and $H(T) = \Pi_{i=d+1}^{n} h_i$.
- $W(T) = \{M \in V(T) \setminus V(H(T))\}$,
- $\mathrm{sat}(T) = \{p \in K[X_1, \ldots, X_n] \mid \exists m \in \mathbb{N},\ H(T)^m p \in \langle T \rangle\}$ the saturated ideal associated with the triangular set T.

Lets consider the particular case of a prime ideal I. In such a situation, the triangular sets decomposition is composed of a unique set T and we have:

$$\overline{W(T)} = V(I).$$

It might sometime happen that a lexicographic Gröbner basis is a triangular set, but it is not always the case. However, one can always extract a triangular set from a lexicographic Gröbner basis as recalled in the next section.

2.2. Computable mathematical objects. The main usable algorithms for the study of polynomial systems are described above. In this section, we will describe briefly the mathematical objects that can be computed in practice. In the case of zero-dimensional systems, one can fully compute the real and complex solutions in an exact way through the Rational Univariate Representation using for univariate polynomials any one of the strategies described above.

Experiments show that one can avoid computing the solutions of systems of large degree with coefficients that are not rational numbers. Under these conditions, systems with about 1000 solutions can be solved. Most of the powerful mathematical results for the study of (real) algebraic varieties suppose that decomposition into irreducible components are computable objects. In practice, such decompositions are difficult to compute and should be, for the moment, considered as not computable for large examples. At the end of this article, we give some examples of use of a new experimental (unpublished) strategy proposed by J.C. Faugère (algorithm F7) for computing the decomposition of any ideal into primes.

As seen before, algorithmic manipulations on ideals are nowadays possible in practice for a large range of examples, mainly through Gröbner bases. Recent algorithmic progress make possible the computation of equi-dimensional decompositions which currently seems to be the maximum of structured information we can compute in practice and with existing software when studying algebraic systems of equations.

Among the few existing solutions one should retain the one proposed in [**Din01**], which combines Gröbner bases and triangular sets computations.

Let us summarize the main related results:

DEFINITION 2.2. A triangular set $T = (t_{d+1}, \ldots, t_n) \subset K[X_1, \ldots, X_n]$ is said to be *regular* if and only if for all $i \in \{d+1, \ldots, n\}$, the initial h_i does not belong to any associated prime ideal of $\text{sat}(t_{d+1}, \ldots, t_{i-1}) \bigcap K[X_1, \ldots, X_{i-1}]$.

A regular triangular set is said to be *separable* if and only if for all $i \in \{d+1, \ldots, n\}$, the partial derivative $\frac{\partial t_i}{\partial X_i}$ does not belong to any associated prime ideal of $\text{sat}(t_{d+1}, \ldots, t_i) \bigcap K[X_1, \ldots, X_i]$.

As mentioned before, the relation between the triangular sets and the variety they represent is not straightforward, but the joint use of Gröbner bases and triangular sets makes it quite easy use in practice to compute equi-dimensional decompositions:

DEFINITION 2.3. Let $G \subset K[X_1, \ldots, X_n]$ be a reduced lexicographical Gröbner basis generating a radical equi-dimensional ideal of dimension d in $K[X_1, \ldots, X_n]$ for the ordering $X_1 < \ldots < X_n$. Let $T = (t_{d+1}, \ldots, t_n)$ be a subset of G such that $\forall g \in G$, $\exists i \in \{d+1, \ldots, n\}$ that satisfies:

((i)) $\text{mvar}(t_i) = \text{mvar}(g)$,
((ii)) $\deg(t_i, \text{mvar}(t_i)) \leq \deg(g, \text{mvar}(t_i))$.

If T is a regular separable triangular set such that $\text{sat}(T) = \langle G \rangle$, then T is said to be extracted from G.

It is not always possible to extract a regular separable triangular set T from a reduced lexicographical Gröbner basis G such that $\text{sat}(T) = \langle G \rangle$ (consider for example $\langle x, y, z \rangle$ for $x < y < z$). In [**Din01**] and [**ARS00**], algorithms are proposed for computing a representation of equidimentional and radical decomposition of any ideal I generated by a set of polynomials. More precisely, one can implement ([**Din01**]) as a subroutine taking as input a polynomial system of equations S and returning a list of couples $\{(G_1, T_1), \ldots, (G_m, T_m)\}$ such that:

- G_i is a reduced lexicographical Gröbner basis $\forall i = 1 \ldots m$,
- T_i is a regular separable triangular set $\forall i = 1 \ldots m$,
- $\text{sat}(T_i) = \langle G_i \rangle$
- $\sqrt{\langle S \rangle} = \bigcap_{i=1}^{m} \langle G_i \rangle$.

2.3. Technical choices. As described in the preceding section, and in the existing software, the following set of functions represents a set of usable and efficient black boxes for the solving algebraic systems of equations:

- roots of zero-dimensional systems and univariate polynomials with rational coefficients,
- ideal representations,
- equi-dimensional decompositions.

Experiments shows that these algorithms may be used for systems of algebraic equations with rational coefficients exclusively (avoid parameters or Puiseux series).

2.4. Choosing the Morse function. Most of the proposed variants of the critical points method use a **coordinate function** or the **distance function** to one point. Lets recall the basic results used in each of these two cases:
Coordinate function. If V is compact and smooth, the set of critical points is finite and intersects every semi-algebraically connected components.
Distance function to one point A. If V is smooth and A generic enough, the set of critical points is finite and intersects every semi-algebraically connected components.

The compactness of the variety is a strong hypothesis from a computational point of view since it is difficult to verify and the smoothness of the variety is assumed in both cases. We then choose the use of the distance function to one generic point.

2.5. The case of a real algebraic set defined by a single equation. In many algorithms, computing a point on each semi-algebraically connected component of a real algebraic set defined by a single equation is a convenient basic black box for computing a point in each semi-algebraically connected component of any real algebraic set (take the sum of squares of polynomials defining the variety).

In this section, we briefly describe a method for performing such a computation.

Let P be any polynomial in $R[X_1, \dots, X_n]$, A any point in R^n and consider the following algebraic set:

$$C(V, A) = \{M \in V | grad_M(P) // \overrightarrow{AM}\}$$

where $V = V(P)$. The set $C(V, A)$ contains the extrema of the distance function $d(A, V)$ and so, meets every semi-algebraically connected component of V, and we can consider having solved the problem if $C(V, A)$ is a finite set of complex points.

Obviously $\sharp C(V, A) = \infty$ if $\sharp\{M \in V | grad_M(P) = \overrightarrow{0}\} = \infty$, ($P$ is a singular polynomial), but we may have $\sharp C(V, A) = \infty$ and $\sharp\{M \in V | grad_M(P) = \overrightarrow{0}\} < \infty$, for example, when A is the center of a circle included in V.

In [**RRS00**], the authors propose several algorithmical solutions based on the computations of Rational Univariate Representations, following the general strategy proposed in [**JH89**].

Testing if $\sharp C(V, A) < \infty$ can easily be done in practice by computing a Gröbner basis for any admissible monomial ordering, so that the case where $\sharp C(V, A) < \infty$ can be solved in a straightforward manner by computing a Rational Univariate Representation of the roots of $C(V, A)$.

When $\sharp C(V, A) = \infty$ two cases occur: either P is singular or A is not generic enough. In [**RRS00**], the authors show that these two cases can be disinguished from each other by computing the dimension of the algebraic set: $S(P) = \{M \in V | grad_M(P) = \overrightarrow{0}\}$. If $S(P)$ is zero-dimensional then there exists a point A such that $\sharp C(V, A) < \infty$. Moreover, such a point can be chosen in a grid $\{0, \dots, d\}^n \in \mathbb{N}^n$ with d sufficiently large (which gives an algorithm).

Thus, the main problem is the case where P is a singular polynomial and $\sharp C(V, A) = \infty = \sharp S(P)$. For such a situation, one have to introduce an infinitesimal ϵ and study the real algebraic set:

$$V_\epsilon = \{M \in \mathbb{C}\langle\epsilon\rangle^n \,|P(M) - \epsilon = 0\}.$$

More precisely, one can compute the points $M \in C(V_\epsilon, A)$ applying the strategy above. Since $P - \epsilon$ is non-singular, $C(V_\epsilon, A)$ is a finite set of points in $\mathbb{C}\langle\epsilon\rangle^n$ and it is sufficient to compute its bounded points ([**RRS00**]). We will not develop this last part since experiments using existing computational tools show that this strategy reaches its limits quite quickly ([**RSS01**]).

One has to point out that the method is very efficient when $\{M \in V | grad_M(P) = \overrightarrow{0}\} < \infty$ since the number of computed points is, in general, quite small and since very powerful tools exists for solving zero dimensional systems. In the general case, the computation of a Rational Univariate Representation of $C(V_\epsilon, A)$ is the bottle-neck since:

- the degrees in ϵ of the result are large;
- working in $\mathbb{Q}(\epsilon)$ prevents one from using various optimizations.

2.6. The general algebraic formulation. In section 2.5 we briefly described an algorithm for computing at least one point in each connected component of a real algebraic set defined by one equation. Experiments ([**RSS01**]) show that the infinitesimal deformation needed for solving the general case (infinitely many singular points) induces hard computations in practice. The objective is now to avoid such a deformation by considering the singular locus independently. The singular locus is an algebraic variety of lower dimension than the variety we want to study. Given any real algebraic set, the basic idea is to compute regular critical points for the chosen Morse function and then to apply the same strategy on the singular locus.

Let $V = V(P_1, \ldots, P_s)$ be an algebraic variety of dimension d and $A \in \mathbb{Q}^n$, and consider the algebraic set:

$$C(V, A) = \{M \in V | \text{rank}(grad_M(P_1), \ldots, grad_M(P_s), \overrightarrow{AM}) \leq n - d\}.$$

If $d(A, V)$ denotes the distance function to one point $A \in R^n$, then:

- $C(V, A)$ may contain the critical points of $d(V, A)$.
- $C(V, A)$ is a priori of smaller dimension than V and contains the singular locus of V.

In practice, due to the lack of efficient tools for computing decompositions into primes, we can not assume that V is irreducible. The main drawback is that the above properties are not, in practice, as generic as we may suppose. Lets describe some *bad* situations.

If $\langle P_1, \ldots, P_s \rangle$ is not radical, then $C(V, A)$ may have the same dimension as V. For example:

$$n = 1 \ , \ V = V(X^2) \Rightarrow C(V, A) = V$$

The condition "$\langle P_1, \ldots, P_s \rangle$ is radical" is not sufficient. In some cases, $C(V, A)$ may not meet every semi-algebraically connected component of V. In the following example:

$$n = 3 \ , \ P_1 = (X_1^2 + X_2^2 - 1)(X_1 - 2) \ , \ P_2 = (X_1 - 2)X_3$$

V is not equi-dimensional and $C(V, A)$ will not meet the component of lowest dimension.

2.6.1. *The main theorem.* The following theorem ([**ARS00**]) shows that if $\langle P_1, \ldots, P_s \rangle$ is radical and if $V(P_1, \ldots, P_s)$ is equi-dimensional, then we have a recursive (with respect to the dimension of the studied varieties) process for computing at zero-dimensional systems whose zero sets intersect each semi-algebraic connected component of V. The two conditions (equi-dimensional and radical) have to be viewed as being minimal algorithmical conditions (in terms of computable objects).

THEOREM 2.4. *Let* $V = V(P_1, \ldots, P_s)$ *be an* **equi-dimensional** *algebraic variety of dimension* d *and* $\langle P_1, \ldots, P_s \rangle$ **radical**. *For* $A \in K^n$*, we define:*

$$C(V, A) = \{M \in V | \text{rank}(grad_M(P_1), \ldots, grad_M(P_s), \overrightarrow{AM}) \leq n - d\}.$$

If D *is a positive integer large enough,* $\exists A \in \{1 \ldots D\}^n$ *such that* $\dim(C(V, A)) < \dim(V)$*, and we have:*

(1) $C(V, A)$ *meets every connected component of* $V \bigcap \mathbb{R}^n$*,*
(2) $C(V, A) = \text{Sing}(V) \bigcup V_0$.

where

- V_0 *is a finite set of points in* $\mathbb{C}^n$*,*
- $\text{Sing}(V) = \{M \in V \mid \text{rank}(grad_M(P_1), \ldots, grad_M(P_s)) < n - d\}$.

Note that the genericity conditions imposed for A are not explicit but can be tested after the computation of $C(V, A)$ simply by testing its dimension.

Theorem 2.4 suggests an algorithm that can be described as follows:

Algorithm 1

- **Input:** A polynomial system S of equations in $K[X_1, \ldots, X_n]$.
- **Output:** A list of zero-dimensional systems whose zero sets intersects every semi-algebraically connected component of $V(S) \bigcap R^n$.
- **External functions:**
  - Minors(S, d, A): compute a polynomial system Q in $K[X_1, \ldots, X_n]$ such that $V(Q) = C(V(S), A)$ where S is a polynomial system of equations in $K[X_1, \ldots, X_n]$ of dimension d.
  - EquiDimDec(S): compute a set $\{S_1, \ldots, S_s\}$ of polynomial systems in $K[X_1, \ldots, X_n]$ such that $V(S_i)$ is equi-dimensional and $< S_i >$ is radical $\forall i = 1 \ldots s$.

(1) list := EquiDimDec(S), result := $\emptyset$,
(2) Choose $A \notin V(S)$.
(3) while list $\neq \emptyset$ do
   - S := first(list), and remove S from list, set d := Dim(S),
   - if $d = 0$ then result := result $\bigcup S$,
   - else
     - (*) Q := Minors(S, d, A) $\bigcup S$ and set u := Dim(Q),
     - if $u = d$ choose another point $A \notin V(S)$ and go to step (*),
     - list := list $\bigcup$ EquiDimDec(Q),
(4) return result.

The algorithm above fits a large part of the objectives:

- all the basic arithmetic operations are done in K (no infinitesimal deformation);
- the needed symbolic preprocessing (equi-dimensional and radical decompositions) are computable quite efficiently in practice;
- the algorithm returns the result without any restriction on the input system.

Unfortunately, the computation of $C(V(Q), A)$ (Minors(Q, d, A)) where $Q = \{Q_1, \ldots, Q_s\} \subset K[X_1, \ldots, X_n]$ is hard to perform in practice. Let introduce some notations to describe precisely this part of the computation. For $A \in C^n$ and $Q = \{Q_1, \dots, Q_s\} \subset K[X_1, \dots, X_n]$, we define, $\forall M \in C^n$, the matrix

$$M_A(Q) = \left[\left[\frac{\partial Q_j}{\partial X_i} \right]_{(i=1\dots n, j=1\dots s)} \middle| \overrightarrow{AM} \right].$$

For $d \in \mathbb{N}$, $0 \leq d < n$, we denote by $\Delta_{A,d}(Q)$ the set of all the minors of order $n-d+1$ in the matrix $M_A(Q)$ such that their last column contains coordinates of AM.

In Algorithm 1, the function Minors(Q, d, A) computes $\Delta_{A,d}(Q)$. Such a computation is difficult since:

- after one recursion step in Algorithm 1, the polynomials in $M_A(Q)$ may be quite large since they are extracted from a Gröbner basis (EquiDimDec(Q)).
- the number of elements in $\Delta_{A,d}(Q)$ is equal to $\binom{s}{n-d}\binom{n}{n-d+1}$.

First experiments show that the computation of $\Delta_{A,d}(Q)$ is the main bottleneck in Algorithm 1.

2.6.2. *Using Triangular Sets.* As mentioned in the previous section, one bottleneck in Algorithm 1 is the computation of a polynomial system Q such that $V(Q) = C(V, A)$, mostly due to the number and size of the minors extracted from the matrices $M_A(Q)$. If G is a lexicographic Gröbner basis, $M_A(G)$ has the following shape:

$$\left[\begin{array}{ccccccccccccccc|c}
\otimes & \otimes & \times & \dots & \times & \otimes & \times & \dots & \times & \otimes & \times & \dots & \times & \\
\vdots & \vdots & \vdots & \vdots & \vdots & \vdots & \vdots & \vdots & \vdots & \vdots & \vdots & \vdots & \vdots & \\
\otimes & \otimes & \times & \dots & \times & \otimes & \times & \dots & \times & \otimes & \times & \dots & \times & \\
0 & \otimes & \times & \dots & \times & \otimes & \times & \dots & \times & \otimes & \times & \dots & \times & \overrightarrow{AM} \\
0 & 0 & 0 & \dots & 0 & \otimes & \times & \dots & \times & \otimes & \times & \dots & \times & \\
\dots & \dots & \dots & \dots & \dots & \dots & \dots & \dots & \dots & \dots & \dots & \dots & \dots & \\
0 & 0 & 0 & \dots & 0 & 0 & 0 & \dots & \times & \otimes & \times & \dots & \times & \\
0 & 0 & 0 & \dots & 0 & 0 & 0 & \dots & 0 & \otimes & \times & \dots & \times &
\end{array}\right]$$

If we consider only the columns containing elements marked $\otimes$, we obtain the matrix $M_A(T)$ where $T = \{t_{d+1}, \dots, t_n\}$ is a triangular set contained in G.

One can show that if T is a regular and separable triangular set such that $sat(T) =< G >$, we can replace $M_A(G)$ by $M_A(T)$ in Algorithm 1, decreasing the number of minors to be computed. More precisely, one can show, that it is sufficient to consider a small subset of these minors. Precisely, if $A = (a_1, \dots a_n)$, $d = \dim(V(Q)) = \dim(V(G))$, consider the restricted list of minors of order $(n-d+1)$ extracted from $\Delta_{A,d}(T)$:

$$\Gamma_A(T) = \{\det(M_A^{(k)}),\ k = 1\dots d\}$$

where

$$M_A^{(k)} = \left[\begin{array}{c|c}
\left[\frac{\partial t_j}{\partial X_k}\right]_{j=d+1\dots n} & X_k - a_k \\
\hline
 & X_{d+1} - a_{d+1} \\
U_T = \left[\frac{\partial t_j}{\partial X_i}\right]_{j=d+1\dots n}^{i=d+1\dots n} & \vdots \\
 & X_n - a_n
\end{array}\right]$$

By studying the zeroes of $\Gamma_A(T)$ instead of those of $\Delta_{A,d}(G)$ we define a variety that contains $C(V(G), A)$. The following theorem shows that we can replace $C(V(G), A)$ by $D(V(G), A) = V(G) \bigcap V(\Gamma_A(T))$ in theorem 2.4:

THEOREM 2.5. *Let G be a lexicographical Gröbner basis and $T = \{t_{d+1}, \ldots, t_n\}$ be a regular separable triangular set such that $T \subset G$ and $sat(T) =< G >$.*
We define:

- $D(V(G), A) = V(G) \bigcap V(\Gamma_A(T))$;
- $d = \dim(G)$;
- $Sep(T) = \prod_{i=d+1}^{n} \frac{\partial t_i}{\partial X_i}$.

If $A \in C^n$ satisfies $\dim(C(V(G), A)) < \dim(V(G))$, then the following relations hold:

(1) $C(V(G), A) \subset D(V(G), A)$,
(2) $(D(V(G), A) \setminus V(Sep(T))) \subset V_0$,
(3) $dim\,(D(V(G), A) \bigcap V(Sep(T))) < \dim(V(G))$.

In particular, $\dim(D(V(G), A)) < \dim(V(G))$ and $D(V(G), A)$ meets every semi-algebraically connected component of $V(G)$.

Note that the condition *T is a regular and separable triangular set such that $T \subset G$ and $sat(T) =< G >$* implies that G is equi-dimensional and radical. Let us now describe the algorithm suggested by theorem 2.5 :

Algorithm 2

- **Input:** A polynomial system S of equations in $K[X_1, \ldots, X_n]$.
- **Output:** A list of zero-dimensional systems whose roots contain at least one point in each semi-algebraically connected component of $V(S) \bigcap R^n$.
- **External functions:**
  - LexTriSetEquiDim(Q): compute a set $\{(G_1, T_1), \ldots, (G_s, T_s)\}$ of couples (G_i, T_i) such that T_i is a triangular system and G_i a lexicographic Gröbner basis such that T_i is regular separable with $sat(T_i) =< G_i >$ and $V(Q) = \bigcup_{i=1}^{s} V(G_i)$.

(1) list := LexTriSetEquiDim(S), result := $\emptyset$,
(2) Choose $A \not\subset V(S)$.
(3) while list $\neq \emptyset$ do
  - (G, T) := first(list), and remove (G, T) from list, set d := Dim(G),
  - if $d = 0$ then result := result $\bigcup G$,
  - else
    - (*) $Q := \Gamma_A(T) \bigcup G$ and set u := Dim(Q)
    - if $u = d$ choose another point $A \not\in V(G)$ and go to step (*).
    - $d := u$; list := list $\bigcup$ LexTriSetEquiDim(Q),

(4) return result.

Algorithms 1 and 2 differ only in the number of minors to be computed and by the strategy used to compute equi-dimensional and radical decompositions.

3. An example

The main parts of Algorithm 1 and 2 are implemented in Maple but use Gb (software developed by J.-C. Faugère) for computing Gröbner bases, and RS (software developed by F. Rouillier) for solving zero-dimensional systems. Separable and regular triangular sets are computed through lexicographic Gröbner bases as proposed in [**Din01**].

We consider the following example:

$$S = \left\{ \begin{array}{l} 2\,a_3 - a_1, \\ 2\,a_4 - a_1, \\ 2\,a_5 - a_2, \\ 2\,a_6 - a_2, \\ -2\,a_7\,a_3 + {a_3}^2 + 2\,a_7\,a_4 - {a_4}^2 + 2\,a_8\,a_5 - {a_5}^2, \\ -2\,a_7\,a_3 + {a_3}^2 + 2\,a_8\,a_6 - {a_6}^2, a_2\,a_9 + a_1\,a_{10} - a_1\,a_2, \\ a_2\,a_{10} - a_1\,a_9, -2\,a_7\,a_3 + {a_3}^2 + 2 \\ a_7\,a_9 - {a_9}^2 + 2\,a_8\,a_{10} - {a_{10}}^2 \end{array} \right.$$

The system S has dimension 3 but the computation of an equi-dimensional decomposition shows that it also has components of dimension 2. Precisely, using the strategy described above, we compute 5 components of dimension 2 and one component of dimension 3 (C_3).

$$C_1 = \left\{ \begin{array}{l} 2\,a_7 - a_3, \\ -2\,a_3 + a_1, \\ a_4 - a_3, \\ 2\,a_8 - a_5, \\ {a_9}^2 + {a_{10}}^2 - 2\,a_{10}\,a_5, \\ a_9\,a_3 - a_{10}\,a_5, \\ a_9\,a_5 + a_{10}\,a_3 - 2\,a_3\,a_5, \\ a_{10}\,{a_3}^2 + a_{10}\,{a_5}^2 - 2\,{a_3}^2 a_5, \\ -2\,a_5 + a_2, \\ a_6 - a_5 \end{array} \right. , \quad C_2 = \left\{ \begin{array}{l} a_1, \\ a_4, \\ 2\,a_8 - a_5, \\ a_9, \\ a_{10}, \\ -2\,a_5 + a_2, \\ a_3, \\ a_6 - a_5 \end{array} \right. ,$$

$$C_3 = \left\{ \begin{array}{l} 2\,a_7\,a_9 + 2\,a_8\,a_{10} - {a_9}^2 - {a_{10}}^2, \\ a_1, \\ a_4, \\ a_2, \\ a_3, \\ a_6, \\ a_5 \end{array} \right. , \quad C_4 = \left\{ \begin{array}{l} a_1, \\ a_4, \\ a_9, \\ a_{10}, \\ a_2, \\ a_3, \\ a_6, \\ a_5 \end{array} \right. , \quad C_5 = \left\{ \begin{array}{l} 2\,a_7 - a_9, \\ a_1, \\ a_4, \\ a_{10}, \\ a_2, \\ a_3, \\ a_6, \\ a_5 \end{array} \right. ,$$

$$C_6 = \left\{ \begin{array}{l} 2\,a_7 - a_3, \\ -2\,a_3 + a_1, \\ a_4 - a_3, \\ a_9, \\ a_{10}, \\ a_2, \\ a_6, \\ a_5 \end{array} \right.$$

Let us examine the component C_1 more closely. Choose $A = [1, 2, \ldots, 10]$.

In algorithm 1, one would have to compute most all the 9×9 minors (Minors(C_1,2,A)) of the (11×11) matrix defining $C(V, A)$.

Let us now consider Algorithm 2.

It is easy to see that a regular and separable triangular set associated with C_1 is:

$$T_1 = \begin{cases} 2\,a_7 - a_3, \\ -2\,a_3 + a_1, \\ a_4 - a_3, \\ 2\,a_8 - a_5, \\ a_9\,a_5 + a_{10}\,a_3 - 2\,a_3\,a_5, \\ a_{10}\,{a_3}^2 + a_{10}\,{a_5}^2 - 2\,{a_3}^2 a_5, \\ -2\,a_5 + a_2, \\ a_6 - a_5 \end{cases}.$$

So, in algorithm 2, one has to compute only 2 determinants in order to define $(D(V(G), A))$. In particular, this means that this first recursion step will generate only 2 new sub-problems of dimension at most 1.

In the next recursion step, the sets $C(V, A)$ will be zero-dimensional so that we will be able to apply dedicated resolution methods, as described above, for computing the sample points.

The author is grateful to Saugata Basu and Laureano Gonzalez-Vega for their help in drafting of this article.

References

[ALM99] P. Aubry, D. Lazard, and M. Moreno Maza, *On the theories of triangular sets*, Journal of Symbolic Computation **28** (1999), 105–124.

[ARS00] P. Aubry, F. Rouillier, and M. Safey, *Real solving for positive dimensional systems*, Tech. Report 2000/009, Laboratoire d'Informatique de Paris VI, 2000.

[Aub99] P. Aubry, *Ensembles triangulaires de polynômes et résolution de systèmes algébriques*, Ph.D. thesis, Université Paris 6, France, 1999.

[BCL82] B. Buchberger, G.-E. Collins, and R. Loos, *Computer algebra symbolic and algebraic computation*, second edition ed., Springer-Verlag, 1982.

[BGHM00] B. Bank, M. Giusti, J. Heintz, and G. MBakop, *Polar varieties and efficient real elimination*, Mathematische Zeitschrift (2000), To appear.

[BPR94] S. Basu, R. Pollack, and M.-F. Roy, *On the combinatorial and algebraic complexity of quantifier elimination*, Proc. 35th IEEE Symp. on Foundations of Computer Science (1994), 632–641.

[BPR96] S. Basu, R. Pollack, and M.F. Roy, *On the combinatorial and algebraic complexity of quantifier elimination*, Journal of Assoc. Comput. Machin. (1996), 1002–1045.

[CA76] G. Collins and A. Akritas, *Polynomial real root isolation using descartes' rule of signs*, SYMSAC, 1976, pp. 272–275.

[Can94] J. Canny, *A toolkit for nonlinear algebra*, Algorithmic foundations of robotics (San Francisco, CA, USA) (K. Goldberg et al., ed.), Workshop on the algorithmic foundations of robotics '94, 17-19 February 1994.

[Col67] G. E. Collins, *Subresultants and reduced polynomial remainder sequences*, Journal of A. C. M. **14** (1967), 128–142.

[Col75] G.E. Collins, *Quantifier elimination for real closed fields by cylindrical algebraic decomposition*, Springer Lecture Notes in Computer Science 33 **33** (1975), 515–532.

[Din01] M. Safey El Din, *Résolution réelle des systèmes polynomiaux en dimension positive*, Ph.D. thesis, Université de Paris VI, 2001.

[Fau99] J.-C. Faugère, *A new efficient algorithm for computing gröbner bases (f4)*, Journal of Pure and Applied Algebra **139** (1999), no. 1-3, 61–88.

[FdSMR98] Jean-Charles Faugère, François Moreau de Saint Martin, and Fabrice Rouillier, *Design of regular nonseparable bidimensional wavelets using Groebner basis techniques*, IEEE SP Transactions Special Issue on Theory and Applications of Filter Banks and Wavelets **46** (1998), no. 4, 845–856.

[FGLM93] J.C. Faugère, P. Gianni, D. Lazard, and T. Mora, *Efficient computation of zero-dimensional gröbner basis by change of ordering*, Journal of Symbolic Computation **16** (1993), no. 4, 329–344.

[GLS01] M. Giusti, G. Lecerf, and B. Salvy, *A gröbner free alternative for solving polynomial systems*, Journal of Complexity **17** (2001), no. 1, 154–211.

[GV88] D. Grigor'ev and N. Vorobjov, *Solving systems of polynomial inequalities in subexponential time*, J. Symbolic Comput. **5** (1988), 37–64.

[GV96] L. Gonzalez-Vega, *Applying quantifier elimination to the birkhoff interpolation problem*, Journal of Symbolic Computation **22** (1996), no. 1, pages 297–310.

[GVLRR89] L. González-Vega, H. Lombardi, T. Recio, and M-F. Roy, *Sturm-habicht sequence*, ISSAC-89 Proceedings, ACM-Press, 1989, pp. 136–146.

[GVLRR94] L. González-Vega, H. Lombardi, T. Recio, and M-F. Roy, *Spécialisation de la suite de sturm*, Informatique théorique et applications **28** (1994), 1–24.

[Hon91] H. Hong, *Comparison of several decision algorithms for the existential theory of the reals*, Tech. report, RISC-Linz, Johannes Kepler University, 1991.

[HRS93] J. Heintz, M.-F. Roy, and P. Solerno, *On the theoritical an practical complexity of the existential theory of reals*, The Computer Journal **36** (1993), no. 5, 427–431.

[JH89] P.S̃olernó J.H̃eintz, M.-F.R̃oy, *On the complexity of semi-algebraic sets*, Proc. IFIP 89, San Francisco, 1989, pp. 293–298.

[Mor97] M. Moreno Maza, *Calculs de pgcd au-dessus des tours d'extensions simples et résolution des systèmes d'équations algébriques*, Ph.D. thesis, Université Paris 6, France, 1997.

[Ren92] J. Renegar, *On the computational complexity and geometry of the first-order theory of the reals*, Journal of Symbolic Computation **13** (1992), 255–352.

[Rou99] F. Rouillier, *Solving zero-dimensional systems through the rational univariate representation*, Journal of Applicable Algebra in Engineering, Communication and Computing **9** (1999), no. 5, 433–461.

[Roy96] M.F. Roy, *Basic algorithms in real algebraic geometry: from Sturm theorem to the existential theory of reals*, Lectures on Real Geometry (F. Broglia, ed.), Expositions in Mathematics, vol. 23, de Gruyter, Berlin, New-York, 1996, pp. 1–67.

[RRS00] F. Rouillier, M.F. Roy, and M. Safey, *Finding at least one point in each connected component of a real algebraic set defined by a single equation*, Journal of Complexity **16** (2000), 716–750.

[RSS01] F. Rouillier, M. Safey, and É. Schost, *Solving the birkhoff interpolation problem via the critical point method: An experimental study*, Automated Deduction in Geometry - Third International Workshop ADG 2000, Zurich Switzerland, September 2000, Revised Papers (J. Richter-Gebert and D. Wang, eds.), Lecture Notes in Artificial Intelligence, no. 2061, Springer, 2001, pp. 26–40.

[RZ01] F. Rouillier and P. Zimmermann, *Efficient isolation of a polynomial real roots*, Tech. Report RR-4113, INRIA, 2001.

[Stu35] C. Sturm, *Mémoire sur la résolution des équations numériques*, Inst. France Sc. Math. Phys. 6 (1835).

SPACES project, LIP6/INRIA Lorraine - Nancy,France

Current address: 8, rue du Capitaine Scott, 75015 Paris

E-mail address: Fabrice.Rouillier@loria.fr

DIMACS Series in Discrete Mathematics
and Theoretical Computer Science
Volume **60**, 2003

Enumerative Real Algebraic Geometry

Frank Sottile

1. Introduction

Consider the following question.

QUESTION 1.1. *Find a priori information about the number of real solutions to a structured system of real polynomial equations*

$$0 = f_1 = f_2 = \cdots = f_N, \qquad \text{where each } f_i \in \mathbb{R}[x_1, \ldots, x_n]\,. \tag{1.1}$$

Two well-defined classes of structured polynomial systems have been studied from this point of view—sparse systems, where the structure is encoded by the monomials in the polynomials f_i—and geometric systems, where the structure comes from geometry. This second class consists of polynomial formulations of enumerative geometric problems, and in this case Question 1.1 is the motivating question of enumerative real algebraic geometry, the subject of this survey. Treating both sparse polynomial systems and enumerative geometry together in the context of Question 1.1 gives useful insight.

Given a system of polynomial equations (1.1) with d complex solutions, we know the following easy facts about its number ρ of real solutions.

(1.2)
$$\begin{aligned}&\text{We have } \rho \leq d, \text{ and}\\ &\text{since } d-\rho \text{ an even integer}, \ \rho \geq \begin{cases} 0 & \text{if } d \text{ is even}\\ 1 & \text{if } d \text{ is odd}\,.\end{cases}\end{aligned}$$

In a surprising number of cases, much better information than this is known.

Research Supported in part by NSF grant DMS-0070494.

2000 *Mathematics Subject Classification.* 14P99, 12D10, 14N10, 14N15, 14M15, 14M25, 14M17.

Structured systems occur in families of systems sharing the same structure. The common structure determines the number d of complex solutions to a general member of the family. We assume throughout that a general system in any family has only simple solutions in that each complex solution occurs without multiplicity.

Given such a family of structured systems whose general member has d complex solutions, perhaps the ultimate answer to our motivating question is to determine exactly which numbers ρ of real solutions can occur and also which systems have a given number of real solutions. Because this level of knowledge may be unattainable, we will be satisfied with less knowledge.

For example, are the trivial bounds given in (1.2) sharp? That is, do there exist systems attaining the maximal and minimal number of real solutions allowed by (1.2)? If these bounds are not sharp, do there exist better sharp bounds? Perhaps we are unable to determine sharp bounds, but can exhibit systems in a family with many (or few) real solutions. This gives lower bounds on the maximum number of real solutions to a system in a family (or upper bounds on the minimum number).

These answers have two parts: bounds and constructions. We shall see that bounds (or other limitations) often come from topological considerations. On the other hand, the constructions often come by deformations from/to a degenerate situation. In enumerative geometry, this is the classical technique of special position, while for sparse systems, it is Viro's method of toric deformations.

In Section 2 we discuss sparse polynomial systems from the point of view of Question 1.1. The heart of this survey begins in Section 3, where we discuss some of the myriad examples of enumerative geometric problems that have been studied from this perspective. In particular, for many enumerative problems the upper bound (1.2) is sharp. In Section 4, we concentrate on enumerative problems from the Schubert calculus, where much work has been done on this question of real solutions. Section 5 is devoted to a conjecture of Shapiro and Shapiro, whose study has led to many recent results in this area. Finally, in Section 6 we describe new ideas of Eremenko and Gabrielov giving lower bounds better than (1.2) for some enumerative problems.

2. Sparse Polynomial Systems

Perhaps the most obvious structure of a system of n polynomials in n variables

$$f_1(x_1,\ldots,x_n) \;=\; f_2(x_1,\ldots,x_n) \;=\; \cdots \;=\; f_n(x_1,\ldots,x_n) \;=\; 0\,, \tag{2.1}$$

is the list of total degrees of the polynomials $f_1, f_2, \ldots, f_n$. For such a system, we have the degree or Bézout upper bound, which is a consequence of the refined Bézout Theorem of Fulton and MacPherson [**18**, §1.23].

THEOREM 2.1 (Bézout Bound). *The system* (2.1), *where the polynomial f_i has total degree $d_i := \deg(f_i)$, has at most $d_1 \cdot d_2 \cdots d_n$ isolated complex solutions.*

The Bézout bound on the number of real solutions is sharp. For example, if

$$f_i \;:=\; (x_i-1)(x_i-2)\cdots(x_i-d_i) \quad i=1,\ldots,n\,, \tag{2.2}$$

then the system (2.1) has $d_1 \cdot d_2 \cdots d_n$ real solutions. The reader is invited to construct systems with the minimum possible number (0 or 1) of real solutions.

A system of polynomial equations with only simple solutions, but with fewer solutions than the Bézout bound is called deficient. For example, fewer monomials in the polynomials lead to fewer solutions. We make this idea more precise. A monomial (or rather its exponent vector) is a point of $\mathbb{N}^n$. The convex hull of the monomials in f is its *Newton polytope*, $New(f)$, a polytope with integral vertices. The terms of f are indexed by the lattice points, $New(f) \cap \mathbb{N}^n$, in the Newton polytope of f.

Figure 1 displays the monomials (dots) and Newton polytopes of the polynomials

$$\begin{aligned} f &= 1 + x - y + xyz^4 , \\ g &= 1 + x - y + 3z - z^3 + 2z^4 . \end{aligned}$$

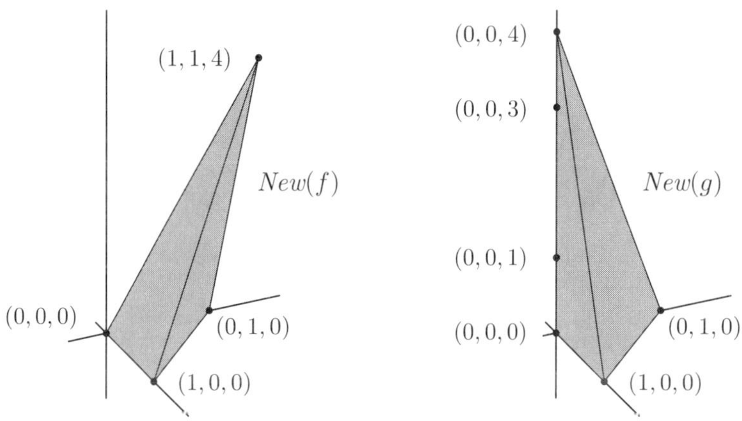

FIGURE 1. Newton Polytopes

Given polytopes $P_1, P_2, \ldots, P_m \subset \mathbb{R}^n$, their Minkowski sum is the pointwise sum

$$P_1 + P_2 + \cdots + P_m = \{p_1 + p_2 + \cdots + p_m \mid p_i \in P_i,\ i = 1, 2, \ldots, m\} .$$

The mixed volume, $MV(P_1, P_2, \ldots, P_m)$, of a collection of m polytopes is

$$\frac{\partial}{\partial t_1}\frac{\partial}{\partial t_2}\cdots\frac{\partial}{\partial t_m}\left[\text{Volume of } t_1P_1 + t_2P_2 + \cdots + t_mP_m\right] .$$

When there are n equal polytopes, the mixed volume is $n!\mathrm{Vol}(P)$, the normalized volume of the common Newton polytope P, and when each P_i is a line segment of length d_i in the ith coordinate (so that $P_i = New(f_i)$, where f_i is the polynomial given in (2.2)), the mixed volume is the Bézout number $d_1 d_2 \cdots d_n$.

Given a list $(P_1, P_2, \ldots, P_n)$ of polytopes in $\mathbb{R}^n$ with vertices in the integral lattice $\mathbb{N}^n$, a *sparse polynomial system* with this structure is a system of polynomials (2.1) with $New(f_i) = P_i$. These sparse systems may have trivial solutions where some coordinates vanish. Thus we only consider solutions in $(\mathbb{C}^\times)^n$. We have the following basic result on the number of solutions to such a sparse system of polynomials.

THEOREM 2.2 (BKK bound). *A sparse polynomial system* (2.1) *with structure* $(P_1, P_2, \ldots, P_n)$ *has at most* $MV(P_1, P_2, \ldots, P_n)$ *isolated solutions in* $(\mathbb{C}^\times)^n$. *If the system is generic given its structure, then it has exactly this number of solutions in* $(\mathbb{C}^\times)^n$.

This result was developed in a series of papers by Kouchnirenko [**43**], Bernstein [**1**], and Khovanskii [**38**]. For simplicity of exposition, we will largely restrict ourselves to the case when the polynomials all have the same Newton polytope P.

Given a polytope P with vertices in the integral lattice, what are the possible numbers of real solutions to systems with that Newton polytope? We shall focus on understanding $\rho(P)$, the maximum number of real solutions to a sparse system with Newton polytope P. The following example serves as an introduction to this question.

EXAMPLE 2.3. Let d be a positive integer, and set

$$\begin{aligned} P_d &:= \text{Convex hull}\,\{(0,0,0),\ (1,0,0),\ (0,1,0)\,,(1,1,d)\}\,, \quad \text{and} \\ Q_d &:= \text{Convex hull}\,\{(0,0,0),\ (1,0,0),\ (0,1,0)\,,(0,0,d)\}\,. \end{aligned}$$

(Figure 1 shows P_4 and Q_4.) These tetrahedrons each have normalized volume d.

A general sparse system with Newton polytope P_d

$$\begin{aligned} A_1xyz^d + B_1x + C_1y + D_1 &= 0 \\ A_2xyz^d + B_2x + C_2y + D_2 &= 0 \\ A_3xyz^d + B_3x + C_3y + D_3 &= 0 \end{aligned}$$

is equivalent to a system of the form

$$z^d = \alpha, \quad x = \beta, \quad y = \gamma,$$

where $\alpha, \beta, \gamma \in \mathbb{R}$. Thus the original system has d complex solutions, but only 0, 1, or 2 real solutions, so $\rho(P_d)$ is 1 or 2, depending upon the parity of d.

A general sparse system with Newton polytope Q_d is 3 polynomials of the form

$$Ax + By + C(z)\,,$$

where $C(z)$ has degree d. This is equivalent to a system of the form

$$f(z) = 0, \quad x = g(z), \quad y = h(z)\,,$$

where $f(z), g(z)$, and $h(z)$ are real polynomials with $\deg(f) = d$ but $\deg(g) = \deg(h) = d - 1$. In this case, the general system again has d complex solutions, but there are systems with any number of real solutions, so $\rho(Q_d) = d$.

2.1. Polyhedral homotopy algorithm. The polyhedral homotopy algorithm of Huber and Sturmfels [**31**] deforms the sparse system (2.1) into a system where the number of solutions is evident. It gives an effective demonstration of the BKK bound and is based upon Sturmfels's generalization [**81**] of Viro's method for constructing real varieties with controlled topology [**85**].

EXAMPLE 2.4. Suppose P is a n-simplex which meets the integral lattice only at its vertices. Translating one vertex to the origin, the others are linearly independent. (Translating corresponds to division by a monomial.) Let M be the $n \times n$ integral matrix whose columns are these vertices. Multiplying M by an invertible integral matrix and taking the resulting columns to be a basis for $\mathbb{R}^n$ corresponds to a multiplicative change of coordinates

$$(x_1, x_2, \ldots, x_n) \longmapsto (y^{m_1}, y^{m_2}, \ldots, y^{m_n})\,,$$

with $y_1, y_2, \ldots, y_n \in (\mathbb{C}^\times)^n$ and $m_1, m_2, \ldots, m_n$ linearily independent integer vectors. Doing this for the Smith normal form of M transforms a polynomial f with Newton polytope P into a polynomial of the form

$$c_0 + c_1 y_1^{d_1} + c_2 y_2^{d_2} + \cdots + c_n y_n^{d_n}\,, \tag{2.3}$$

where $d_1 \mid d_2 \mid \cdots \mid d_n$ and $d_1 d_2 \cdots d_n = n!\mathrm{Vol}(P)$, the normalized volume of P. (These d_i are the invariant factors of the integral matrix M.) A system consisting of n general polynomials of the form (2.3) is equivalent to one of the form

$$y_1^{d_1} = \alpha_1\,, \quad y_2^{d_2} = \alpha_2\,, \quad \ldots\,, \quad y_2^{d_n} = \alpha_n\,.$$

Thus a general system whose Newton polytope is P has $n!\mathrm{Vol}(P)$ simple complex solutions.

The combinatorial structure underlying the homotopy algorithm of Huber and Sturmfels is that of a regular triangulation, which is a special case of a regular subdivision. A regular subdivision P_w of a lattice polytope P is given by a lifting function

$$w\colon P \cap \mathbb{N}^n \longrightarrow \mathbb{Q}_{\geq 0}$$

as follows. Set

$$Q \;:=\; \text{Convex hull}\,\{(a, w(a)) \mid a \in P \cap \mathbb{N}^n\} \;\subset\; \mathbb{R}^{n+1}$$

This lifted polytope Q has distinguished lower facets, those facets whose inward pointing normal vector has positive last coordinate. Forgetting the last coordinate projects these lower facets into $\mathbb{R}^n$ (and hence P). Their totality gives the *regular subdivision* P_w of P. When all the lattice points in $P \cap \mathbb{N}^n$ lift to vertices of lower facets of Q, and these lower facets are all simplices, then P_w is a *regular triangulation* of P.

In Figure 2 the triangulation on the left is regular and the triangulation on the right is not regular. Consider a hypothetical lifting function w for the triangulation on the right. We assume that w takes the value 0 at the three interior vertices. The clockwise neighbour of any vertex of the big triangle must be lifted higher than that vertex. (Consider the figure they form with the parallel edge of the interior triangle.) Since the edge of the big triangle is lifted to a concave broken path, this implies that each vertex is lower than its clockwise neighbour, which is impossible, except in some M.C. Escher woodcuts.

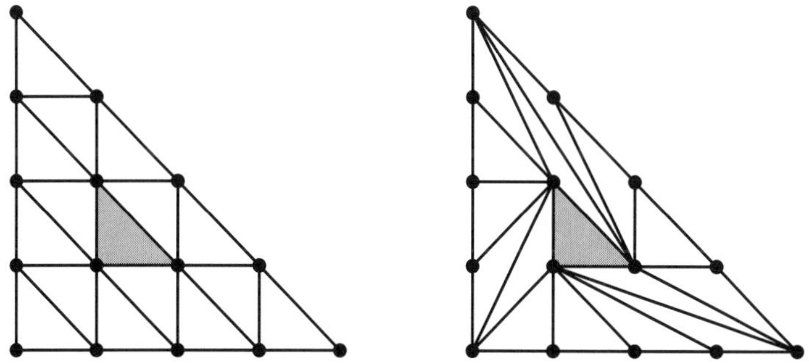

FIGURE 2. Regular and non-regular triangulations

Crucial to this algorithm are polynomial systems derived from the original system and the regular subdivision P_w of P by a lifting function w. Given a polynomial f with Newton polytope P and a face F of the subdivision P_w, consider the sum of terms of f with exponent vector in F, restricting f to the face F. Doing this for each polynomial f_i in our original system, we obtain the facial subsystem of (2.1) given by F.

We continue with the algorithm of Huber and Sturmfels. Given a lifting function $w\colon P\cap\mathbb{N}^n \to \mathbb{Q}_{\geq 0}$ and a polynomial f with Newton polytope P

$$f(x) \;=\; \sum_{a\in P\cap\mathbb{N}^n} c_a x^a\,,$$

we multiply the term with exponent a by $t^{w(a)}$ to obtain

$$f(x;t) \;:=\; \sum_{a\in P\cap\mathbb{N}^n} c_a x^a t^{w(a)}\,.$$

Modifying all polynomials in the original sparse system in this way gives the *lifted system*, a family of sparse systems depending upon the parameter t.

$$f_1(x;t) \;=\; f_2(x;t) \;=\; \cdots \;=\; f_n(x;t) \;=\; 0\,, \tag{2.4}$$

Solutions to this system are algebraic functions $t\mapsto x(t)$ in the parameter t. In a neighborhood of 0 in the complex plane, each branch is expressed as a Puiseaux series

$$\begin{aligned} x(t) &= (x_1(t), x_2(t),\ldots,x_n(t))\,, \qquad \text{with}\\ x_i(t) &= t^{u_i}y_i + \text{ higher order terms in } t\,, \end{aligned}$$

where the y_i are non-zero constants and the exponents $u_i\in\mathbb{Q}$. Substituting this expression into (2.4), we obtain

$$0 \;=\; \sum_{a\in P\cap\mathbb{N}^n} c_a y^a t^{u\cdot a + w(a)} + \text{ higher order terms in } t\,.$$

The exponent of t is the dot product $(u,1)\cdot(a,w(a))$. Thus the terms of lowest order in t correspond to points $(a,w(a))$ in the lifted polytope where the exponent vector $(u,1)$ achieves its minimum—a face in the lower hull of the lifted polytope. Let F be the corresponding face of the regular subdivision P_w. Removing the common factor $t^{u\cdot a+w(a)}$ from these lowest order terms gives the restriction of f to the face F. Thus the coefficients y_i of the leading terms of the Puiseaux expansion are solutions to the facial subsystem of the original system (2.1) given by the face F selected by the initial Puiseaux exponents.

Suppose the original system is general in the sense that each of its facial subsystems given by faces F of the regular subdivision has solutions only if $\mathrm{Vol}(F)>0$. Consider substituting the Puiseaux series into the polynomial system (2.4) and then taking the limit as t approaches 0. The resulting polynomial system for the leading coefficients y of the Puiseaux series are solutions to the corresponding facial system. By previous assumption, this has no solutions unless F is a facet of the subdivision P_w. In that case, the inward pointing normal vector $(u,1)$ to the corresponding facet of the lifted polytope gives the initial exponent u of the Puiseaux expansion.

The full Puiseaux series can be reconstructed from its initial terms (see [**31**] for details) giving the 1-1 correspondence

$$\left\{\begin{array}{l}\text{Solutions to facial subsystems}\\ \text{of (2.1) given by facets of } P_w\end{array}\right\} \Longleftrightarrow \left\{\begin{array}{l}\text{Branches of the algebraic func-}\\ \text{tion } t \mapsto x(t) \text{ near } t=0\end{array}\right\}$$

This is the number of solutions to (2.4) for general $t \in \mathbb{C}$, so the number of solutions to the original sparse system equals the number of solutions to all the facial subsystems given by facets of P_w.

Now suppose the regular subdivision P_w is a regular triangulation and each facial subsystem is general. Then the facets F of P_w are simplices with no interior vertices. Since general sparse systems whose Newton polytope is such a simplex have exactly $n!\mathrm{Vol}(F)$ solutions, the number of solutions to facial subsystems given by facets of P_w is exactly the sum of the normalized volumes of all facets of P_w, which is the normalized volume of P.

2.2. Real solutions to sparse polynomial systems. How many real solutions are there to a sparse polynomial system (2.1) with common Newton polytope P? By Example 2.4, if P is the convex hull of $v_0, v_1, \ldots, v_n$, a simplex with no interior lattice points, then a general system (2.1) is equivalent to the system of binomials

$$y_1^{d_1} = \alpha_1, \quad y_2^{d_2} = \alpha_2, \quad \ldots, \quad y_n^{d_n} = \alpha_n\,, \tag{2.5}$$

with each $a_i \neq 0$. Here, $d_1 \mid d_2 \mid \cdots \mid d_n$ are the invariant factors of the matrix whose ith column is $v_i - v_0$ and $d_1 d_2 \cdots d_n$ is the normalized volume of P. Following Sturmfels [**80**], let $e(P)$ be the number of these invariant factors which are even. If $e(P) = 0$, so that d is odd, then P is an odd cell.

PROPOSITION 2.5. *The polynomial system* (2.5) *has* $2^{e(P)}$ *real solutions if* $a_i >$ 0 *whenever* d_i *is even, and* 0 *real solutions otherwise. In particular, if* P *is an odd cell, then there is one real solution.*

THEOREM 2.6 (Sturmfels [**80**, Corollary 2.3]). *The maximum number* $\rho(P)$ *of real solutions to a sparse polynomial system* (2.1) *with common Newton polytope* P *satisfies*

$$\#\text{ odd cells in } P_w \;\leq\; \rho(P)\,,$$

for any regular triangulation P_w *of* P.

PROOF. In the limit as $t \to 0$, the lifted system (2.4) given by a regular triangulation P_w of P becomes the disjunction of facial subsystems of (2.1), one for each facet F of P_w. Thus the number of real solutions in the limit is a constructive lower bound for $\rho(P)$.

By Proposition 2.5, the number of odd cells in P_w is a lower bound on the number of real solutions to the facial subsystem given by the facets of P_w. □

Proposition 2.5 also gives an upper bound on the limiting number of real solutions ρ of the lifted system (2.4) as $t \to 0$ (due to Sturmfels [**80**, Theorem 2.2].)

$$\rho \leq \sum_{F \text{ a facet of } P_w} 2^{e(F)}\,. \tag{2.6}$$

More sophisticated accounting of the possible signs of the coefficients of facial subsystems improves this bound. This accounting is accomplished in [**54**] and [**35**], leading to a combinatorial upper bound for such limiting systems. Itenberg and

Roy [**35**] show there is a system (2.1) for which this upper bound is attained, and thus obtain

$$\text{Combinatorial upper bound for limiting systems} \;\leq\; \rho(P)\,.$$

Itenberg and Roy then conjectured that this combinatorial bound was in fact the global bound, that is, they conjectured that the maximal number of real solutions occurs in limiting systems. They also gave a similar bound for $\rho_+(P)$, the number of solutions with positive coordinates. This was too optimistic, for Li and Wang [**48**] found a remarkably simple counterexample to this conjecture of Itenberg and Roy[†]

$$y - x - 1 \;=\; 200 - 100y^3 + 900x^3 - x^3y^3 \;=\; 0\,.$$

This system has 3 positive solutions

$$(0.317659, 1.317659), \quad (.659995, 1.659995), \quad \text{and} \quad (8.12058, 9.12058)\,,$$

but the combinatorial upper bound is 2. Thus systems with the maximal number of real solutions cannot in general be constructed with these limiting techniques.

We still have more questions than answers. Among the questions are:

(1) Improve this lower bound for $\rho(P)$ (or for $\rho_+(P)$) of Itenberg and Roy.
(2) Find new general methods to construct systems with many real solutions.
(3) Determine which polytopes P have the property that $\rho(P) < n!\mathrm{Vol}(P)$, that is, not all the solutions can be real.

2.3. Fewnomial bounds. These definitions, constructions, and results apply to what have come to be known as fewnomial systems. A fewnomial is a polynomial f with *few monomials*—the monomials of f are members of some set $\mathcal{A}$ not necessarily equal to the lattice points within its convex hull. In particular, the results of Section 2.2 give lower bounds on the maximum number of real solutions $\rho(\mathcal{A})$ to a system of fewnomials whose monomials are from $\mathcal{A}$. Here, we use a regular triangulation Δ_w of the point set $\mathcal{A}$ induced from a lifting function $w\colon \mathcal{A} \to \mathbb{Q}$.

When $n = 1$, consider the binomial (a fewnomial) system

$$x^d - 1 \;=\; 0\,.$$

This has d complex solutions. Similarly, the number of complex solutions to a general fewnomial system is equal to its BKK bound. The above binomial system has either 1 or 2 real solutions, and so we see that the number of real solutions to a fewnomial system should be less than its BKK bound. The question is: How much less?

Khovanskii [**38, 39**] established a very general result concerning systems where each f_i is a *polynomial* function of the monomials x^a for $a \in \mathcal{A}$. He proves that the number of real solutions to such a system is at most $2^n 2^{\binom{N}{2}}(n+1)^N$, where $N = \#\mathcal{A}$, the number of monomials in $\mathcal{A}$. When the polynomial functions are linear, they are polynomials with monomials from $\mathcal{A}$, and hence we have Khovanskii's fewnomial bound.

$$\rho(\mathcal{A}) \;\leq\; 2^n \cdot 2^{\binom{N}{2}}(n+1)^N\,.$$

While this bound seems outrageously large, it does not depend upon the volume of the convex hull of $\mathcal{A}$, but rather on the algebraic complexity—the ambient dimension n and the size N of $\mathcal{A}$. That such a bound exists at all was revolutionary.

[†]Strictly speaking, this counterexample is to their more general conjecture concerning *mixed systems* which are systems of polynomials with possibly different Newton polytopes.

We compare this complexity bound to the combinatorial upper bound (2.6) on the number of real solutions to a lifted fewnomial system (2.4) in the limit as $t \longrightarrow 0$. The invariant $e(F)$ of a facet of the regular triangulation Δ_w of $\mathcal{A}$ is at most n. Thus

$$\rho \;\leq\; 2^n \cdot \#\text{facets of } \Delta_w \;\left(\leq\; 2^n \cdot \binom{N}{n+1}\right),$$

as a facet of Δ_w involves $n+1$ points of $\mathcal{A}$. This bound is typically much lower than Khovanskii's bound. For example, consider two trinomials in two variables. Here $n = 2$, and after multiplying each equation by a suitable monomial, we have $\#\mathcal{A} = 5$. Thus we have Khovanskii's fewnomial bound

$$\rho(\mathcal{A}) \;\leq\; 2^2 \cdot 2^{\binom{5}{2}} \cdot 3^5 \;=\; 995{,}328\,.$$

In contrast, a triangulation of 5 points in the plane has at most 5 simplices

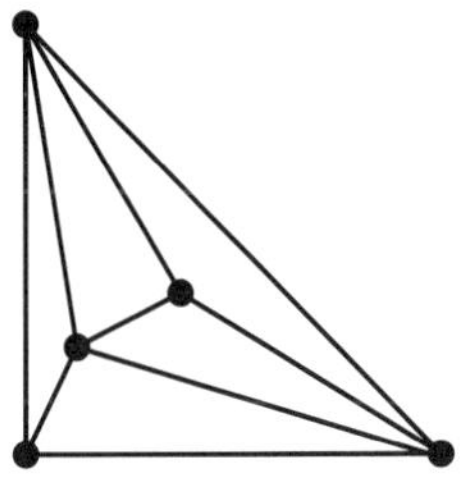

and so the bound ρ for limiting lifted systems is

$$\rho \;\leq\; 2^2 \cdot 5 \;=\; 20\,.$$

It may be more feasible to look for bounds on $\rho_+(\mathcal{A})$, the number of solutions with positive coordinates. Kouchnirenko made† the following conjecture.

CONJECTURE 2.7 (Kouchnirenko (see [**5**, p. 300])). *A system of n polynomials in n variables with only simple solutions whose ith polynomial has m_i monomials has at most*

$$(m_1 - 1)(m_2 - 1)\cdots(m_n - 1)$$

solutions with positive coordinates.

For a system of two trinomials in 2 variables, Kouchnirenko's conjecture asserts that $\rho_+ \leq 4$. In 2000, Haas [**24**] gave the counterexample

$$\begin{aligned} x^{108} + 1.1y^{54} - 1.1y &= 0 \\ y^{108} + 1.1x^{54} - 1.1x &= 0, \end{aligned}$$

a system of two trinomials with 5 solutions having positive coordinates. Although Kouchnirenko's conjecture is false, the question remains: Is the true value for $\rho_+(\mathcal{A})$ (or $\rho(\mathcal{A})$) closer to Khovanskii's bound or to the number in Kouchnirenko's conjecture? Recent work suggests that it is the latter.

THEOREM 2.8 (Li, Rojas, and Wang [**47**]). *A system of 2 polynomials in 2 variables*

$$f_1(x,y) \;=\; f_2(x,y) \;=\; 0\,,$$

---

†Apparently, Kouchnirenko did not believe this bound. Nonetheless, the conjecture and its attribution have passed into folklore.

where f_1 has 3 terms and f_2 has m terms, and every solution is isolated, has at most

$$2 \cdot (2^{m-1} - 1),$$

solutions with positive coordinates.

When $m = 3$, so we have two trinomials in 2 variables, this gives a bound of 6. There is presently no known example with more that 5 positive solutions. Does Theorem 2.8 give the correct value of 6, or is Haas's construction with 5 solutions the best possible?

REMARK 2.9. Khovanskii's bound for the number of real solutions with positive coordinates holds also for systems of power functions, where the exponents of monomials are arbitrary real numbers. One might suppose that this added generality is the source of the large size of his bound and its apparent lack of tightness for polynomial systems. However, Napoletani [**53**] has shown that the complexity bounds are the same for both polynomials and for power functions.

3. Enumerative Real Algebraic Geometry

In his treatise on enumerative geometry, Schubert [**62**] declared enumerative geometry to be concerned with all questions of the following form: How many geometric figures of a fixed type satisfy certain conditions? This includes problems as diverse as the number of lines on a cubic surface (27) and the number of lines meeting four fixed lines in 3-space (see Section 4 for the answer). These are archetypes for two distinct classes of enumerative geometric problems. For the purpose of this survey, we ignore the first class (except in Section 3.1) and concentrate on the second. Specifically, we consider conditions imposed by geometric figures that may be moved independently into general position.

Enumerative geometry (in the broad sense) had a great flourishing in the 19th century in the hands of Chasles, Schubert, Halphen, Zeuthen, and others. (The survey of Kleiman [**41**] is a good historical source.) At that time, it had long been recognized that it was necessary to work over the complex numbers to ensure the existence of solutions. We know of only a handful of cases where the number of real solutions was considered. (We discuss some in Section 3.1.) Asking how many solutions can be real is the motivating question of enumerative real algebraic geometry and an analog of Question 1.1.

QUESTION 3.1. *In a given enumerative geometric problem, if the general figures are chosen to be real, how many of the solution figures can be real?*

For example, how many of the lines on a real cubic surface can be real? (Answer: all 27.) How many of the lines meeting four given real lines can be real? (Answer: all can.) This question was posed by Fulton [**19**, p. 55]: "The question of how many solutions of real equations can be real is still very much open, particularly for enumerative problems."

This problem is fundamentally hard. Of the geometric figures satisfying real conditions, some will be real while the rest will occur in complex conjugate pairs, and the number which are real will depend subtly on the configuration of the conditions. Despite this difficulty, this is an important question often asked in applications.

One fruitful variant has been whether it is possible that *all* solution figures can be real. We call an enumerative problem *fully real* if there are general real conditions for which all solution figures are real. That is, if the upper bound of (1.2) is sharp. In light of the situation for sparse polynomial systems, it is quite surprising that there are no known enumerative problems which are not fully real. For this it is important that the conditions are imposed by general figures. A related question is whether the opposite situation can occur: Is it possible to have no (or only one) real solutions? We give some examples in Section 4.2.4. We shall see in Section 6 that there are many enumerative problems whose number of real solutions is always at least 2.

In the above passage, Fulton [**19**] goes on to ask: "For example, how many of the 3264 conics tangent to five general conics can be real?" He answered this question in the affirmative in 1986, but did not publish that result. Later, Ronga, Tognoli, and Vust [**60**] gave a careful argument that all 3264 can be real. This example is very striking, both for the number, 3264, and because this problem of conics has long been an important testing ground for ideas in enumerative geometry.

One difficulty with enumerative real algebraic geometry is that there are few techniques or theorems with a wide range of applicability. Sometimes a direct calculation suffices (Sections 3.3 and 3.4) or more commonly real solutions are constructed by deforming from a special configuration, as in the homotopy algorithm of Huber and Sturmfels in Section 2.1. Attempts to formalize this method include 'Schubert induction' [**74**]; this is presented in the proof of Theorem 4.6, and used to establish the other results of Section 4. Another formalization is the notion of 'real effective algebraic equivalence' [**69**]. That (together with Theorem 4.4) gives such results as [**69**, Theorem 18]. The enumerative problem of $\binom{2n-2}{n-1}n^{2n-3}$ codimension 2 planes meeting $2n-2$ rational normal curves is fully real and [**67**, Corollary 5]: The enumerative problem of 11,010,048 2-planes in $\mathbb{P}^5$ meeting 9 general Veronese surfaces is fully real.

We devote the rest of this section to a description of some enumerative problems in the context of Question 3.1.

3.1. Enumerative problems not involving general conditions. We give a brief tour of some enumerative problems that do not involve general conditions, but nonetheless raise some interesting questions regarding real solutions.

Schläfli (see the survey of Coxeter [**11**]) showed there are 4 possibilities for the number of real lines on a real cubic surface: 27, 15, 7, or 3. Recent developments in enumerative geometry (mirror symmetry) have led to the solution of a large class of similar enumerative problems involving, among other things, the number of rational curves of a fixed degree on a Calabi-Yau threefold. (See the book of Cox and Katz [**10**].) For example, on a general quintic hypersurface in $\mathbb{P}^4$ there are 2875 lines [**26**], 609,250 conics [**36**], and 371,206,375 twisted cubics. The number of twisted cubics and higher degree rational curves was computed in the seminal paper of Candelas, de la Ossa, Green, and Parkes [**6**]. How many of the curves can be real in problems of this type? For example, how many real lines can there be on a real quintic hypersurface in $\mathbb{P}^4$?

A real homogeneous polynomial $f(x)$ is positive semi-definite (psd) if $f(x) \geq 0$ whenever x is real. Hilbert [**27**] proved that a psd ternary quartic is a sum of three squares of real quadratic forms. In fact, a general quartic is a sum of three squares of *complex* quadratic forms in 63 different ways [**86**]. Powers and Reznick [**56**] studied the question of how many ways one may represent a ternary quartic as a real sum or difference of three squares. In every instance, they found that 15 of the 63 ways involved real quadratic forms. Is it true that a general psd quartic is a sum or a differrence of three real squares in exactly 15 different ways? In how many ways is it a sum of three squares?†

A general plane curve C of degree d has $3d(d-2)$ flexes. These are the points on C where the Hessian determinant of the form defining C vanishes. Since the Hessian determinant has degree $3(d-2)$, we expect there to be $3d(d-2)$ such points. This involves intersecting the curve with its Hessian curve, and *not* with a general curve of degree $3(d-2)$. A real smooth plane cubic has 3 of its 9 flexes real. Zeuthen [**89**] found that at most 8 of the 24 flexes of a real plane quartic can be real. An example of a plane quartic with 8 real flexes is provided by the Hilbert quartic [**28**], which is defined by

$$(x^2+2y^2-z^2)(2x^2+y^2-z^2)+z^4/100 \;=\; 0\,.$$

We display this curve in Figure 3, marking the flexes with dots. Klein [**42**] later

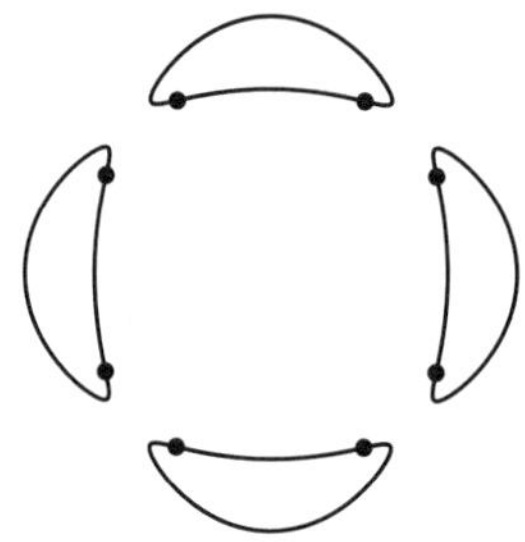

FIGURE 3. Hilbert's quartic: a plane quartic with 8 flexes

showed that a general real plane curve has at most 1/3 of its flexes real.

Harnack [**25**] proved that a smooth real algebraic curve of genus g has at most $g+1$ topological components, and he constructed real algebraic curves of genus g with $g+1$ components. In particular, a plane curve of degree d has genus $g=(d-1)(d-2)/2$ and there are real plane curves of degree d with $g+1$ components. (An example is provided by Hilbert's quartic, which has genus 3.) Finer topological questions than enumerating the components leads to (part of) Hilbert's 16th problem [**29**], which asks for the determination of the topological types of smooth projectively embedded real algebraic varieties.

A variant concerns rational plane curves of degree d. A general rational plane curve of degree d has $3(d-2)$ flexes and $g=(d-1)(d-2)/2$ nodes. Theorem 4.2 implies that there exist real rational plane curves of degree d with all $3(d-2)$ flexes real, which we call maximally inflected curves. See Section 5.1 for the connection. Such curves have at most $g-d+2$ of their nodes real, and there exist curves with the extreme values of 0 and of $g-d+2$ real nodes [**37**]. For example, a rational quartic ($d=4$) has 6 flexes and $g=3$ nodes. If all 6 flexes are real, then at most

†With Powers, Reznick, and Scheiderer, we have shown that the answers to these two questions are 15 and 8, respectively [**57**].

one node is real. Figure 4 shows maximally inflected quartics with and 0 and 1 nodes. The flexes are marked by dots.

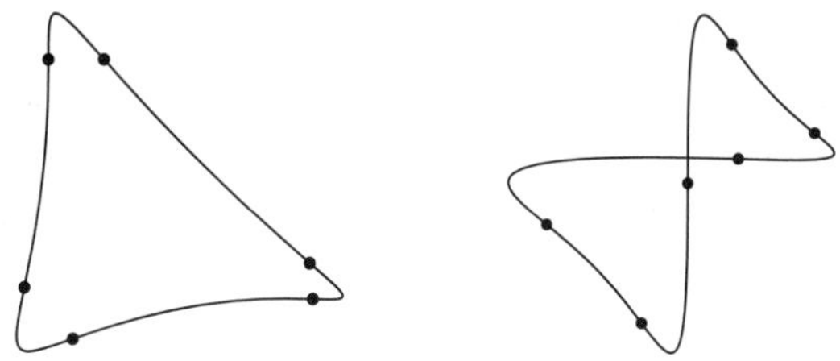

FIGURE 4. Rational quartics with 6 real flexes

Recently, Huisman asked and answered a new question about real curves. A component X of a real algebraic curve is a *psuedoline* if its homology class $[X]$ in $H^1(\mathbb{P}^n_{\mathbb{R}}, \mathbb{Z}/2\mathbb{Z})$ is non-zero, and an *oval* otherwise.

QUESTION 3.2. *Given a smooth (irreducible over $\mathbb{C}$) real algebraic curve C in $\mathbb{P}^n$ of genus g and degree c, how many real hypersurfaces of degree d are tangent to at least s components of C with order of tangency at least m?*

Let ν be that number, when there are finitely many such hypersurfaces.

THEOREM 3.3 (Huisman [**33**, Theorem 3.1]). *When $s = g$ and $gm = cd$, and the restriction*

$$H^0(\mathbb{P}^n, \mathcal{O}(d)) \longrightarrow H^0(C, \mathcal{O}(d))$$

is an isomorphism, then ν is finite. Moreover, ν is non-zero if and only if

(1) *Both m and d are odd and C consists of exactly g psuedolines, or*
(2) *m is even and either d is even or all components of C are ovals.*

In case (1), $\nu = m^g$, and in case (2),

$$\nu = \begin{cases} (g+1)m^g & \textit{if } C \textit{ has } g+1 \textit{ components} \\ m^g & \textit{if } C \textit{ has } g \textit{ components.} \end{cases}$$

It is notable that this problem can only be stated over the real numbers.

3.2. The Stewart-Gough platform. The position of a rigid body in $\mathbb{R}^3$ has 6 degrees of freedom. This is exploited in robotics, giving rise to the Stewart-Gough platform [**23, 79**]. Specifically, suppose we have 6 fixed points $A_1, A_2, \ldots, A_6$ in space and 6 points $B_1, B_2, \ldots, B_6$ on a rigid body B (the framework of Figure 5). The body is controlled by varying each distance l_i between the fixed point A_i and the point B_i on B. This may be accomplished by attaching rigid actuators to spherical joints located at the points B_i, or by suspending the platform from cables.

Given a position of the body B in space, the distances $l_1, l_2, \ldots, l_6$ are uniquely determined. A fundamental problem is the inverse question: Given a platform (positions of the A_i fixed and the relative positions of the B_i specified) and a sextuple of distances $l_1, l_2, \ldots, l_6$, what is the position of the platform in space?

It had long been understood that several positions were possible for a given sextuple of lengths. An early work in 1897 showed there could be as many as 16 different positions [**4**]. This leads to the following enumerative problem.

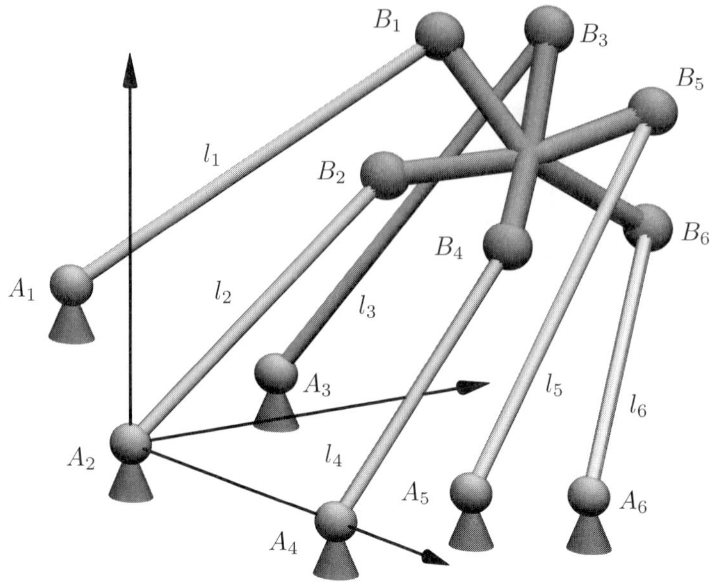

FIGURE 5. A Stewart Platform

QUESTION 3.4. *For a given Stewart platform, how many (complex) positions are there for a generic choice of the distances $l_1, l_2, \ldots, l_6$? How many of these can be real?*

In the early 1990's, several approaches (numerical experimentation [**58**], intersection theory [**61**], Gröbner bases [**46**], resultants [**51**], and algebra [**52**]) each showed that there are 40 complex positions of a general Stewart platform. The obviously practical question of how many positions could be real remained open until 1998, when Dietmaier introduced a novel method to find a value of the distances $l_1, l_2, \ldots, l_6$ with all 40 positions real.

THEOREM 3.5 (Dietmaier [**14**]). *All* 40 *positions can be real.*

Dietmaier's method will find future applications to other problems of this kind. He began with a formulation of the problem as a system of equations depending upon the distances $l_1, l_2, \ldots, l_6$. An initial solution for a given instance of the distances gave 6 real solutions and 17 pairs of complex conjugate solutions. He then used an ingenious algorithm to vary the distances in search of a configuration with all 40 solutions real.

This algorithm systematically varies the distances with the intention of increasing the number of real solutions. It proceeds in two stages. In the first stage, a pair of complex conjugate solutions $(x, \overline{x})$ are driven together, eventually creating a double solution, while at the same time the existing real solutions are kept bounded away from one another. At the formation of a double (necessarily real) solution, the distances are further incremented to create two new nearby real solutions (x_1, x_2), which are then driven apart in the second stage. This procedure is repeated again with another pair of complex conjugate solutions, and *et cetera*. Figure 6 ilustrates the two stages.

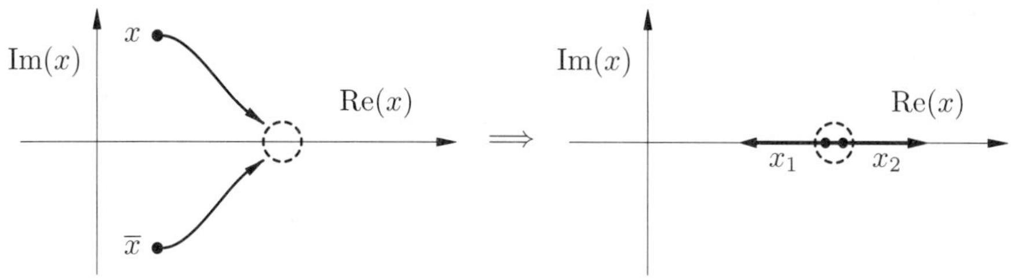

FIGURE 6. The two stages of Dietmaier's algorithm.

In each stage, Dietmaier accomplishes the given task (eg. colliding conjugate solutions) by linearizing the system at the current solutions and then solving a linear program for the optimal increment of the distances for the given goal. Changing the distances, he uses Newton's method beginning with the current solutions to find soutions for the new set of distances, and then repeats this procedure until the goal is acheived (eg. the conjugate pair collides). This is an application of numerical homotopy continuation [**84**].

While there is no guarantee that this method will even successfully collide two conjugate solutions, Dietmaier uses it to find a sextuple of distances with all 40 solutions real. While at each step the solutions are only numerical approximations to the actual solutions, the condition number N guarantees the existence of a genuine solution within $1/N$ of each approximate solution. Since the approximate real solutions were separated by more than $2/N$, the requirement that non-real solutions occur in complex conjugate pairs forced these genuine solutions to be real.

3.3. Real rational cubics through 8 points in $\mathbb{P}^2_{\mathbb{R}}$. There are 12 singular (rational) cubic curves containing 8 general points in the plane.

THEOREM 3.6 (Degtyarev and Kharlamov [**12**, Proposition 4.7.3]). *Given* 8 *general points in* $\mathbb{P}^2_{\mathbb{R}}$*, there are at least* 8 *singular real cubics containing them, and there are choices of the* 8 *points for which all* 12 *singular cubics are real.*

PROOF. Since a cubic equation in the plane has 10 coefficients, the space of cubics is identified with $\mathbb{P}^9$. The condition for a plane cubic to contain a given point is linear in these coefficients. Given 8 general points, these linear equations are independent and so there is a pencil ($\mathbb{P}^1$) of cubics containing 8 general points in $\mathbb{P}^2$.

Two cubics P and Q in this pencil meet transversally in 9 points. Since curves in the pencil are given by $aP+bQ$ for $[a,b]\in\mathbb{P}^1$, any two curves in the pencil meet transversally in these 9 points. Let Z be $\mathbb{P}^2$ blown up at these same 9 points. We have a map

$$f\,:\,Z\longrightarrow\mathbb{P}^1\,,$$

where $f^{-1}([a,b])$ is the cubic curve defined by the polynomial $aP+bQ$.

Consider the Euler characteristic $\chi(Z)$ of Z first over $\mathbb{C}$ and then over $\mathbb{R}$. Blowing up a smooth point on a surface replaces it with a $\mathbb{P}^1_{\mathbb{C}}$ and thus increases the Euler characteristic by 1. Since $\chi(\mathbb{P}^2_{\mathbb{C}})=3$, we see that $\chi(Z)=12$. The general fiber of f is a smooth plane cubic which is homeomorphic to $S^1\times S^1$, and so has

Euler characteristric 0. Thus only the singular fibers of f contribute to the Euler characteristic of Z. Assume that the 8 points are in general position so there are only nodal cubics in the pencil. A nodal cubic has Euler characteristic 1. Thus there are 12 singular fibers of f and hence 12 singular cubics meeting 8 general points in $\mathbb{P}^2_{\mathbb{C}}$.

Consider now the Euler characteristic of $Z_{\mathbb{R}}$. Blowing up a smooth point on a real surface replaces the point by $\mathbb{P}^1_{\mathbb{R}} = S^1$, and hence decreases the Euler characterstic by 1. Since $\chi(\mathbb{P}^2_{\mathbb{R}}) = 1$, we have $\chi(Z_{\mathbb{R}}) = 1 - 9 = -8$. A nonsingular real cubic is homeomorphic to either one or two disjoint copies of S^1, and hence has Euler characteristic 0. Again the Euler characteristic of $Z_{\mathbb{R}}$ is carried by its singular fibers. There are two nodal real cubics; either the node has two real branches or two complex conjugate branches so that the singular point is isolated. Call these curves real nodal and complex nodal, respectively. They are displayed in Figure 7. The real nodal curve is homoemorphic to a figure 8 and has Euler characteristic

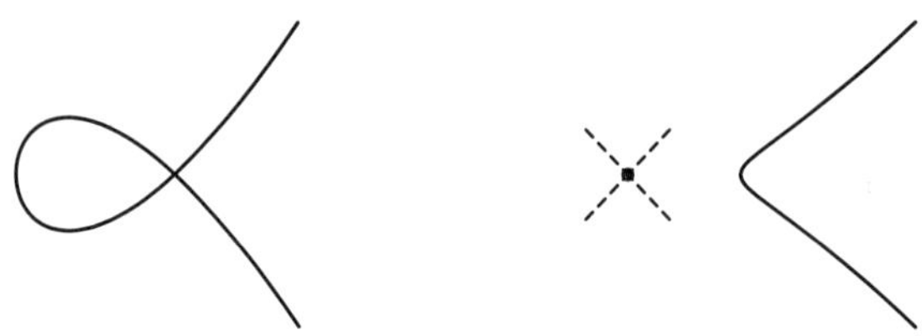

FIGURE 7. A real nodal and complex nodal curve

-1, while the complex nodal curve is the union of a S^1 with a point and so has Euler characteristic 1.

Among the singular fibers, we have

$$\begin{aligned} -8 &= \#\text{complex nodal} - \#\text{real nodal}, \quad \text{with} \\ 12 &\geq \#\text{complex nodal} + \#\text{real nodal}. \end{aligned}$$

Thus there are at least 8 real nodal curves containing 8 general points in $\mathbb{P}^2_{\mathbb{R}}$. The pencil of cubics containing the 2 complex nodal cubics of Figure 8 has 10 real nodal cubics. Thus there are 12 real rational cubics containing any 8 of the 12 points in Figure 8.

□

REMARK 3.7. This classical problem of 12 plane cubics containing 8 points generalizes to the problem of enumerating rational plane curves of degree d containing $3d - 1$ points. Let N_d be the number of such curves, which satisfies the recursion [**20**]

$$N_d = \sum_{d_1+d_2=d} N_{d_1} N_{d_2} \left[d_1^2 d_2^2 \binom{3d-4}{3d_1-2} - d_1^3 d_2 \binom{3d-4}{3d_1-1} \right].$$

The values $N_1 = N_2 = 1$ are trivially fully real, and we have just seen that $N_3 = 12$ is fully real. The next case of $N_4 = 620$ (computed by Zeuthen [**88**]) seems quite challenging.

$$\frac{(y+9x-28)^2}{4} = (x-1)x^2$$

$$\frac{(x+10y-28)^2}{4} = y(y+1)^2$$

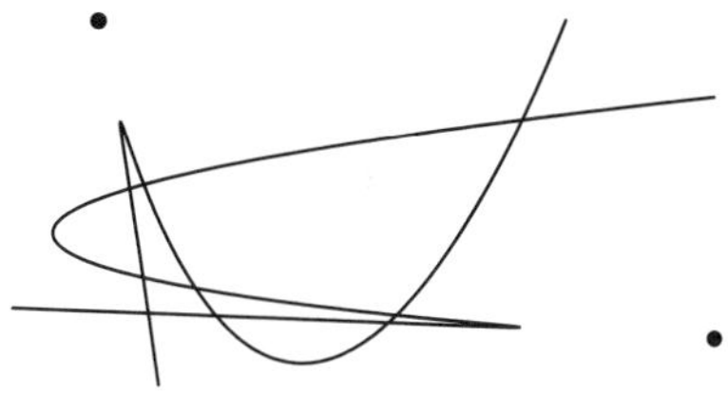

FIGURE 8. Complex nodal curves meeting in 9 points

REMARK 3.8. The most interesting feature of Theorem 3.6 is the existence of a lower bound on the number of real solutions, which is a new phenomenon. In Section 6 we shall see evidence that this may be a pervasive feature of this field.

3.4. Common tangent lines to $2n-2$ spheres in $\mathbb{R}^n$. Consider the following.

QUESTION 3.9. *How many common tangent lines are there to $2n-2$ spheres in $\mathbb{R}^n$?*

For example, when $n=3$, how many common tangent lines are there to four spheres in $\mathbb{R}^3$? Despite its simplicity, this question does not seem to have been asked classically, but rather arose in discrete and computational geometry[†]. The case $n=3$ was solved by Macdonald, Pach, and Theobald [**49**] and the general case more recently[§] [**78**].

THEOREM 3.10. *$2n-2$ general spheres in $\mathbb{R}^n$ ($n \geq 3$) have $3 \cdot 2^{n-1}$ complex common tangent lines, and there are $2n-2$ such spheres with all common tangent lines real.*

Figure 9 displays a configuration of 4 spheres in $\mathbb{R}^3$ with 12 real common tangents. The number $2n-2$ is the dimension of the space of lines in $\mathbb{R}^n$ and is necessary for there to be finitely many common tangents.

Represent a line in $\mathbb{R}^n$ by a point p on the line and its direction vector $v \in \mathbb{P}^{n-1}$. Imposing the condition

$$p \cdot v = 0\,, \tag{3.1}$$

makes this representation unique. Here $\cdot$ is the usual Euclidean dot product. Write v^2 for $v \cdot v$. A line (p, v) is tangent to the sphere with center c and radius r when

$$v^2p^2 - 2v^2 p \cdot c + v^2c^2 - (v \cdot c)^2 - r^2v^2 = 0\,.$$

Suppose we have $2n-2$ spheres with centers $c_1, c_2, \ldots, c_{2n-2}$ and corresponding radii $r_1, r_2, \ldots, r_{2n-2}$. Without any loss of generality, we may assume that the last

---

[†]The question of the maximal number of (real) common tangents to 4 balls was first formulated by David Larman [**45**] at DIMACS in 1990.

[§]This was solved during the DIMACS Workshop on Algorithmic and Quantitative Aspects of Real Algebraic Geometry in Mathematics and Computer Science.

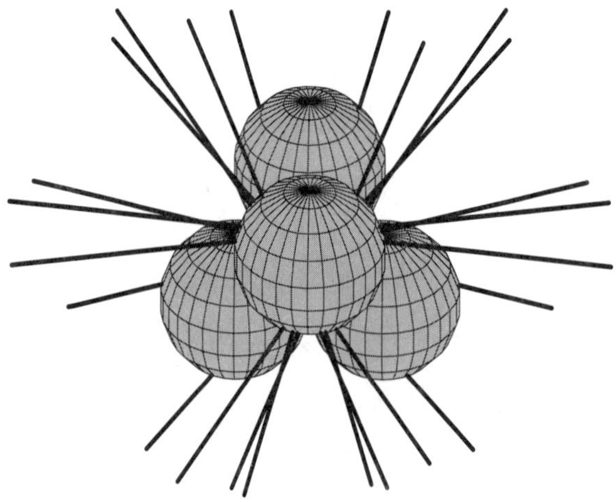

FIGURE 9. Four spheres with 12 common tangents

sphere is centered at the origin and has radius r. Then its equation is

$$v^2p^2 - v^2r^2 \;=\; 0\,. \tag{3.2}$$

Subtracting this from the equations for the other spheres, we obtain the equations

$$2v^2p\cdot c_i \;=\; v^2c_i^2 - (v\cdot c_i)^2 - v^2(r_i^2 - r^2) \tag{3.3}$$

for $i = 1, 2, \ldots, 2n-3$.

These last equations are linear in p. If the centers $c_1, c_2, \ldots, c_n$ are linearly independent (which they are, by our assumption on generality), then we use the corresponding equations to solve for v^2p as a homogeneous quadratic in v. Substituting this into the equations (3.1) and (3.2) gives a cubic and a quartic in v, and substituting the expression for v^2p into (3.3) for $i = n+1, n+2, \ldots, 2n-3$ gives $n-3$ quadratics in v. By Bézout's Theorem, if there are finitely many complex solutions to these equations, their number is bounded by $3\cdot 4\cdot 2^{n-3} = 3\cdot 2^{n-1}$.

This upper bound is attained with all solutions real. Suppose that the spheres all have the same radius, r, and the first four have centers

$$\begin{aligned}
c_1 &:= (\;\;1,\;\;\;1,\;\;\;1,\; 0,\ldots,0)\,,\\
c_2 &:= (\;\;1,-1,-1,\; 0,\ldots,0)\,,\\
c_3 &:= (-1,\;\;\;1,-1,\; 0,\ldots,0)\,,\\
c_4 &:= (-1,-1,\;\;\;1,\; 0,\ldots,0)\,,
\end{aligned}$$

and subsequent centers are at the points $\pm ae_j$ for $j = 4, 5, \ldots, n$, where $e_1, e_2, \ldots, e_n$ is the standard basis for $\mathbb{R}^n$. Let $\gamma := a^2(n-1)/(a^2+n-3)$, which is positive.

THEOREM 3.11 ([**78**, Theorem 5]). *When*

$$a\,r\,(r^2-3)\,(3-\gamma)\,(a^2-2)\,(r^2-\gamma)\,\left((3-\gamma)^2 + 4\gamma - 4r^2\right) \;\neq\; 0\,,$$

there are exactly $3\cdot 2^{n-1}$ *complex lines tangent to the spheres. If we have*

(a) $\dfrac{1}{4}\,(3-\gamma)^2 + \gamma \;>\; r^2 \;>\; \gamma$ *and*

(b) $\dfrac{n-1}{n-4}+2 > a^2 > 2$,

then all the $3 \cdot 2^{n-1}$ *lines are in fact real.*

Theorem 3.10 is false when $n = 2$. There are 4 lines tangent to 2 circles in the plane, and all may be real. The argument given for Theorem 3.10 fails because the centers of the spheres do not affinely span $\mathbb{R}^2$. This case of $n = 2$ does generalize, though.

THEOREM 3.12 (Megyesi [**50**]). *Four unit spheres in* $\mathbb{R}^3$ *whose centers are coplanar but otherwise general have* 12 *common complex tangents. At most* 8 *of these* 12 *are real.*

REMARK 3.13. This problem of common tangents to 4 spheres with equal radii and coplanar centers gives an example of an enumerative geometric problem that is not fully real. We do not feel this contradicts the observation that there are no enumerative problems not known to be fully real, as the spheres are not sufficiently general.

4. Schubert Calculus

The classical Schubert calculus of enumerative geometry is concerned with questions of enumerating linear subspaces of a vector space or projective space that satisfy incidence conditions imposed by other linear subspaces. A non-trivial instance is the question posed at the beginning of Section 3.

QUESTION 4.1. *How many lines in space meet four general lines* ℓ_1, ℓ_2, ℓ_3, *and* ℓ_4*?*

Three pairwise skew lines ℓ_1, ℓ_2, and ℓ_3 lie on a unique smooth quadric surface Q. There are two families of lines that foliate Q—one family includes ℓ_1, ℓ_2, and ℓ_3 and the other consists of the lines meeting each of ℓ_1, ℓ_2, and ℓ_3. The fourth line ℓ_4 meets Q in two points, and each of these points determines a line in the second family. Thus there are 2 lines μ_1, μ_2 in space that meet general lines ℓ_1, ℓ_2, ℓ_3, and ℓ_4. Figure 10 shows this configuration. Figure 10 also shows how the two lines can

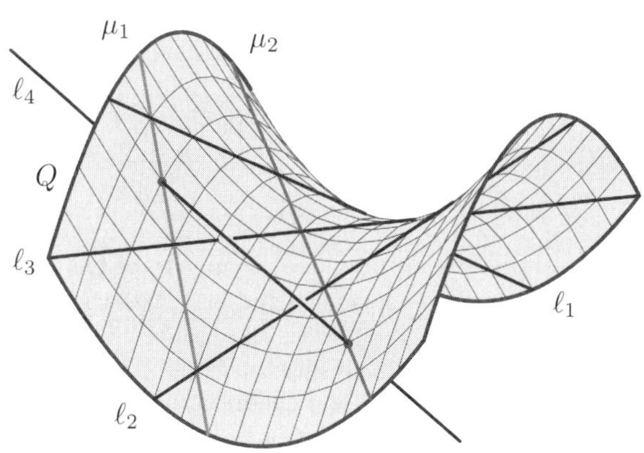

FIGURE 10. The two lines meeting four general lines in space.

be real—if ℓ_4 meets Q in two real points. (The two lines are complex when ℓ_4 meets Q in two complex conjugate points.)

The classical Schubert calculus is a vast generalization of this problem of four lines. In the 1980's Robert Speiser suggested to Fulton that the classical Schubert calculus may be a good testing ground for Question 3.1. This was also considered by Chiavacci and Escamilla-Castillo [**8**]. We will discuss increasingly more general versions of the Schubert calculus, and the status of Question 3.1 for each.

Consider first more general problems involving lines. The space of lines in $\mathbb{P}^n$ is a smooth projective variety of dimension $2n-2$ called the Grassmannian of lines in $\mathbb{P}^n$. The set of lines meeting a linear subspace L of dimension $n-1-l$ has codimension l in the Grassmannian. Thus given general linear subspaces $L_1, L_2, \ldots, L_s$ of $\mathbb{P}^n$ with $\dim L_i = n-1-l_i$ where $l_1+l_2+\cdots+l_s = 2n-2$, we expect (and there are indeed) finitely many lines in $\mathbb{P}^n$ meeting each linear subspace $L_1, L_2, \ldots, L_s$. Schubert [**64**] discovered algorithms for computing this number $d(l_1, l_2, \ldots, l_s)$ of lines. For example, if each $l_i = 1$, so that $s = 2n-2$, then this number is the nth Catalan number†

$$C_n \;=\; \frac{1}{n}\binom{2n-2}{n-1}. \tag{4.1}$$

Enumerative problems of lines in $\mathbb{P}^n$ meeting general linear subspaces furnished the first infinite family of non-trivial enumerative problems known to be fully real.

THEOREM 4.2 ([**68**, Theorem C]). *Given positive integers $l_1, l_2, \ldots, l_s$ with $l_1 + l_2 + \cdots + l_s = 2n-2$, there exist linear subspaces $L_1, L_2, \ldots, L_s$ of $\mathbb{RP}^n$ with $\dim L_i = n+1-l_i$ such that there are exactly $d(l_1, l_2, \ldots, l_s)$ complex lines meeting each subspace L_i, and each of these lines are real.*

4.1. The special Schubert calculus. More generally, we may ask how many linear subspaces of a fixed dimension meet general linear subspaces. We formulate this question in terms of linear subspaces of a vector space.

The set of k-dimensional subspaces (k-planes) of an n-dimensional vector space forms the *Grassmannian of k-planes in n-space*, $\mathrm{Gr}(k,n)$, a smooth projective variety of dimension $k(n-k)$. Those k-planes meeting a linear subspace L of dimension $n-k+1-l$ non-trivially (that is, the intersection has positive dimension) form the special Schubert subvariety $\Omega(L)$ of $\mathrm{Gr}(k,n)$ which has codimension l. The special Schubert calculus is concerned with the following question.

QUESTION 4.3. *Given general linear subspaces $L_1, L_2, \ldots, L_s$ of $\mathbb{C}^n$ with $\dim L_i = n-k+1-l_i$ where $l_1+l_2+\cdots+l_s = k(n-k)$, how many k-planes K meet each subspace L_i non-trivially, that is, satisfy*

$$K \cap L_i \;\neq\; \{0\} \qquad i = 1, 2, \ldots, s\,? \tag{4.2}$$

The condition (4.2) is expressed in the global geometry of $\mathrm{Gr}(k,n)$ as the number of points in the intersection of the special Schubert varieties

$$\Omega(L_1) \cap \Omega(L_2) \cap \cdots \cap \Omega(L_s)\,, \tag{4.3}$$

when the intersection is transverse. (A general theorem of Kleiman [**40**] guarantees transversality when the L_i are in general position, and also implies transversality for the other intersections considered in this section.)

†This indexing of the Catalan numbers is shifted from that of some other authors.

There are algorithms due to Schubert [**65**] (when each $l_i = 1$) and Pieri [**55**] to compute the expected number of solutions. When each $l_i = 1$, Schubert [**63**] showed that the number of solutions is equal to

$$d(n,k) \;:=\; \frac{1!\,2!\,\cdots\,(k-1)!\cdot[k\,(n-k)]!}{(n-k)!\,(n-k+1)!\cdots(n-1)!}\,.$$

A line in $\mathbb{P}^n$ is a 2-plane in $(n+1)$-space and two linear subspaces in $\mathbb{P}^n$ meet if and only if the corresponding linear subspaces in $(n+1)$-space have a non-trivial intersection. Thus the problem of lines in projective space corresponds to the case $k = 2$ of the special Schubert calculus. While the geometric problem generalizes easily from $k = 2$ to arbitrary values of k, the proof of Theorem 4.2 does not. There is, however, a relatively simple argument that this special Schubert calculus is fully real.

THEOREM 4.4 ([**71**, Theorem 1]). *Suppose $n > k > 0$ and $l_1, l_2, \ldots, l_s$ are positive integers with $l_1 + l_2 + \cdots + l_s = k(n-k)$. Then there are linear subspaces $L_1, L_2, \ldots, L_s$ of $\mathbb{R}^n$ in general position with* $\dim L_i = n - k + 1 - l_i$ *such that each of the a priori complex k-planes K satisfying* (4.2) *are in fact real.*

We present an elementary proof of this result in the important special case when each $l_i = 1$ so that the conditions are simple, meaning each $\Omega(L_i)$ has codimension 1. This proof generalizes to show that some other classes of enumerative problems in the Schubert calculus are fully real (see Sections 4.2.2 and 4.2.3). This generalization constructs sufficiently many real solutions using a limiting argument, as in Section 2.2. Just as the arguments of Section 2.2 were linked to the homotopy algorithms of Huber and Sturmfels, the proof of Theorem 4.4 leads to numerical homotopy methods for solving these problems [**30, 32**].

We develop further geometric properties of Grassmann varieties. The kth exterior power of the embedding $K \to \mathbb{C}^n$ of a k-plane K into $\mathbb{C}^n$ gives the embedding

$$\mathbb{C} \simeq \wedge^k K \longrightarrow \wedge^k \mathbb{C}^n\,, \tag{4.4}$$

whose image is a line in $\wedge^k\mathbb{C}^n$ and thus a point in the projective space $\mathbb{P}(\wedge^k\mathbb{C}^n) \simeq \mathbb{P}^{\binom{n}{k}-1}$. This point determines the k-plane K uniquely. The *Plücker embedding* is the resulting projective embedding of the Grassmannian

$$\mathrm{Gr}(k,n) \longrightarrow \mathbb{P}^{\binom{n}{k}-1}\,.$$

The $\binom{n}{k}$ homogeneous Plücker coordinates for the Grassmannian in this embedding are realized concretely as follows. Represent a k-plane K as the row space of a $k \times n$ matrix, also written K. A maximal minor of $K = (a_{ij})$ is the determinant of a $k \times k$ submatrix of K: Given a choice of columns $\alpha\colon 1 \le \alpha_1 < \alpha_2 < \cdots < \alpha_k \le n$, set

$$p_\alpha(K) \;:=\; \det \begin{bmatrix} a_{1,\alpha_1} & a_{1,\alpha_2} & \cdots & a_{1,\alpha_k} \\ a_{2,\alpha_1} & a_{2,\alpha_2} & \cdots & a_{2,\alpha_k} \\ \vdots & \vdots & \ddots & \vdots \\ a_{k,\alpha_1} & a_{k,\alpha_2} & \cdots & a_{k,\alpha_k} \end{bmatrix}\,.$$

The vector $(p_\alpha(K))$ of $\binom{n}{k}$ maximal minors of K defines the map (4.4) giving *Plücker coordinates* for K. Let $\binom{[n]}{k}$ be the collection of these indices of Plücker coordinates.

The indices $\binom{[n]}{k}$ have a natural Bruhat order

$$\beta \le \alpha \iff \beta_j \le \alpha_j \quad \text{for} \quad j = 1, 2, \ldots, k\,.$$

The Schubert variety Ω_α is

$$(4.5)\qquad \Omega_\alpha \;=\; \{K \in \mathrm{Gr}(k,n) \mid p_\beta(K) = 0 \text{ for } \beta \not\leq \alpha\}\,.$$

This has dimension $|\alpha| := \sum_j (\alpha_j - j)$.

The relevance of the Plücker embedding to Question 4.3 when $l_i = 1$ is seen as follows. Let L be a $(n-k)$-plane, represented as the row space of a $(n-k)$ by n matrix, also written L. Then a general k-plane K meets L non-trivially if and only if

$$\det \begin{bmatrix} K \\ L \end{bmatrix} \;=\; 0\,.$$

Laplace expansion along the rows of K gives

$$(4.6)\qquad 0 \;=\; \det \begin{bmatrix} K \\ L \end{bmatrix} \;=\; \sum_{\alpha \in \binom{[n]}{k}} p_\alpha(K) L_\alpha\,,$$

where L_α is the appropriately signed minor of L given by the columns complementary to α. Hence the set $\Omega(L)$ of k-planes that meet the $(n-k)$-plane L non-trivially is a hyperplane section of the Grassmannian in its Plücker embedding.

Thus the set of k-planes meeting $k(n-k)$ general $(n-k)$-planes non-trivially is a complementary linear section of the Grassmannian, and so the number $d(k,n)$ of such k-planes is the degree of the Grassmannian in its Plücker embedding. More generally, if $\alpha \in \binom{[n]}{k}$ and $L_1, L_2, \ldots, L_{|\alpha|}$ are general $(n-k)$-planes, then the number of points in the intersection

$$(4.7)\qquad \Omega_\alpha \cap \Omega(L_1) \cap \Omega(L_2) \cap \cdots \cap \Omega(L_{|\alpha|})$$

is the degree $d(\alpha)$ of the Schubert variety Ω_α, which we now compute.

An intersection $X \cap Y$ is *generically transverse* if X and Y meet transversally along an open subset of every component of $X \cap Y$. When $\beta, \alpha \in \binom{[n]}{k}$ satisfy $\beta < \alpha$ but there is no index γ with $\beta < \gamma < \alpha$, then we write $\beta \lessdot \alpha$. The following fact is elementary and due to Schubert.

THEOREM 4.5. *Let $\alpha \in \binom{[n]}{k}$ and set H_α to be the hyperplane defined by $p_\alpha = 0$. Then*

$$\Omega_\alpha \cap H_\alpha \;=\; \bigcup_{\beta \lessdot \alpha} \Omega_\beta\,,$$

and the intersection is generically transverse.

In fact this intersection is transverse along $\Omega_\beta^\circ := \Omega_\beta - \bigcup_{\delta < \beta} \Omega_\delta$. We obtain the recursion for the degree $d(\alpha)$ of the Schubert variety Ω_α

$$d(\alpha) \;=\; \sum_{\beta \lessdot \alpha} d(\beta)\,.$$

Since the minimal Schubert variety is a point (which has degree 1), this gives a conceptual formula for $d(\alpha)$. Let $\hat{0} = (1, 2, \ldots, k)$ be the minimal element in the Bruhat order.

$$(4.8)\quad d(\alpha) \;=\; \text{the number of saturated chains in the Bruhat order from } \hat{0} \text{ to } \alpha\,.$$

Figure 11 displays both the Bruhat order for $k = 3$ and $n = 6$ (on the left) and the degrees of the corresponding Schubert varieties (on the right).

Consider the action of the non-zero real numbers $\mathbb{R}^\times$ on $\mathbb{R}^n$

$$(4.9)\qquad t.e_j \;:=\; t^j \cdot e_j\,,$$

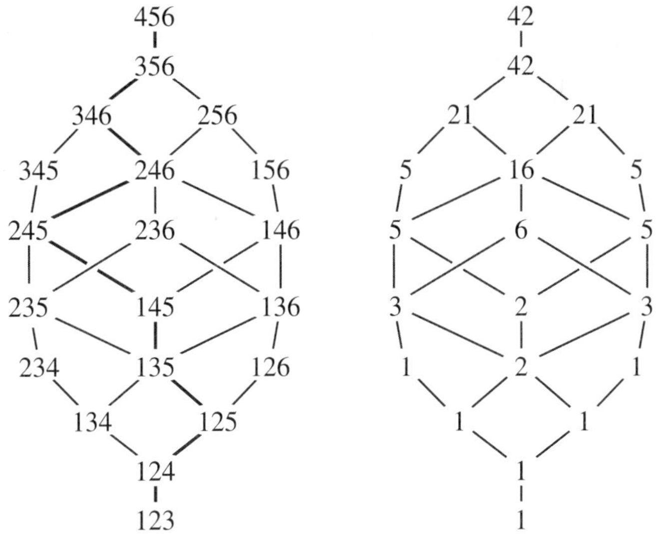

FIGURE 11. Bruhat order and degrees of Schubert varieties, $k = 3$, $n = 6$

where $t \in \mathbb{R}^\times$ and $e_1, e_2, \ldots, e_n$ is a basis for $\mathbb{R}^n$ (corresponding to the rows of the $n \times n$ identity matrix). Let L be a $(n-k)$-plane. By (4.6), the equation for a k-plane K to meet $t.L$ non-trivially is

$$0 = \sum_\beta t^{\binom{n+1}{2}-|\beta|} L_\beta p_\beta(K)\,.$$

For $K \in \Omega_\alpha$, the sum is over $\beta \leq \alpha$ by (4.5). Removing the common factor $t^{\binom{n+1}{2}-|\alpha|}$ gives

$$0 = \sum_{\beta\leq\alpha} t^{|\alpha|-|\beta|} L_\beta p_\beta(K)\,. \tag{4.10}$$

The case $l_i = 1$ of Theorem 4.4 is implied by the case $\alpha = (n-k+1, \ldots, n-1, n) = \hat{1}$ ($\Omega_{\hat{1}} = \mathrm{Gr}(k,n)$) of the following theorem.

THEOREM 4.6 ([**76**, Theorem 4.2]). *Let $L \subset \mathbb{R}^n$ be a $(n-k)$-plane, none of whose Plücker coordinates vanishes. Then there exist real numbers $t_1, t_2, \ldots, t_{k(n-k)} \in \mathbb{R}^\times$ such that for every $\alpha \in \binom{[n]}{k}$ the intersection*

$$\Omega_\alpha \cap \Omega(t_1.L) \cap \Omega(t_2.L) \cap \cdots \cap \Omega(t_{|\alpha|}.L)$$

is transverse (so it contains $d(\alpha)$ points) with all points real.

PROOF. We induct on m to construct numbers $t_1, t_2, \ldots, t_{k(n-k)} \in \mathbb{R}^\times$ having the property that, for all $\alpha \in \binom{[n]}{k}$ with $|\alpha| = m$,

$$\Omega_\alpha \cap \Omega(t_1.L) \cap \Omega(t_2.L) \cap \cdots \cap \Omega(t_m.L)$$

is transverse (over $\mathbb{C}$) and each of its $d(\alpha)$ points are real.

The case $m = 0$ is trivial, as $|\alpha| = 0$ implies that $\alpha = \hat{0}$ and $\Omega_{\hat{0}}$ is a single (real) point in $\mathrm{Gr}(k,n)$. Suppose we have constructed $t_1, t_2, \ldots, t_m \in \mathbb{R}^\times$ with the above properties.

Let $\alpha \in \binom{[n]}{k}$ with $|\alpha| = m+1$ and consider

$$(4.11)\quad \left(\sum_{\beta \lessdot \alpha} \Omega_\beta\right) \cap (\Omega(t_1.L) \cap \cdots \cap \Omega(t_m.L)) \;=\; \sum_{\beta \lessdot \alpha} \Omega_\beta \cap \Omega(t_1.L) \cap \cdots \cap \Omega(t_m.L)\,.$$

Each summand in the second sum is transverse (over $\mathbb{C}$) and consists of $d(\beta)$ real points. The intersection on the left will be transverse and consist of $d(\alpha) = \sum_{\beta \lessdot \alpha} d(\beta)$ real points only if no two summands in the second sum share a point in common.

If two summands β, γ with $\beta \neq \gamma$ share a point, then

$$(4.12)\qquad \Omega_\beta \cap \Omega_\gamma \cap \Omega(t_1.L) \cap \Omega(t_2.L) \cap \cdots \cap \Omega(t_m.L)$$

is non-empty. By (4.5), $\Omega_\beta \cap \Omega_\gamma$ is contained in a union of Schubert varieties of dimension less than m. By (4.10) the condition for a fixed k-plane $K \in \Omega_\delta$ to meet $t.L$ non-trivially is a polynomial of degree at most $|\delta|$ in t. Thus a fixed k-plane $K \in \Omega_\delta$ lies in at most $|\delta|$ distinct $(n-k)$-planes in the family $t.L$. Hence the intersection (4.12) is empty, and so the summands in the second sum of (4.11) are disjoint. This argument also shows that the points in the summand indexed by β lie in $\Omega_\beta^\circ = \Omega_\beta - \bigcup_{\delta < \beta} \Omega_\delta$.

For $t \in \mathbb{R}$, let $Z_t \subset \Omega_\alpha$ be the set of k-planes K satisfying the polynomial (4.10). For $t \neq 0$, we have $Z_t = \Omega_\alpha \cap \Omega(t.L)$. Since L has no vanishing Plücker coordinates, the constant term of that polynomial is $p_\alpha(K)$, and so by Theorem 4.5,

$$Z_0 \;=\; \Omega_\alpha \cap H_\alpha \;=\; \sum_{\beta \lessdot \alpha} \Omega_\beta\,.$$

Thus Z_0 meets

$$(4.13)\qquad \Omega(t_1.L) \cap \Omega(t_2.L) \cap \cdots \cap \Omega(t_m.L)$$

transversally (over $\mathbb{C}$) in $d(\alpha)$ real points. We see that there exists $\epsilon_\alpha > 0$ such that if $0 < t \leq \epsilon_\alpha$, then Z_t meets (4.13) transversally (over $\mathbb{C}$) in $d(\alpha)$ real points. Since the intersection defining Z_0 is transverse along each Ω_β° for $\beta \lessdot \alpha$, we may assume that for $0 < t \leq \epsilon_\alpha$,

$$\Omega_\alpha \cap \Omega(t.L) \cap \Omega(t_1.L) \cap \cdots \cap \Omega(t_m.L)$$

is transverse (over $\mathbb{C}$) and consists of $d(\alpha)$ real points.

Let $t_{m+1} = \min\{\epsilon_\alpha \colon |\alpha| = m+1\}$. Then for any $\alpha \in \binom{[n]}{k}$ with $|\alpha| = m+1$,

$$\Omega_\alpha \cap \Omega(t_1.L) \cap \cdots \cap \Omega(t_m.L) \cap \Omega(t_{m+1}.L)$$

is transverse (over $\mathbb{C}$) and consists of $d(\alpha)$ real points. □

4.2. Further extensions of the Schubert calculus. This special Schubert calculus admits further extensions, some of which are known to be fully real. Interestingly, some problems involving the Lagrangian Grassmannian 4.2.4 are known to be *fully unreal*, a new phenomenon. Further investigation is encouraged; in particular, the Lagrangian Schubert calculus may yield the first class of enumerative which are *not* fully real.

4.2.1. *General Schubert calculus.* A *flag* in n-space is a sequence of linear subspaces $F_\bullet\colon F_1 \subset F_2 \subset \cdots \subset F_n$ with $\dim F_i = i$. Given $\alpha \in \binom{[n]}{k}$, the Schubert condition of type α on a k-plane K imposed by the flag $F_\bullet$ is

$$\dim K \cap F_{n+1-\alpha_j} \geq k+1-j \qquad \text{for} \quad j = 1, 2, \ldots, n\,. \tag{4.14}$$

The Schubert variety $X_\alpha F_\bullet \subset \mathrm{Gr}(k,n)$ is the set of all k-planes K satisfying (4.14).

We relate this to the definitions of Section 4.1. Let $e_1, e_2, \ldots, e_n$ be a basis for $\mathbb{R}^n$ and for its complexification $\mathbb{C}^n$. Defining F_i to be the linear span of $e_1, e_2, \ldots, e_i$ gives the flag $F_\bullet$. Then the Schubert variety $X_\alpha F_\bullet$ is $\Omega_{\alpha^\vee}$, where $\alpha^\vee_j = n{+}1{-}\alpha_{k+1-j}$ for each j, and so the codimension of $X_\alpha F_\bullet$ is $|\alpha|$. A special Schubert condition is when $\alpha = (1, \ldots, k{-}1, k{+}l)$ so that $X_\alpha F_\bullet = \Omega(F_{n-k+1-l})$.

The general problem of the classical Schubert calculus of enumerative geometry asks, given $\alpha^1, \alpha^2, \ldots, \alpha^s \in \binom{[n]}{k}$ with $|\alpha^1| + |\alpha^2| + \cdots + |\alpha^s| = k(n-k)$ and general flags $F_\bullet{}^1, F_\bullet{}^2, \ldots, F_\bullet{}^s \subset \mathbb{C}^n$, how many points are there in the intersection[†]

$$X_{\alpha^1} F_\bullet{}^1 \cap X_{\alpha^2} F_\bullet{}^2 \cap \cdots \cap X_{\alpha^s} F_\bullet{}^s\,? \tag{4.15}$$

There are algorithms due to Pieri [**55**] and Giambelli [**21**] to compute these numbers. Other than the case when the α^i are indices of special Schubert varieties, it remains open whether the general Schubert calculus is fully real. (See [**73**] and [**70**] for some cases.)[§]

4.2.2. *Quantum Schubert calculus.* The space $\mathcal{M}^q_{k,n}$ of degree q maps $M\colon \mathbb{P}^1 \to \mathrm{Gr}(k,n)$ is a smooth quasi-projective variety [**9**]. A point $t \in \mathbb{P}^1$ and a Schubert variety $X_\alpha F_\bullet$ together impose a quantum Schubert condition on maps $M \in \mathcal{M}^q_{k,n}$,

$$M(t) \;\in\; X_\alpha F_\bullet\,.$$

The set of such maps has codimension $|\alpha|$ in $\mathcal{M}^q_{k,n}$. The quantum Schubert calculus of enumerative geometry asks the following question.

QUESTION 4.7. *Given $\alpha^1, \alpha^2, \ldots, \alpha^s \in \binom{[n]}{k}$ with $|\alpha^1| + |\alpha^2| + \cdots + |\alpha^s| = \dim \mathcal{M}^q_{k,n} = qn + k(n-k)$, how many maps $M \in \mathcal{M}^q_{k,n}$ satisfy*

$$M(t_i) \;\in\; X_{\alpha^i} F_\bullet{}^i \qquad \textit{for } i = 1, 2, \ldots, s\,,$$

where $t_1, t_2, \ldots, t_s \in \mathbb{P}^1$ are general and $F_\bullet{}^1, F_\bullet{}^2, \ldots, F_\bullet{}^s$ are general flags?

Algorithms to compute this number were proposed by Vafa [**82**] and Intriligator [**34**] and proven by Siebert and Tian [**66**] and by Bertram [**3**]. A simple quantum Schubert condition defines a subvariety of codimension 1,

$$M(t) \;\cap\; L \;\neq\; \{0\}\,, \tag{4.16}$$

where L is a $(n{-}k)$-plane. Let $d(q;k,n)$ be the number of maps $M \in \mathcal{M}^q_{k,n}$ satisfying $\dim \mathcal{M}^q_{k,n}$-many general simple quantum Schubert conditions (4.16). A combinatorial formula for this number was given by Ravi, Rosenthal, and Wang [**59**]. For a survey of this particular enumerative problem and its importance to linear systems theory, see [**75**].

[†] In this survey flags are general when the corresponding intersection is transverse.

[§] While this survey was in review, Ravi Vakil communicated to the author a proof that the classical Schubert calculus is fully real [**83**].

THEOREM 4.8 ([**72**, Theorem 1.1]). *Let $q \geq 0$ and $n > k > 0$ and set $N := \dim \mathcal{M}_{k,n}$. Then there exist $t_1, t_2, \ldots, t_N \in \mathbb{RP}^1$ and $(n-k)$-planes $L_1, L_2, \ldots, L_N \subset \mathbb{R}^n$ such that there are exactly $d(q;k,n)$ maps $M \in \mathcal{M}^q_{k,n}(\mathbb{C})$ satisfying*

$$M(t_i) \cap L_i \neq \{0\} \qquad \text{for } i = 1, 2, \ldots, N\,,$$

and all of them are real.

As with the classical Schubert calculus, the question of whether the general quantum Schubert calculus is fully real remains open.

4.2.3. *Schubert calculus of flags.* Let $\mathbf{a} := 0 < a_1 < \cdots < a_r < a_{r+1} = n$ be a sequence of integers. The manifold of partial flags in n-space (or the flag manifold) $\mathbb{F}\ell_{\mathbf{a}}$ is the collection of partial flags of subspaces

$$E_\bullet \;:\; E_{a_1} \subset E_{a_2} \subset \cdots \subset E_{a_r}$$

in n-space, where $\dim E_{a_i} = a_i$. A complete flag $F_\bullet$ is when $\mathbf{a} = 1, 2, \ldots, n-1$.

The Schubert varieties of $\mathbb{F}\ell_{\mathbf{a}}$ are indexed by permutations $w \in \mathcal{S}_n$, the symmetric group on n letters, whose descents only occur at positions in $\mathbf{a}$. That is, $w(i) > w(i+1)$ implies that $i \in \{a_1, a_2, \ldots, a_r\}$. Let $I_{\mathbf{a}}$ be this set of indices. For a complete flag $F_\bullet$ and $w \in I_{\mathbf{a}}$, $\mathbb{F}\ell_{\mathbf{a}}$ has a Schubert variety

$$X_w F_\bullet \;:=\; \big\{E_\bullet \in \mathbb{F}\ell_{\mathbf{a}} \mid \dim E_{a_j} \cap F_i \geq \#\{l \leq a_j \mid w(l) \leq i\}\big\}\,.$$

The codimension of $X_w F_\bullet$ is $\ell(w) := \#\{i < j \mid w(i) > w(j)\}$. We state the general question in the Schubert calculus of enumerative geometry for flags.

QUESTION 4.9. *Given permutations $w_1, w_2, \ldots, w_s \in I_{\mathbf{a}}$ with $\ell(w_1) + \ell(w_2) + \cdots + \ell(w_s) = \dim \mathbb{F}\ell_{\mathbf{a}}$ and general flags $F_\bullet{}^1, F_\bullet{}^2, \ldots, F_\bullet{}^s$, what is the number of points in the intersection*

$$X_{w_1} F_\bullet{}^1 \cap X_{w_2} F_\bullet{}^2 \cap \cdots \cap X_{w_s} F_\bullet{}^s\ ?$$

There are algorithms [**2, 13**] for computing this number and the numbers for the Lagrangian Schubert calculus in Section 4.2.4. When $w = (i, i+1)$, $X_w F_\bullet$ is the simple Schubert variety, written $X_i F_\bullet$,

$$X_i F_\bullet \;=\; X_{(i,i+1)} F_\bullet \;=\; \{E_\bullet \in \mathbb{F}\ell_n \mid E_i \cap F_{n-i} \neq \{0\}\}\,. \tag{4.17}$$

THEOREM 4.10 ([**74**, Corollary 2.2]). *Given a list $i_1, i_2, \ldots, i_N$ ($N = \dim \mathbb{F}\ell_{\mathbf{a}}$) of numbers with $i_j \in \mathbf{a}$, there exist real flags $F_\bullet{}^1, F_\bullet{}^2, \ldots, F_\bullet{}^N$ such that the intersection of simple Schubert varieties*

$$X_{i_1} F_\bullet{}^1 \cap X_{i_2} F_\bullet{}^2 \cap \cdots \cap X_{i_N} F_\bullet{}^N$$

is transverse and consists only of real flags.

The case of $\mathbf{a} = 2 < n-2$ of this theorem was proven earlier [**69**, Theorem 13]. It remains open whether the general Schubert calculus of flags is fully real.

4.2.4. *Unreality in the Lagrangian Schubert calculus.* Let V be a $2n$-dimensional vector space equipped with the alternating bilinear form

$$\Big\langle \sum x_i e_i\,,\ \sum y_j e_j \Big\rangle \;:=\; \sum_{i=1}^{n} x_i y_{2n+1-i} - y_i x_{2n+1-i}\,. \tag{4.18}$$

A subspace $H \subset V$ is *isotropic* if the restriction of the form to H is identically zero, $\langle H, H\rangle \equiv 0$. The dimension of an isotropic subspace is at most n and *Lagrangian subspaces* are isotropic subspaces with this maximal dimension. The Lagrangian

Grassmannian $LG(n)$ is the set of Lagrangian subspaces of V, an algebraic manifold of dimension $\binom{n+1}{2}$.

A flag $F_\bullet$ in V is *isotropic* if F_n is Lagrangian and $\langle F_i, F_{2n-i}\rangle \equiv 0$ for all $i = 1, 2, \ldots, 2n-1$. Given an isotropic flag $F_\bullet$, the Lagrangian Grassmannian has Schubert varieties $\Psi_\lambda F_\bullet$ indexed by decreasing sequences $\lambda\colon n \geq \lambda_1 > \lambda_2 > \cdots > \lambda_l > 0$ of positive integers, called strict partitions. (Here l can be any integer between 0 and n). The codimension of $\Psi_\lambda F_\bullet$ is $|\lambda| = \lambda_1 + \lambda_2 + \cdots + \lambda_l$. The Lagrangian Schubert calculus asks for the number of points in a transverse zero-dimensional intersection of Schubert varieties.

The *simple Schubert variety* $\Psi_1 F_\bullet$ consists of those Lagrangian subspaces meeting the Lagrangian subspace F_n non-trivially. The simple Lagrangian Schubert calculus is fully unreal.

THEOREM 4.11. *There exist isotropic real flags* $F_\bullet{}^1, F_\bullet{}^2, \ldots, F_\bullet{}^{\binom{n+1}{2}}$ *such that the intersection of Schubert varieties*

$$\Psi_1 F_\bullet{}^1 \cap \Psi_1 F_\bullet{}^2 \cap \cdots \cap \Psi_1 F_\bullet{}^{\binom{n+1}{2}}$$

is transverse with **no** *real points.*

REMARK 4.12. In [**74**, Theorem 4.2] flags were given so that the intersection had no real points, and it was not known if that intersection was transverse. Perturbing those flags slightly (so that the intersection becomes transverse) gives the above result.

We do not know if these (or many other) enumerative problems in the Lagrangian Schubert calculus are fully real. Experimentation[†] suggests that the situation is complicated. Briefly, while many other enumerative problems in the Lagrangian Schubert calculus are fully unreal, there are a few which are fully real. For example, there exist 2 real isotropic 2-planes and 2 real isotropic 3-planes such that all 4 Lagrangian subspaces meeting each of these are real (see Theorem 5.13).

The problems of Theorem 4.11 may give examples of enumerative problems that are not fully real. Experimental evidence suggests however that this may be unlikely. For example, (case $n = 3$ of Theorem 4.11) there are 16 Lagrangian subspaces in $\mathbb{C}^6$ having non-trivial intersection with 6 general Lagrangian subspaces. We computed 30,000 random instances of this enumerative problem, and found two examples of 6 real Lagrangian subspaces such that all 16 Lagrangian subspaces meeting them are real. Table 1 shows the number of these 30,000 systems having a given number of real solutions.

Number of real solutions	0	2	4	6	8	10	12	14	16
Frequency	4983	8176	9314	5027	1978	445	67	8	2

TABLE 1. Frequency with given number of real solutions.

[†]This will be reported in the forthcoming article [**77**].

5. The Conjecture of Shapiro and Shapiro

The results of Section 4 were inspired by a remarkable conjecture of Boris Shapiro and Michael Shapiro. Let $\gamma\colon \mathbb{R} \to \mathbb{R}^n$ be the rational normal curve

$$\gamma(t) \;=\; (1,\, t,\, t^2,\, \ldots,\, t^{n-1}) \;=\; \sum_{i=1}^{n} t^{i-1} e_i\,.$$

For $t \in \mathbb{R}$, define the flag $F_\bullet(t)$ by any of the three equivalent ways

$$(5.1)\quad F_i(t) \;=\; \text{linear span of } \gamma(t),\ \frac{d}{dt}\gamma(t),\ \ldots,\ \frac{d^{i-1}}{dt^{i-1}}\gamma(t)\,,$$

$$(5.2)\qquad = \text{row space} \begin{bmatrix} 1 & t & t^2 & & \cdots & & t^{n-1} \\ 0 & 1 & 2t & & \cdots & & (n{-}1)t^{n-2} \\ \vdots & & \ddots & \ddots & & & \vdots \\ 0 & \cdots & 0 & 1 & it & \cdots & \binom{n-1}{i-1}t^{n-i} \end{bmatrix},$$

$$(5.3)\qquad = i\text{-plane osculating rational normal curve } \gamma \text{ at } \gamma(t)\,.$$

This makes sense for $t \in \mathbb{C}$ and is extended to $t = \infty \in \mathbb{P}^1$ by setting $F_i(\infty)$ to be the row space of the last i rows of the $n \times n$ identity matrix.

Conjecture 5.1 (Shapiro-Shapiro). *Let $\alpha^1, \alpha^2, \ldots, \alpha^s \in \binom{[n]}{k}$ be such that $|\alpha^1| + |\alpha^2| + \ldots + |\alpha^s| = k(n-k)$. Then, for every distinct $t_1, t_2, \ldots, t_s \in \mathbb{R}$, the intersection of Schubert varieties*

$$(5.4)\qquad X_{\alpha^1}F_\bullet(t_1) \cap X_{\alpha^2}F_\bullet(t_2) \cap \cdots \cap X_{\alpha^s}F_\bullet(t_s)$$

is (a) *transverse, and* (b) *consists only of real points.*

Eisenbud's and Harris's dimensional transversality result [**15**, Theorem 2.3] guarantees that the intersection (5.4) is zero-dimensional. Not only does Conjecture 5.1 state that the classical Schubert calculus is fully real, but it also proposes flags witnessing this full reality. This conjecture has been central to subsequent developments in the real Schubert calculus and it has direct connections to other parts of mathematics, including linear systems theory and linear series on $\mathbb{P}^1$ (see Remark 5.8). The article [**73**] and the web page [**70**] give a more complete discussion.

One aspect of this conjecture which we relate is the following.

Theorem 5.2 ([**73**, Theorem 3.3]). *For a given k and n, the general case of Conjecture 5.1 follows from the special case when each Schubert condition is simple, that is, when each $\alpha^i = (1, 2, \ldots, k{-}1, k{+}1)$.*

Consider these osculating flags $F_\bullet(t)$ in more detail. If the ith row of the matrix (5.2) is multiplied by t^i (which does not affect its row space when $t \neq 0$), then the entry in position i,j is $\binom{j-1}{i-1}t^j$, and so we have

$$F_i(t) \;=\; t.F_i(0)\,,$$

where $t.F_i(0)$ is given by the action (4.9) of $\mathbb{R}^\times$ on $\mathbb{R}^n$. The αth Plücker coordinate of $F_i(0)$ is

$$(5.5)\qquad p_\alpha(F_i(0)) \;=\; \prod_{j<l} \frac{\alpha_j - \alpha_l}{j - l}\,,$$

which is non-vanishing. Thus Theorem 4.6 has the following corollary.

THEOREM 5.3 ([**71**, Theorem 1]). *There exist* $t_1, t_2, \ldots, t_{k(n-k)} \in \mathbb{R}$ *such that there are exactly* $d(k,n)$ k*-planes meeting each* $(n-k)$*-plane* $F_{n-k}(t_i)$ *non-trivially, and all are real. Equivalently, if* $\alpha = (1,2,\ldots,k-1,k+1)$ *so that* $|\alpha| = 1$*, then the intersection of Schubert varieties*

$$X_\alpha F_\bullet(t_1) \cap X_\alpha F_\bullet(t_2) \cap \cdots \cap X_\alpha F_\bullet(t_{k(n-k)}) \tag{5.6}$$

is transverse with all points real.

This establishes a weak form of Conjecture 5.1 for simple Schubert conditions, replacing the quantifier *for all* $t_i \in \mathbb{R}$ by *there exists* $t_i \in \mathbb{R}$.

If the parameters t_i in (5.6) vary, then the number of real points in that intersection could change, but only if two points first collide (prior to spawning a complex conjugate pair of solutions). This is the reverse of the progression in Dietmaier's algorithm, as displayed in Figure 6. This situation cannot occur if the intersection remains transverse. Together with Theorems 5.2 and 5.3 we deduce the following result.

THEOREM 5.4 ([**71**, Theorem 6]). *Part* (a) *of Conjecture 5.1 for simple conditions implies part* (b) *for arbitrary Schubert conditions.*

If $t > 0$, then by (5.5) the Plücker coordinates of $F_i(t)$ are strictly positive. An upper triangular $n \times n$ matrix g is *totally positive* if every $i \times i$ subdeterminant of g is non-negative, and vanishes only if that subdeterminant vanishes on all upper triangular matrices. For example, when $t > 0$ and $i = n$, the matrix (5.2) is totally positive. Write $\Gamma(t)$ for this matrix. It has the form $e^{t\eta}$ where η is the principal nilpotent matrix $\Gamma'(0)$. Observe that if $t_1 < t_2 < \cdots < t_s$, then

$$F_\bullet(t_i) \;=\; \Gamma(t_i - t_{i-1}) \cdot F_\bullet(t_{i-1})$$

Conjecture 5.1 has a more general version involving totally positive matrices.

CONJECTURE 5.5 (Shapiro-Shapiro [**73**, Conjecture 4.1]). *Let* $\alpha^1, \alpha^2, \ldots, \alpha^s \in \binom{[n]}{k}$ *be such that* $|\alpha^1| + |\alpha^2| + \ldots + |\alpha^s| = k(n-k)$ *and suppose* $g_2, g_3, \ldots, g_s$ *are totally positive matrices. Given any real flag* ${F_\bullet}^1$*, define* ${F_\bullet}^i$ *for* $i > 1$ *by* ${F_\bullet}^i := g_i \cdot {F_\bullet}^{i-1}$*, the intersection of Schubert varieties*

$$X_{\alpha^1}{F_\bullet}^1 \cap X_{\alpha^2}{F_\bullet}^2 \cap \cdots \cap X_{\alpha^s}{F_\bullet}^s$$

is transverse with all points real.

There is some experimental evidence for this version of Conjecture 5.1. Subsequent conjectures involving osculating flags have versions involving totally positive matrices. We leave their statements to the reader, they will be explored more fully in [**77**].

5.1. Rational functions with real critical points. By far the strongest evidence for Conjecture 5.1 is that it is true when k or $n-k$ is equal to 2.

THEOREM 5.6 (Eremenko and Gabrielov [**16**]). *Conjecture 5.1 is true when one of* k *or* $n-k$ *is* 2.

This is a consequence of a theorem about rational functions with real critical points. A rational function is an algebraic map $\varphi\colon \mathbb{P}^1 \to \mathbb{P}^1$. Two rational functions φ_1 and φ_2 are equivalent if $\varphi_1 = g \circ \varphi_2$, where g is a fractional linear transformation.

THEOREM 5.7 ([**16**]). *If all the critical points of a rational function are real, then it is equivalent to a real rational function.*

Consider the composition

$$\mathbb{P}^1 \longrightarrow \mathbb{P}^d \dashrightarrow \mathbb{P}^1\,, \tag{5.7}$$

where the first map is the rational normal curve

$$\gamma\,:\,[s,t]\;\longmapsto\;[s^d,\,s^{d-1}t,\,\ldots,\,st^{d-1},\,t^d]\,,$$

and the second is a linear projection

$$[x_0,x_1,\ldots,x_d]\;\longmapsto\;[\Lambda_1(x),\Lambda_2(x)]\,,$$

where Λ_1 and Λ_2 are independent linear forms. Let $E\subset\mathbb{P}^d$ be the center of this projection, the linear subspace where $\Lambda_1=\Lambda_2=0$. When E is disjoint from the rational normal curve, this composition defines a rational function of degree d, and all such rational functions occur in this manner. In fact, equivalence classes of rational maps are exactly those maps with the same center of projection. A rational function φ has a critical point at $t\in\mathbb{P}^1$ if $d\varphi$ vanishes at t. If we consider the composition (5.7), this implies that the center E meets the line tangent to the rational normal curve $\gamma(\mathbb{P}^1)$ at $\gamma(t)$.

Goldberg [**22**] asked (and answered) the question: how many equivalence classes of rational functions of degree d have a given set of $2d-2$ critical points? Reasoning as above, she reduced this to the problem of determining the number of codimension 2 planes E in $\mathbb{P}^d$ meet $2d-2$ given tangents to the rational normal curve. Formulating this in the dual projective space, we recover the problem of the Introduction to Section 4: Determine the number of lines in $\mathbb{P}^d$ meeting $2d-2$ general codimension 2 planes. The answer is the dth Catalan number, $C_d=\frac{1}{d}\binom{2d-1}{d-1}$ (4.1).

Consider the above description in the affine cone over $\mathbb{P}^d$. The centers $E\in\mathrm{Gr}(d-1,d+1)$ giving a rational function of degree d with critical point at $t\in\mathbb{C}\subset\mathbb{P}^1$ are points in the Schubert variety $X_\alpha F_\bullet(t)$, where α is the simple Schubert condition of Theorem 5.2 and $F_\bullet(t)$ is the flag of subspaces osculating the rational normal curve $\gamma(\mathbb{P}^1)$ at $\gamma(t)$. An equivalence class of rational functions contains a real rational function when the common center E is real. In this way, Theorem 5.7 implies Conjecture 5.1 when $n-k=2$ and each $|\alpha^i|=1$. By Theorem 5.2, this implies the full conjecture when $n-k=2$. Working in the dual space, we deduce Conjecture 5.1 when $k=2$.

Gabrielov and Eremenko prove Theorem 5.7 by showing there exist C_d distinct real rational functions with critical points at a given set of $2d-2$ real numbers. Let $\varphi\colon\mathbb{P}^1\to\mathbb{P}^1$ be a real rational function of degree d with only real critical points. Then $\varphi^{-1}(\mathbb{P}^1_\mathbb{R})\subset\mathbb{P}^1_\mathbb{C}$ is an embedded graph containing $\mathbb{P}^1_\mathbb{R}$ which is stable under complex conjugation and whose vertices have valence 4 and are at the critical points. There are exactly C_d isotopy classes of such graphs. For each isotopy class Γ and collection of $2d-2$ distinct real points, they essentially construct a rational function φ of degree d having these critical points with $\varphi^{-1}(\mathbb{P}^1_\mathbb{R})\in\Gamma$. A key point involves degenerate rational maps with fewer than $2d-2$ critical points. It may be interesting to relate this to the degenerations in the proof of Theorem 4.6.

REMARK 5.8. If we project the rational normal curve from a center E of codimension $m+1$, then the image of the projection is $\mathbb{P}^m$ and the composition is a parameterized rational curve in $\mathbb{P}^m$. Post-composition by an element of $\mathrm{Aut}(\mathbb{P}^m)$

defines an equivalence relation on such maps and equivalence classes are determined by the centers of projection. These centers are codimension $m+1$ linear series of degree d on $\mathbb{P}^1$. Such a linear series is *ramified* at a point $t \in \mathbb{P}^1$ when the center E meets the m-plane osculating the rational normal curve. A rational curve/linear series is *maximally inflected* if the ramification is at real points. We just considered the case $d=1$ of rational functions with real critical points, and the case $d=2$ was discussed in Section 3.1. The existence of maximally inflected curves with simple ramification is a consequence of Theorem 5.3, and Conjecture 5.1 predicts this is a rich class of real curves in $\mathbb{P}^m$. This connection between linear series and the Schubert calculus originated in work of Castelnuovo [**7**].

5.2. Generalizations of Conjecture 5.1. The Grassmannian, flag manifolds, and Lagrangian Grassmannian are examples of flag varieties G/P where G is a reductive algebraic group and P a parabolic subgroup. These flag varieties have Schubert varieties and the most general form of the Schubert calculus involves zero-dimensional intersections of these Schubert varieties. Likewise, these flag varieties have real forms (given by split† forms of G and P) and there is a generalization of Conjecture 5.1 for these real forms. This generalization is false, but in a very interesting way. We describe what is known about this general conjecture for the flag manifolds and the Lagrangian Grassmannian, and give conjectures describing what we believe to be true.

5.2.1. *The manifold of partial flags.* Let $\mathbf{a} := 0 < a_1 < a_2 < \cdots < a_r < a_{r+1} = n$ be a sequence of integers. The straightforward generalization (which was its original form and which is false) of Conjecture 5.1 for the flag manifold $\mathbb{F}\ell_{\mathbf{a}}$ declares that a zero-dimensional intersection of Schubert varieties given by flags osculating the real rational normal curve consists only of real points.

For $\alpha \in \binom{[n]}{a_i}$, we have the *Grassmannian Schubert variety*

$$Y_{\alpha,i}F_\bullet := \{E_\bullet \in \mathbb{F}\ell_{\mathbf{a}} \mid E_{a_i} \in X_\alpha F_\bullet \subset \mathrm{Gr}(a_i, n)\}.$$

For example, if $|\alpha| = 1$, then $Y_{\alpha,i}F_\bullet$ is the simple Schubert variety $X_iF_\bullet$ of Section 4.2.3.

EXAMPLE 5.9. Let $\mathbf{a} = 2 < 3 < 5 = n$ so that $\mathbb{F}\ell_{\mathbf{a}}$ is the manifold of partial flags $E_2 \subset E_3$ in 5-space. This flag manifold has two simple Schubert varieties

$$\begin{aligned} X_2F_\bullet &= \{E_2 \subset E_3 \mid E_2 \cap F_3 \neq \{0\}\} \\ X_3F_\bullet &= \{E_2 \subset E_3 \mid E_3 \cap F_2 \neq \{0\}\} \end{aligned}$$

A calculation [**74**, Example 2.5] shows that

$$\begin{aligned} X_2F_\bullet(-8) \cap X_3F_\bullet(-4) \cap X_2F_\bullet(-2) \cap X_3F_\bullet(-1) \cap \\ X_2F_\bullet(1) \cap X_3F_\bullet(2) \cap X_2F_\bullet(4) \cap X_3F_\bullet(8) \end{aligned} \tag{5.8}$$

is transverse with *none* of its 12 points real.

Thus the straightforward generalization of Conjecture 5.1 is completely false. On the other hand, if $t_1 < t_2 < \cdots < t_8$ are any of the 24,310 subsets of eight numbers from

$$\{-6, -5, -4, -3, -2, -1, 1, 2, 3, 5, 7, 11, 13, 17, 19, 23, 29\},$$

†Split is a technical term: $\mathbb{R}^\times \subset \mathbb{C}^\times$ is a split form of GL_1, but $S^1 \subset \mathbb{C}^\times$ is not.

then the intersection

$$X_2F_\bullet(t_1) \cap X_2F_\bullet(t_2) \cap X_2F_\bullet(t_3) \cap X_2F_\bullet(t_4) \cap X_3F_\bullet(t_5) \cap X_3F_\bullet(t_6) \cap X_3F_\bullet(t_7) \cap X_3F_\bullet(t_8) \tag{5.9}$$

is transverse with *all* of its 12 points real [**74**, Example 2.5].

We have used more than 2×10^7 seconds of CPU time investigating this problem of intersections of Schubert varieties in manifolds of partial flags given by flags osculating the rational normal curve, and a picture is emerging of what to expect, at least for Grassmannian Schubert varieties.

Suppose we have a list of indices of Grassmannian Schubert varieties

$$(\alpha^1, i_1),\ (\alpha^2, i_2),\ \ldots,\ (\alpha^s, i_s), \qquad \text{with} \quad \alpha^j \in \binom{[n]}{i_j}$$

where $i_j \in \{a_1, \ldots, a_r\}$ and $|\alpha^1| + |\alpha^2| + \cdots + |\alpha^s| = \dim \mathbb{F}\ell_{\mathbf{a}}$. Call such a list *Grassmannian Schubert data* for $\mathbb{F}\ell_{\mathbf{a}}$. Consider an intersection of Grassmannian Schubert varieties

$$Y_{\alpha^1,i_1}F_\bullet(t_1) \cap Y_{\alpha^2,i_2}F_\bullet(t_2) \cap \cdots \cap Y_{\alpha^s,i_s}F_\bullet(t_s)\,, \tag{5.10}$$

where $t_1 < t_2 < \cdots < t_s$ are distinct real numbers and $F_\bullet(t)$ is the flag osculating the rational normal curve γ at $\gamma(t)$.

CONJECTURE 5.10. *Let* $\mathbf{a} = 1 < a_1 < a_2 < \cdots < a_r < a_{r+1} = n$ *and*

$$(\alpha^1, i_1),\ (\alpha^2, i_2),\ \ldots,\ (\alpha^s, i_s), \qquad \textit{with} \quad \alpha^j \in \binom{[n]}{i_j}$$

be Grassmannian Schubert data for $\mathbb{F}\ell_{\mathbf{a}}$.

(1) *For every choice of real numbers* $t_1 < t_2 < \cdots < t_s$*, the intersection* (5.10) *is* (a) *transverse with* (b) *with all points of intersection real if the indices* i_j *are in order:* $i_1 \leq i_2 \leq \cdots \leq i_s$ *or else* $i_1 \geq i_2 \geq \cdots \geq i_s$.

(2) *If the indices* i_j *are not in order, nor is any cyclic permutation of the indices, then there exist real numbers* $t_1 < t_2 < \cdots < t_s$ *such that the intersection* (5.10) *is transverse with* **all** *points real, and there exist real numbers* $t_1 < t_2 < \cdots < t_s$ *such that the intersection is transverse with* **not all** *points real.*

In particular, enumerative problems involving Grassmannian Schubert varieties on $\mathbb{F}\ell_{\mathbf{a}}$ *are fully real.*

REMARK 5.11.

(1) When each $|\alpha^i| = 1$, there are real numbers $t_1 < t_2 < \ldots < t_s$ such that the intersection (5.10) is transverse with all points real [**74**, Corollary 2.2]. Thus the number of real points in the intersection (5.10) is expected to vary when the the indices i_j are not in order.

(2) When $r = 1$, $\mathbb{F}\ell_{\mathbf{a}}$ is the Grassmannian $\mathrm{Gr}(a_1, n)$ and the condition on the indices being ordered is empty, so this case of Conjecture 5.10 reduces to Conjecture 5.1.

(3) There is considerable evidence for this conjecture when $r = 2$. Part (2) is true for every set of Grassmannian Schubert data in $\mathbb{F}\ell_{2<3}\mathbb{C}^5$ and all

except one such set in $\mathbb{F}\ell_{2<4}\mathbb{C}^6$†. Many instances of these same enumerative problems with ordered partitions have been computed, and in each instance of (1) the intersection (5.10) is transverse with all points real.

(4) As for Conjecture 5.1, (a) implies (b) in Part (1) of Conjecture 5.10.

(5) We have tested no instances of the intersection (5.10) with $r > 2$, so the truth may differ from the exact statement of Conjecture 5.10.

(6) Conjecture 5.10 has nothing to say when the Schubert data are not Grassmannian.

5.2.2. *The Lagrangian Grassmannian.* As in Section 4.2.4, let V be a $2n$-dimensional vector space equipped with the alternating form $\langle\cdot,\cdot\rangle$ defined by (4.18). Set

$$\gamma(t) := \left(1,\, t,\, \frac{t^2}{2}, \ldots, \frac{t^n}{n!}, -\frac{t^{n+1}}{(n+1)!}, \frac{t^{n+2}}{(n+2)!}, \ldots, (-1)^{n-1}\frac{t^{2n-1}}{(2n-1)!}\right). \tag{5.11}$$

The flag $F_\bullet(t)$ of subspaces osculating this rational normal curve is isotropic for all $t \in \mathbb{C}$.

Given a strict partition $\lambda\colon \lambda_1 > \lambda_2 > \cdots > \lambda_l > 0$ and an isotropic flag $F_\bullet$, set

$$\Psi_\lambda F_\bullet := \{H \in LG(n) \mid H \cap F_{n+1-\lambda_i} \geq i \text{ for } i = 1, \ldots, l\}.$$

The codimension of this Schubert variety is $|\lambda| = \lambda_1 + \lambda_2 + \cdots + \lambda_l$. For example,

$$\Psi_a F_\bullet = \{H \mid H \cap F_{n+1-a} \neq \{0\}\}.$$

With these definitions, we state the version of Theorem 4.11 proven in [**74**].

THEOREM 5.12 ([**74**, Theorem 4.2]). *Given a strict partition λ, let $N := \binom{n+1}{2} - |\lambda| = \dim \Psi_\lambda F_\bullet$. If $N > 1$, then there exist distinct real numbers $t_1, t_2, \ldots, t_N$ such that*

$$\Psi_\lambda F_\bullet(0) \cap \Psi_1 F_\bullet(t_1) \cap \Psi_1 F_\bullet(t_2) \cap \cdots \cap \Psi_1 F_\bullet(t_N)$$

is zero-dimensional with **no** *points real.*

Thus the generalization of Conjecture 5.1 is badly false for the Lagrangian Grassmannian. On the other hand it holds for some enumerative problems in the Lagrangian Grassmannian.

THEOREM 5.13 ([**77**]). *For any distinct real numbers t_1, t_2, t_3, t_4, the intersection of Schubert varieties in $LG(3)$*

$$\Psi_1 F_\bullet(t_1) \cap \Psi_1 F_\bullet(t_2) \cap \Psi_2 F_\bullet(t_3) \cap \Psi_2 F_\bullet(t_4)$$

is transverse with **all** *points real.*

These two results illustrate a dichotomy that is emerging from experimentation. Call a list of strict partitions $\lambda^1, \lambda^2, \ldots, \lambda^s$ with $|\lambda^1| + |\lambda^2| + \cdots + |\lambda^s| = \binom{n+1}{2}$ Lagrangian Schubert data. In every instance we have computed of a zero-dimensional intersection of Lagrangian Schubert varieties whose flags osculate the rational normal curve (5.11), the intersection has been transverse with either all points real or no points real. Most interestingly, the outcome—all real or no real—has depended only upon the Lagrangian Schubert data of the intersection.

†The one exception is due to our inability to compute any instances.

CONJECTURE 5.14. *Given Lagrangian Schubert data* $\lambda^1, \lambda^2, \ldots, \lambda^s$ *and distinct real numbers* $t_1, t_2, \ldots, t_s$, *the intersection*

$$\Psi_{\lambda^1} F_\bullet(t_1) \cap \Psi_2 \lambda^2 F_\bullet(t_2) \cap \cdots \cap \Psi_{\lambda^s} F_\bullet(t_s)$$

is transverse with either

(a) *all points real, or*

(b) *no points real,*

and the outcome (a) *or* (b) *depends only upon the list* $\lambda^1, \lambda^2, \ldots, \lambda^s$.

We do not have a good idea what distinguishes the Lagrangian Schubert data giving all points real from the data giving no points real. Further experimentation is needed.

5.2.3. *Further generalizations of Conjecture 5.1.* The status of the generalizations of Conjecture 5.1 to other flag varieties is almost completely unknown. There is one flag variety, the Orthogonal Grassmannian, for which much is known. In particular, the analog of Theorem 5.3 (involving codimension 1 Schubert varieties given by isotropic flags osculating the rational normal curve) holds for the Orthogonal Grassmannian [**74**, Corollary 3.3]. There has also been a significant amount of computer experimentation testing cases of the obvious generalization of Conjecture 5.1 for the orthogonal Grassmannian, and in each, all points of intersection were found to be real. Lastly, we remark that we do not know of any reasonable version of Conjecture 5.1 for the quantum Schubert calculus.

6. Lower Bounds in the Schubert calculus

In Section 3.3, we saw that of the 12 rational cubics meeting 8 real points in the plane, at least 8 were real. This was the first instance of a non-trivial lower bound on the number of real solutions to a problem in enumerative geometry. Recent work of Eremenko and Gabrielov shows this phenomenon is pervasive in the Schubert calculus.

Recall from Section 4.1 that the k-planes in $\mathbb{C}^n$ meeting $k(n-k)$ general $(n-k)$-planes non-trivially is a complementary dimensional linear section of the Grassmannian,

$$\Lambda \cap \mathrm{Gr}(k,n)\,,$$

where Λ has codimension $k(n-k)$ in Plücker space. The number of such k-planes is the degree of $\mathrm{Gr}(k,n)$. This is also the degree of the linear projection

$$\pi_E \;:\; \mathrm{Gr}(k,n) \hookrightarrow \mathbb{P}^{\binom{n}{k}-1} \dashrightarrow \mathbb{P}^{k(n-k)}$$

with center of projection a plane E with codimension $k(n-k)+1$ disjoint from $\mathrm{Gr}(k,n)$. The connection between these two definitions of degree is that when $E \subset \Lambda$, $\pi_E(\Lambda)$ is a point $x \in \mathbb{P}^{k(n-k)}$ and

$$\Lambda \cap \mathrm{Gr}(k,n) \;=\; \pi_E^{-1}(x)\,.$$

Since complex manifolds are canonically oriented, the degree of such a linear projection is just the number of points in the inverse image of a regular value x.

An important such linear projection is the Wronski map. Let $L(t)$ be the $(n-k)$-plane osculating the rational normal curve (5.1). By (4.6) and the discussion following Theorem 5.2, the equation for a k-plane K to meet $L(t)$ non-trivially is

$$\sum t^{\binom{n+1}{2}-\binom{n-k+1}{2}-|\alpha|} L_\alpha p_\alpha(K)\,, \tag{6.1}$$

where $p_\alpha(K)$ is the αth Plücker coordinate of K and L_α is the appropriately signed Plücker coordinate of $L(0)$ complementary to α. The association of a k-plane K to the polynomial (6.1) is the *Wronski map*

$$\pi_W \; : \; \mathrm{Gr}(k,n) \;\longrightarrow\; \mathbb{P}^{k(n-k)}\,,$$

where $\mathbb{P}^{k(n-k)}$ is the space of polynomials of degree at most $k(n-k)$, modulo scalars. This is a linear projection as the coefficients in (6.1) are linear in the Plücker coordinates. If f has distinct roots $t_1, t_2, \ldots, t_{k(n-k)}$, then $\pi_W^{-1}(f)$ is the set of k-planes meeting each of $L(t_1), L(t_2), \ldots, L(t_{k(n-k)})$ non-trivially. Observe that $\pi_W^{-1}(f)$ is given by real linear equations on the Grassmannian if and only if f has real coefficients, which includes the case when the t_i are all real (the situation of the Shapiro Conjecture 5.1).

To see why this is called the Wronski map, consider $\mathrm{Gr}(k,n)$ as the set of k-planes in the space of polynomials of degree at most $n-1$. Given a k-plane

$$K \;=\; \text{linear span}\{f_1, f_2, \ldots, f_k\}\,,$$

the Wronski determinant of K is

$$W(K) \;=\; \det \begin{bmatrix} f_1 & f_2 & \cdots & f_k \\ f_1' & f_2' & \cdots & f_k' \\ \vdots & \vdots & \ddots & \vdots \\ f_1^{(k-1)} & f_2^{(k-1)} & \cdots & f_k^{(k-1)} \end{bmatrix}\,,$$

a polynomial of degree at most $k(n-k)$, well-defined modulo scalars. Under a choice of coordinates given by the coefficients of a polynomial, $W(K) = \pi_W(K)$.

Since real manifolds are not necessarily orientable, the degree of a map is a $\mathbb{Z}/2\mathbb{Z}$-valued invariant. However, Kronecker [**44**] defined the degree of a regular map $\mathbb{P}^2_{\mathbb{R}} \to \mathbb{P}^2_{\mathbb{R}}$, which he called the *characteristic*, and his definition makes sense for many maps $f\colon X \to Y$ of (not necessarily orientable) compact manifolds.

First suppose that X is oriented. For a regular value $y \in Y$ of f, define

$$\operatorname{char} f \;:=\; \left| \sum_{x\in f^{-1}(y)} \operatorname{sgn}\det df_x \right|\,,$$

using local coordinates in X consistent with its orientation and any local coordinate near $y \in Y$. The sum is well-defined up to multiplication by ± 1. This number is independent of choices, if Y is connected.

When X may not be orientable, let $\widetilde{X}$ be the space of orientations of X, which is canonically oriented, and similarly let $\widetilde{Y}$ be the space of orientations of Y. Then $\widetilde{X} \to X$ and $\widetilde{Y} \to Y$ are 2 to 1 coverings of X and Y, with covering group $\mathbb{Z}/2\mathbb{Z}$. The map f is *orientable* if it has a lift $\widetilde{f}\colon \widetilde{X} \to \widetilde{Y}$ that is equivariant with respect to the covering group $\mathbb{Z}/2\mathbb{Z}$. Define the characteristic of an orientable map f to be $\operatorname{char}\widetilde{f}$. This is well-defined, even if $\widetilde{Y}$ consists of 2 components. This characteristic satisfies a fundamental inequality.

PROPOSITION 6.1. *Let $f\colon X \to Y$ be an orientable map. Then, for every regular value $y \in Y$,*

$$\# f^{-1}(y) \;\geq\; \operatorname{char} f\,.$$

We call this characteristic the *real degree* of the map f.

Consider this notion for the Wronski map of the real Grassmannian $\mathrm{Gr}_{\mathbb{R}}(k,n)$. The Grassmannian has dimension $k(n-k)$ and is orientable if and only if n is even. Eremenko and Gabrielov [**17**] compute the degree of the Wronski map. To state their result, we introduce some additional combinatorics. Recall the interpretation (4.8)

$$\text{degree } \mathrm{Gr}(k,n) \;=\; \#\text{ chains in Bruhat order from } \hat{0} \text{ to } \hat{1}\,,$$

where $\hat{1} = (n-k+1,\ldots,n-1,n)$ is the top element in the Bruhat order. We introduce a statistic on these chains. Each cover $\alpha \lessdot \beta$ in the Bruhat order has a unique index i with

$$\alpha_i \;<\; \beta_i \;=\; \alpha_i+1 \qquad \text{but} \qquad \alpha_j \;=\; \beta_j \quad \text{for } j \neq i\,.$$

The word $w(\text{ч})$ of a chain $\text{ч}\colon \hat{0} \lessdot \alpha^1 \lessdot \alpha^2 \lessdot \cdots \lessdot \alpha^{k(n-k)} = \hat{1}$ is the sequence of indices $(i_1, i_2, \ldots, i_{k(n-k)})$ of these indices for covers in ч. An inversion in such a word is a pair $j < l$ with $i_j < i_l$, and the weight, $\omega(\text{ч})$ is the number of inversions in the word of the chain ч. For instance the chain in the Bruhat order of $\mathrm{Gr}(3,6)$ highlighted in Figure 11

$$123 \lessdot 124 \lessdot 125 \lessdot 135 \lessdot 145 \lessdot 245 \lessdot 246 \lessdot 346 \lessdot 356 \lessdot 456\,,$$

has word 332213121 and length 5. Define the inversion polynomial

$$I(k,n)(q) \;:=\; \sum q^{\omega(\text{ч})}\,,$$

the sum over all chains ч in the Bruhat order from $\hat{0}$ to $\hat{1}$.

THEOREM 6.2 (Eremenko and Gabrielov [**17**, Theorem 1]). *The characteristic of the Wronski map* $\pi_W\colon \mathrm{Gr}_{\mathbb{R}}(k,n) \to \mathbb{P}_{\mathbb{R}}^{k(n-k)}$ *is* $|I(k,n)(-1)|$.

They prove this by an induction reminiscent of that used in the proof of Theorem 4.6. In the course of their proof, they construct a polynomial f with all roots real having $d(k,n)$ (=degree of the Grassmannian) real points in $\pi_W^{-1}(f)$ (one for each chain in the Bruhat order) and show that

$$\det d\pi_W \;=\; (-1)^{\omega(\text{ч})}\,,$$

at the point in $\pi_W^{-1}(f)$ corresponding to the chain ч.

White [**87**] studied the statistic $|I(k,n)(-1)|$ and showed that it equals zero if and only if n is even, and that $|I(2,2n)(-1)| = C_n$. The main result of Eremenko and Gabrielov is the following, which gives a lower bound for the number of real solutions in some cases of the Shapiro conjecture.

COROLLARY 6.3 (Eremenko and Gabrielov [**17**, Corollary 2]). *Let* $L_1, L_2, \ldots, L_{k(n-k)}$ *be codimension* $(n-k)$*-planes in* $\mathbb{R}^n$ *osculating the rational normal curve at* $k(n-k)$ *general real points. Then number* ρ *of real* k*-planes* K *meeting each* L_i *non-trivially satisfies*

$$|I(k,n)(-1)| \;\leq\; \rho \;\leq\; I(n,k)(1)\,.$$

In particular, when n *is odd this number* ρ *is non-zero.*

As we remarked before, this lower bound holds if the osculating planes are no longer required to be real, but that the set $\{t_1, t_2, \ldots, t_{k(n-k)}\}$ of points of osculation are real (the roots of a real polynomial).

Consider now a linear projection π_E different from the Wronski map. If we move the center E, the degree of π_E does not change as long as E does not meet the Grassmannian. Thus if E is in the same component of the space of codimension $k(n-k)+1$ planes not meeting the Grassmannian as is the center of the Wronski projection π_W, then Corollary 6.3 applies to $\pi_E^{-1}(x)$ for real x.

These new ideas of Eremenko and Gabrielov, particularly their notion of real degree, greatly increase our understanding of the real Schubert calculus. We emphasize that this is an important start in the search for lower bounds to other enumerative problems. Not all enumerative problems involve a linear projection of a submanifold of projective space. In fact, of the other enumerative problems we have considered in the Schubert calculus, only those of Theorem 4.11 on the Lagrangian Grassmannian involve linear projections, and they have a lower bound of 0. There has been very little experimental work investigating lower bounds. We describe some of it in the final section.

6.1. Lower bounds in the Schubert calculus for flags? Conjecture 5.10 speculates that a zero-dimensional intersection of Grassmannian Schubert varieties in a flag manifold $\mathbb{F}\ell_{\mathbf{a}}$ has only real points, when the Schubert varieties are given by flags osculating the rational normal curve at points and when the indices of the Grassmannian Schubet data are ordered. If the indices cannot be ordered, then we conjectured (and have found experimentally) that there is a selection of osculating flags with all points of intersection real, as well as a selection with not all points real.

Table 6.1 shows the results of computing 160,000 instances of the intersection

Necklace	Number of Real Solutions						
	0	2	4	6	8	10	12
22223333	0	0	0	0	0	0	20000
22322333	0	0	9	1677	3835	6247	8232
22233233	0	0	67	3015	6683	4822	5413
22332233	0	0	136	1533	7045	5261	6025
22323323	0	0	303	2090	6014	7690	3903
22323233	0	37	1944	4367	6160	4634	2858
22232333	0	195	1476	1776	3628	4546	8379
23232323	251	929	5740	3168	5420	2828	1662

of 4 simple Schubert varieties $X_2F_\bullet(t_i)$ and 4 simple Schubert varieties $X_3F_\bullet(t_i)$ in the manifold $\mathbb{F}\ell_{2<3}$ of partial flags $E_2 \subset E_3$ in 5-space. Here the numbers $t_1 < t_2 < \cdots < t_8$ were 20,000 random subsets of 8 numbers between 1 and 80, chosen using Maple's random number generator. For each choice, we considered the 8 possibilities of orderings of the indices, and for each system we computed the number of real solutions out of 12. Observe that the apparent lower bound on the number of real solutions depends on ordering of the indices of the Schubert data. Other calculations we have done reinforce this observation.

We conclude by reminding the reader of the observation in Section 3 that the number of real solutions to a problem in enumerative goemetry depends subtly on the configuration of the conditions.

Acknowledgements

We thank the following people: Alexandre Eremenko, William Fulton, Andrei Gabrielov, Viatcheslav Kharlamov, T.-Y. Li, Ricky Pollack, Felice Ronga, Marie-Françoise Roy, Bernd Sturmfels, and Thorsten Theobald.

References

[1] D. N. Bernstein, *The number of roots of a system of equations*, Funct. Anal. Appl., 9 (1975), pp. 183–185.
[2] I. N. Bernstein, I. M. Gelfand, and S. I. Gelfand, *Schubert cells and cohomology of the spaces G/P*, Russian Mathematical Surveys, 28 (1973), pp. 1–26.
[3] A. Bertram, *Quantum Schubert calculus*, Adv. Math., 128 (1997), pp. 289–305.
[4] R. Bricard, *Mémoire sur la théorie de l'octaèdre articulé*, J. Math. Pures et Appliquées, 3 (1897), pp. 113–150.
[5] P. Burgisser, M. Clausen, and M. Shohrollahi, *Algebraic Complexity Theory*, Springer-Verlag, 1997.
[6] P. Candelas, X. C. de la Ossa, P. S. Green, and L. Parkes, *A pair of Calabi-Yau manifolds as an exactly soluble superconformal theory*, Nuclear Phys. B, 359 (1991), pp. 21–74.
[7] G. Castelnuovo, *Numero delle involuzioni razionali gaicenti sopra una curva di dato genere*, Rendi. R. Accad. Lineci, 4 (1889), pp. 130–133.
[8] R. Chiavacci and J. Escamilla-Castillo, *Schubert calculus and enumerative problems*, Bollettino Un. Math. Ital., 7 (1988), pp. 119–126.
[9] J. Clark, *The consistent selection of local coordinates in linear system identification*, in Proc. Joint Automatic Control Conference, 1976, pp. 576–580.
[10] D. A. Cox and S. Katz, *Mirror symmetry and algebraic geometry*, American Mathematical Society, Providence, RI, 1999.
[11] H. S. M. Coxeter, *The twenty-seven lines on the cubic surface*, in Convexity and its applications, Birkhäuser, Basel, 1983, pp. 111–119.
[12] A. I. Degtyarev and V. M. Kharlamov, *Topological properties of real algebraic varieties: Rokhlin's way*, Uspekhi Mat. Nauk, 55 (2000), pp. 129–212.
[13] M. Demazure, *Désingularization des variétés de Schubert généralisées*, Ann. Sc. E. N. S. (4), 7 (1974), pp. 53–88.
[14] P. Dietmaier, *The Stewart-Gough platform of general geometry can have 40 real postures*, in Advances in Robot Kinematics: Analysis and Control, Kluwer Academic Publishers, 1998, pp. 1–10.
[15] D. Eisenbud and J. Harris, *Divisors on general curves and cuspidal rational curves*, Invent. Math., 74 (1983), pp. 371–418.
[16] A. Eremenko and A. Gabrielov, *Rational functions with real critical points and B. and M. Shapiro conjecture in real enumerative geometry*. Annals of Math., 155 (2002), pp. 105–129.
[17] ———, *Degrees of real Wronski maps*. Discr. Comput. Geom., 28 (2002), pp. 331–347.
[18] W. Fulton, *Intersection Theory*, no. 2 in Ergebnisse der Math., Springer-Verlag, 1984.
[19] ———, *Introduction to Intersection Theory in Algebraic Geometry*, CBMS 54, AMS, 1996. second edition.
[20] W. Fulton and R. Pandharipande, *Notes on stable maps and quantum cohomology*, in Algebraic Geometry, Santa Cruz 1995, J. Kollár, R. Lazarsfeld, and D. Morrison, eds., vol. 62, Part 2 of Proc. Sympos. Pure Math., Amer. Math. Soc., 1997, pp. 45–96.
[21] G. Giambelli, *Rizoluzione del problema degli spazi secanti*, Memor. Real. Acc. di Torino, 52 (1902), pp. 171–193.
[22] L. Goldberg, *Catalan numbers and branched coverings by the Riemann sphere*, Adv. Math., 85 (1991), pp. 129–144.
[23] V. Gough, *Contribution to discussion papers on research in automobile stability and control in tyre performance*, 1957.
[24] B. Haas, *A counterexample to Kouchnirenko's conjecture*. B Beiträge zur Algebra und Geometrie, 43 (2002), pp. 1-8.

[25] A. HARNACK, *Über die Vieltheiligkeit der eben algebraischen Kurven*, Math. Ann., 10 (1876), pp. 189–198.

[26] J. HARRIS, *Galois groups of enumerative problems*, Duke Math. J., 46 (1979), pp. 685–724.

[27] D. HILBERT, *Über die Darstellung definiter Formen als Summe von Formen-quadraten*, Math. Ann., 32 (1888), pp. 342–350.

[28] D. HILBERT, *Über die reellen Züge algebraischen Curven*, Math. Ann., 38 (1891), pp. 115–138.

[29] ———, *Sur les problèmes futurs des mathématiques*, in Proceedings of the Second International Congress of Mathematicians, Paris 1900, Gauthier-Villars, 1902, pp. 58–114.

[30] B. HUBER, F. SOTTILE, AND B. STURMFELS, *Numerical Schubert calculus*, J. Symb. Comp., 26 (1998), pp. 767–788.

[31] B. HUBER AND B. STURMFELS, *A polyhedral method for solving sparse polynomial systems*, Math. Comp., 64 (1995), pp. 1541–1555.

[32] B. HUBER AND J. VERSCHELDE, *Pieri homotopies for problems in enumerative geometry applied to pole placement in linear systems control*, SIAM J. Control and Optim., 38 (2000), pp. 1265–1287.

[33] J. HUISMAN, *On the number of real hypersurfaces hypertangent to a given real space curve*. Illinois J. Math. 46 (2002), pp. 145–153.

[34] K. INTRILIGATOR, *Fusion residues*, Mod. Phys. Lett. A, 6 (1991), pp. 3543–3556.

[35] I. ITENBERG AND M.-F. ROY, *Multivariate Descartes' rule*, Beiträge zur Algebra und Geometrie, 37 (1996), pp. 337–346.

[36] S. KATZ, *On the finiteness of rational curves on quintic threefolds*, Compositio Math., 60 (1986), pp. 151–162.

[37] V. KHARLAMOV AND F. SOTTILE, *Maximally inflected real rational curves*. `math.AG/0206268`. 2002.

[38] A. KHOVANSKII, *Newton polyhedra and the genus of complete intersections*, Funct. Anal. Appl., 12 (1978), pp. 38–46.

[39] ———, *Fewnomials*, Trans. of Math. Monographs, 88, AMS, 1991.

[40] S. L. KLEIMAN, *The transversality of a general translate*, Compositio Math., 28 (1974), pp. 287–297.

[41] S. L. KLEIMAN, *Problem 15. Rigorous foundation of Schubert's enumerative calculus*, in Mathematical Developments arising from Hilbert Problems, vol. 28 of Proc. Sympos. Pure Math., Amer. Math. Soc., 1976, pp. 445–482.

[42] F. KLEIN, *Eine neue Relation zwischen den Singularitäten einer algebraischen Kurve*, Math. Ann., 10 (1876), pp. 199–209.

[43] A. KOUCHNIRENKO, *A Newton polyhedron and the number of solutions of a system of k equations in k unknowns*, Usp. Math. Nauk., 30 (1975), pp. 266–267.

[44] L. KRONECKER, *Leopold Kronecker's Werke*, Chelsea, NY, 1968.

[45] D. LARMAN. Problem posed in the Problem Session of the DIMACS Workshop on Arrangements, Rutgers University, New Brunswick, NJ, USA, 1990.

[46] D. LAZARD, *Generalized Stewart platform: how to compute with rigid motions?*, in IMACS-SC'93, 1993.

[47] T. Y. LI, J. M. ROJAS, AND X. WANG, *Counting Real Connected Components of Trinomial Curve Intersections and m-nomial Hypersurfaces*. 2002.

[48] T. Y. LI AND X. WANG, *On multivariate Descartes' rule—a counterexample*, Beiträge Algebra Geom., 39 (1998), pp. 1–5.

[49] I. MACDONALD, J. PACH, AND T. THEOBALD, *Common tangents to four unit balls in* $\mathbb{R}^3$, Discrete Comput. Geom., 26 (2001), pp. 1–17.

[50] G. MEGYESI, *Lines tangent to four unit spheres with affinely dependent centres*, Discrete Comput. Geom., 26 (2001), pp. 493–497.

[51] B. MOURRAIN, *The 40 generic positions of a parallel robot*, in ISSAC'93, Kiev, ACM Press, 1993, pp. 173–182.

[52] ———, *Enumeration problems in geometry, robotics, and vision*, in Algorithms in Algebraic Geometry and Applications (MEGA-94, Santander), L. González-Vega and T. Recio, eds., vol. 143 of Progress in Math., Birkhäuser, 1996, pp. 285–306.

[53] D. NAPOLETANI, *A power function approach to Kouchnirenko's conjceture*. Symbolic computation: solving equations in algebra, geometry, and engineering (South Hadley, MA, 2000), 99–106, Contemp. Math., 286, Amer. Math. Soc., Providence, RI, 2001.

[54] P. Pedersen and B. Sturmfels, *Mixed monomial bases*, in Algorithms in Algebraic Geometry and Applications, L. González-Vega and T. Recio, eds., vol. 143 of Progress in Mathematics, Proceedings, MEGA-94, Birkhäuser, 1996, pp. 307–316.

[55] M. Pieri, *Sul problema degli spazi secanti*, Rend. Reale Ist. lombardo, 26 (1893), pp. 534–546.

[56] V. Powers and B. Reznick, *Notes towards a constructive proof of Hilbert's theorem on ternary quartics*, in Quadratic forms and their applications (Dublin, 1999), Amer. Math. Soc., Providence, RI, 2000, pp. 209–227.

[57] V. Powers, B. Reznick, C. Scheiderer, and F. Sottile, *A New Proof of Hilbert's Theorem on Ternary Quartics*. In preparation.

[58] M. Raghavan, *The Stewart platform of general geometry has 40 configurations*, in ASME Design and Automation Conf., vol. 32-2, 1991, pp. 397–402.

[59] M. Ravi, J. Rosenthal, and X. Wang, *Degree of the generalized Plücker embedding of a quot scheme and quantum cohomology*, Math. Ann., 311 (1998), pp. 11–26.

[60] F. Ronga, A. Tognoli, and T. Vust, *The number of conics tangent to 5 given conics: the real case*, Rev. Mat. Univ. Complut. Madrid, 10 (1997), pp. 391–421.

[61] F. Ronga and T. Vust, *Stewart platforms without computer?*, in Real Analytic and Algebraic Geometry, Proceedings of the International Conference, (Trento, 1992), Walter de Gruyter, 1995, pp. 196–212.

[62] H. Schubert, *Kalkul der abzählenden Geometrie*, Springer-Verlag, 1879. reprinted with an introduction by S. Kleiman, 1979.

[63] ———, *Anzahl-Bestimmungen für lineare Räume beliebiger Dimension*, Acta. Math., 8 (1886), pp. 97–118.

[64] ———, *Die n-dimensionalen Verallgemeinerungen der fundamentalen Anzahlen unseres Raume*, Math. Ann., 26 (1886), pp. 26–51. (dated 1884).

[65] ———, *Losüng des Charakteritiken-Problems für lineare Räume beliebiger Dimension*, Mittheil. Math. Ges. Hamburg, (1886), pp. 135–155. (dated 1885).

[66] B. Siebert and G. Tian, *On quantum cohomology rings of Fano manifolds and a formula of Vafa and Intrilligator*, Asian J. Math., 1 (1997), pp. 679–695.

[67] F. Sottile, *Enumerative geometry for real varieties*, in Algebraic Geometry, Santa Cruz 1995, J. Kollár, R. Lazarsfeld, and D. Morrison, eds., vol. 62, Part 1 of Proc. Sympos. Pure Math., Amer. Math. Soc., 1997, pp. 435–447.

[68] ———, *Enumerative geometry for the real Grassmannian of lines in projective space*, Duke Math. J., 87 (1997), pp. 59–85.

[69] ———, *Real enumerative geometry and effective algebraic equivalence*, J. Pure Appl. Alg., 117 & 118 (1997), pp. 601–615. Proc., MEGA'96.

[70] ———, *The conjecture of Shapiro and Shapiro*. An archive of computations and computer algebra scripts, `www.math.umass.edu/~sottile/pages/shapiro`, 1999.

[71] ———, *The special Schubert calculus is real*, ERA of the AMS, 5 (1999), pp. 35–39.

[72] ———, *Real rational curves in Grassmannians*, J. Amer. Math. Soc., 13 (2000), pp. 333–341.

[73] ———, *Real Schubert calculus: Polynomial systems and a conjecture of Shapiro and Shapiro*, Exper. Math., 9 (2000), pp. 161–182.

[74] ———, *Some real and unreal enumerative geometry for flag manifolds*, Mich. Math. J., 48 (2000), pp. 573–592. Special Issue in Honor of Wm. Fulton.

[75] ———, *Rational curves in Grassmannians: systems theory, reality, and transversality*, in Advances in Algebraic Geometry Motivated by Physics, E. Previato, ed., vol. 276 of Contemp. Math., AMS, 2001, pp. 9–42.

[76] ———, *Elementary transversality in the Schubert calculus for arbitrary characteristic*. `math.AG/0010319`, 2001.

[77] ———, *Some real and unreal conjectures on enumerative geometry for flag manifolds*. In preparation, 2001.

[78] F. Sottile and T. Theobald, *Common tangent lines to $2n-2$ spheres in $\mathbb{R}^n$*. Trans. Amer. Math. Soc., 354 (2002), 4815-4829.

[79] D. Stewart, *A platform with 6 degree of freedom*, Proc. of the Institution of Mechanical Engineers, 180 (1965-66), pp. 371–386.

[80] B. Sturmfels, *On the number of real roots of a sparse polynomial system*, in Hamiltonian and gradient flows, algorithms and control, vol. 3 of Fields Inst. Commun., American Mathematical Society, Providence, 1994, pp. 137–143.

[81] ———, *Viro's theorem for complete intersections*, Ann. della Scuola Nor. Sup. Pisa, Sc. Fis. Mat., Ser IV, XXI (1994), pp. 377–386.

[82] C. VAFA, *Topological mirrors and quantum rings*, in Essays on Mirror Manifolds, S.-T. Yau, ed., International Press, Hong Kong, 1992, pp. 96–119.

[83] R. VAKIL, *Schubert induction.* In preparation.

[84] J. VERSCHELDE, *Polyhedral homotopies for dense, sparse, and determinantal systems.* MSRI preprint 1999-041, 1999.

[85] O. Y. VIRO, *Real algebraic plane curves: Constructions with controlled topology*, Algebra i Analiz, 1 (1989), pp. 1–73. Translation in Leningrad Math. J., 1 (1990), pp. 1059–1134.

[86] C. WALL, *Is every quartic a conic of conics?*, Math. Proc. Cambridge Philos. Soc., 109 (1991), pp. 419–424.

[87] D. WHITE, *Sign-balanced posets.* J. Combin. Theory Ser. A 95 (2001), no. 1, 1–38.

[88] H. G. ZEUTHEN, *Almindelige Egenskaber ved Systemer af plane Kurver*, Danske Videnskabernes Selskabs Skrifter, Naturvidenskabelig og Mathematisk, Afd. **10** Bd. IV, (1873), pp. 286–393.

[89] ———, *Sur les différentes formes des courbes planes du quartriéme order*, Math. Ann., 7 (1874), pp. 410–432.

DEPARTMENT OF MATHEMATICS, UNIVERSITY OF MASSACHUSETTS, LEDERLE GRADUATE RESEARCH TOWER, AMHERST, MASSACHUSETTS, 01003, USA

E-mail address: sottile@math.umass.edu

URL: http://www.math.umass.edu/~sottile

DIMACS Series in Discrete Mathematics
and Theoretical Computer Science
Volume **60**, 2003

Combinatorial Roadmaps in Configuration Spaces of Simple Planar Polygons

Ileana Streinu

Abstract. One-degree-of-freedom mechanisms induced by minimum pseudo-triangulations with one convex hull edge removed have been recently introduced by the author to solve a family of non-colliding motion planning problems for planar robot arms (open or closed polygonal chains). They induce canonical roadmaps in configuration spaces of simple planar polygons with fixed edge lengths.

While the combinatorial part is well understood, the search for efficient solutions to the algebraic components of the algorithm is posing a number of interesting questions, some of which are addressed in this paper. A list of open problems and further research topics on pointed pseudo-triangulations and related structures motivated by this work is appended.

1. Introduction

In this paper we address aspects of a computational algebraic nature arising in the Pseudo-Triangulation Roadmap Algorithm introduced by the author in [**20**] to solve the Carpenter's Rule Problem. We formulate the problems and discuss several possible solutions, together with the theoretical and computational challenges they induce and related open questions and conjectures.

The **Carpenter's Rule Problem.** Consider a simple planar polygonal chain (linkage) with fixed edge lengths (*robot arm*). Orient it so that the interior lies to the left when walking along the polygon in the positive direction. The edges (*bars*) are allowed to move freely around the vertices (*joints*). We want to avoid collisions between the bars while moving the linkage continuously from an initial to a final configuration with the same orientation. It suffices to show that we can move from any position to a convex polygon position. Then, to move between any two configurations, take one path in reverse (it is easy to move between two distinct convex positions). That this is always possible was shown by Connelly, Demaine

1991 *Mathematics Subject Classification.* Primary 68U05, 65D18; Secondary 52C25, 52C40, 52C30, 52C45, 05C10, 13P10.

Key words and phrases. Computational Geometry, Rigidity Theory, Computational Algebraic Geometry.

Research supported by NSF RUI grant CCR-0105507 and NSF-DARPA CARGO CCR-0138374.

and Rote [**8**], based on Rote's ground-breaking idea of using *expansive infinitesimal motions* to avoid collisions.

For finding algorithmically a path in configuration space between any two compatible positions, the author has proposed in [**20**] a combinatorial approach: the path consists of a finite number of arcs, each being the unique trajectory of a one-degree-of-freedom mechanism induced by a *pointed* (or *minimum*) *pseudo-triangulation.*

Expansive Motions and Pseudo-Triangulation Mechanisms. A one-degree-of-freedom (1DOF) mechanism is a bar-and-joint framework whose configuration space is a one-dimensional curve. The mechanism is said to be *(infinitesimally) expansive* at some point in its configuration space, if all the pairwise interdistances between its vertices simultaneously increase or decrease (including the possibility of some staying the same), when the mechanism is moved infinitesimally along its unique trajectory in configuration space. Expansive motions guarantee that no collisions will occur. It is shown in [**20**] that pseudo-triangulations with one convex hull edge removed (defined in section 2) are (infinitesimally) expansive one-degree-of-freedom mechanisms.

The infinitesimally expansive motions form a cone ([**8**] and [**19**]), and linear programming can be used to find a set of infinitesimal velocities of the moving points. Pseudo-triangulations with a convex hull edge removed correspond to canonical basic feasible solutions found by such a linear program. They can be computed very efficiently geometrically, without using linear programming. Our algorithmic approach ([**20**]) is to start with a mechanism induced by a pseudo-triangulation and move it until a special event occurs (alignment of two edges). This invalidates the expansive property, should the motion continue. Therefore, at that point, the mechanism is locally reconfigured to get a new pseudo-triangulation and the motion continues. After a finite number of such steps, the convex position is attained.

Results. In this paper we investigate the algebraic components of this algorithm. Two general problems which await efficient and accurate numerical solutions are discussed: the **Simulation of Motion** for a pseudo-triangulation mechanism, and the **Detection of the Next Event.** We exhibit some special cases which are satisfactorily solved. These depend on a certain inductive construction for pseudo-triangulations introduced in [**20**] and modelled after the Henneberg constructions of [**14**] for generically minimally rigid graphs (see [**12**], [**23**], [**21**]).

Organization. The paper is organized as follows. To make the paper self-contained, in section 2 we introduce the necessary concepts and give an overview of the pseudo-triangulation roadmap algorithm of [**20**]. In section 3 we state the two problems addressed in this paper and formulate them algebraically. In section 4 we discuss some of the drawbacks of using standard techniques from computational algebraic geometry in solving them. In section 5 we include proofs and heuristics based on rigid-components and in section 6 we describe heuristics based on Henneberg constructions. Open problems, conjectures and research topics (both combinatorial and algebraic) related to this approach are included in almost every section, and additional ones are gathered in section 7.

2. Definitions and Preliminaries

This largely self-contained section introduces the basic terminology, definitions, problems and relevant previous results.

References. For rigidity theory terminology and classical results, we refer the reader to [**18**], [**23**], [**24**] and [**12**]. In particular, rigidity, first-order and generic rigidity, as well as classical results on 2-dimensional rigidity such as Laman's theorem and the Henneberg constructions are to be found there. For computational algebraic geometry terminology and results, see [**9**] and [**10**]: Gröbner bases and the homotopy method are described there.

Basic concepts, notations and abbreviations. Our setting is the Euclidian plane. We denote by $\mathcal{P} = \{p_1, \cdots, p_n\} \subset R^2$ a finite set of planar points. When it is a function of a time parameter t we may denote it as $P(t) = \{p_1(t), \cdots, p_n(t)\}$. We abbreviate *counter-clockwise* as *ccw* and *one-degree-of-freedom mechanism* as *1DOF mechanism.*

Basic Definitions. A *pseudo-triangle* is a simple planar polygon with exactly three convex vertices (its *corners*). These are joined by three inner convex polygonal chains called its *side chains.* See Figure 1.

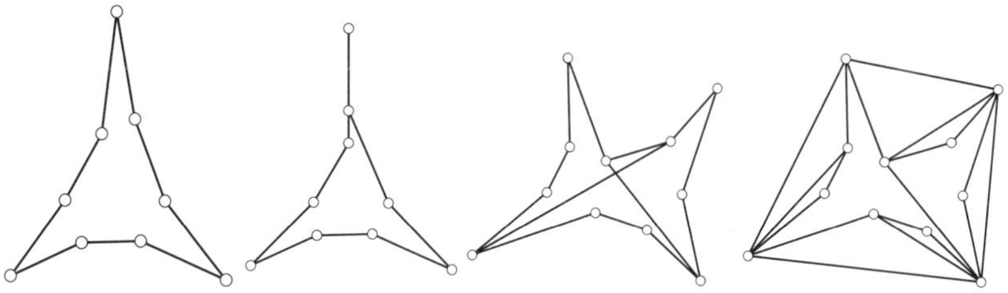

FIGURE 1. A pseudo-triangle, a semi-simple pseudo-triangle, a flip in a pseudo-quadrilateral and a pointed (minimum) pseudo-triangulation.

A set of vectors $v_1, \cdots, v_k \in R^2$ is *pointed* if there is no linear combination with not-all-zero positive coefficients summing them up to zero. Such vectors lie in a half plane. Equivalently, among the k angles spanned by pairs of cyclically consecutive vectors, one is *strictly* larger than π (reflex). [1]

A *framework* (G, L) is a graph $G = (V, E)$, $V = \{1, \cdots, n\}$, together with a given set of edge lengths $L = \{l_{ij} | ij \in E\}$. A *configuration* or *realization* of (G, L) is a set of points $P = \{p_1, \cdots, p_n\}$ such that if the vertices of G are mapped to P, $i \in V \mapsto p_i \in P$, then the length of the line segment $\overline{p_i p_j}$ equals l_{ij} if $ij \in E$. The notation $G(P)$, referring to both G and P, is used to denote the realization of G on the set of points P. Questions of realizability of a particular set of edge lengths will not be of interest to us here, since we already start with a set of realizable lengths via an initial embedding $G(P)$ of the graph on a planar point set.

The *configuration space* of (G, L) is the set of all its possible realizations. If non empty, this space can be factored to the reorientation of the plane, as each realization comes with a mirror image which is itself realizable. It also contains a 3-dimensional subspace corresponding to rigid plane transformations (translations and rotations). We will later make specific choices that will induce this factorization and then we will be concerned only with questions of realizability in the quotient space. For the particular case of planar polygons, if we factor out the rigid motions, it is well known that their configuration spaces are $(n-3)$-dimensional manifolds,

[1]The terminology used in [**20**] for the same concept was *acyclic* for reasons related to the oriented matroids motivation.

with the sole exception when there exists a configuration where all the vertices lie on a line, see [**15**].

An embedding is *planar* or *non-crossing* if any pair of non-incident edges (ij and kl, $i, j \notin \{k, l\}$) are disjoint as line segments in the embedding ($\overline{p_i p_j} \cap \overline{p_k p_l} = \emptyset$). In the particular case when G is a path or a cycle the embedding is a planar polygonal path or a polygon. Instead of "non-crossing", a polygon will be called *simple*. A pseudo-triangle is a special case of a simple polygon. A *semi-simple pseudo-triangle* may have two overlapping edges adjacent to one of the corners, i.e. one or more of the internal convex angles of the corners may be zero. See Fig. 1. [2]

Pointed Pseudo-Triangulations. A *pseudo-triangulation* is a tilling of the convex hull of a planar point set with pseudo-triangles. In other words it is an embedded planar graph whose outer face is the complement of the convex hull and each interior face is a pseudo-triangle. A *minimum pseudo-triangulation* uses the minimum number of edges. A *pointed pseudo-triangulation* has an angle larger than π adjacent to every vertex, i.e. the edge vectors around each vertex form a pointed set. See Fig. 1. Minimum and pointed pseudo-triangulations turn out to be the same object. They have remarkable combinatorial and rigidity theoretic properties, see [**20**]. For the sake of completeness, we reproduce here the relevant ones, which will be used implicitly throughout the paper.

THEOREM 2.1. [**20**] *Let $G(P)$ be a graph embedded on a set of n points, satisfying any one of the following 6 properties. Then it satisfies all the others.*

(1) *pointed pseudo-triangulation*
(2) *minimum pseudo-triangulation*
(3) *maximally non-crossing and pointed*
(4) *pseudo-triangulation with $2n-3$ edges*
(5) *non-crossing, pointed with $2n-3$ edges*
(6) *admits a non-crossing and pointed Henneberg construction (described below)*

From now on, we will work only with pointed pseudo-triangulations and refer to them, for conciseness, simply as *pseudo-triangulations*. A pseudo-triangulation has $2n-3$ edges and since both planarity and pointedness are hereditary properties, each subset of k vertices spans at most $2k-3$ edges. This hereditary $(2n-3)$-property characterizes *generically minimally rigid graphs* and is known as the *Laman property* ([**16**], see [**18**]). Graphs satisfying Laman's property are called *Laman graphs.* Minimality means that removing any edge causes the (generic) framework to move, hence it becomes a mechanism. Laman graphs are infinitesimally rigid in *almost all* possible embeddings, called *generic embeddings.*

In general, removing $k \geq 1$ edges from a minimally infinitesimally rigid graph creates an infinitesimal mechanism with k degrees of freedom, i.e. k is the dimension of its space of infinitesimal motions. The configuration space of a 1DOF mechanism is a curve obtained by integrating the infinitesimal motions and is *in general* a manifold. It may have several connected components. On each component, one may choose a parametrization and move in exactly two directions on the closed curve. The distance between at least one pair of points $p_i p_j$ changes: it either increases or decreases strictly. A mechanism is *(locally or infinitesimally) expansive* for some position in its configuration space if all the interdistances between pairs of

[2]Semi-simple pseudo-triangles appear in the Pseudo-Triangulation Roadmap Algorithm 2.3 at the beginning of each step except the first.

points simultaneously increase (or stay the same), or simultaneously decrease (or stay the same) (infinitesimally). If this property holds at a generic position, then it holds on an open neighborhood: the mechanism is *expansive* as it moves along that portion of its configuration space.

THEOREM 2.2. [**20**] *Pseudo-triangulations are always minimally infinitesimally rigid (and hence rigid). A pseudo-triangulation with a convex hull edge removed is an (infinitesimally) expansive 1DOF mechanism. As it moves on its component in configuration space, it remains expansive as long as the underlying combinatorial pseudo-triangulation structure doesn't change.*

The definition of a combinatorial pseudo-triangulation will be given a bit later in this section.

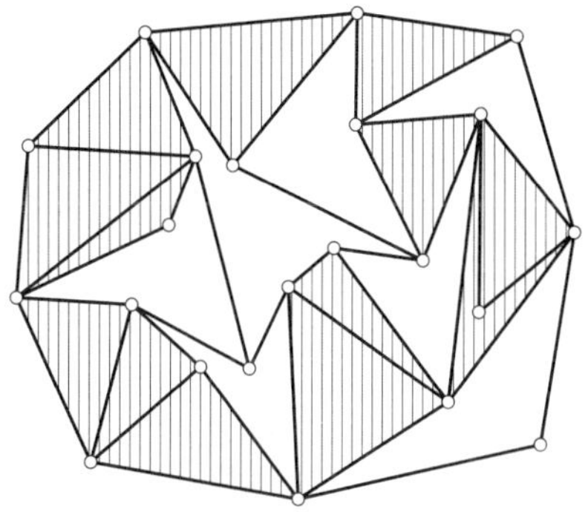

FIGURE 2. The rigid components of a pseudo-triangulation mechanism.

We will refer to a pseudo-triangulation with a convex hull edge removed as a *pseudo-triangulation mechanism.* Like any Laman graph with an edge removed, a pseudo-triangulation mechanism decomposes into *maximally rigid components*, or *r-components*: maximal subsets of some k vertices spanning exactly $2k-3$ edges. See [**12**]. This *r-decomposition* has a nice structure, [**20**]: each component is a pseudo-triangulation of a convex subset of points (i.e. a subset whose convex hull does not include points outside the set). The interesting components have at least 3 vertices, although we can view the remaining edges also as r-components. These are bitangents between various subcomponents, in the sense used in [**17**] for pseudo-triangulations of the free space among convex obstacles. Figure 2 gives an example of such an r-decomposition.

Theorem 2.1 (3) implies that every set of pointed edges can be extended to a pointed pseudo-triangulation. In particular, the edges of a simple planar polygon form a pointed set of edges. Any extension of it to a pseudo-triangulation, which will be referred to from now on as a *pseudo-triangulation of the polygon*, has the property that the underlying graph contains a Hamiltonian cycle (the polygon). In general, define a *Hamiltonian pseudo-triangulation* to be a pseudo-triangulation whose underlying graph is Hamiltonian. This is to be distinguished from what is usually referred to in the literature as a (pointed) *polygon pseudo-triangulation*, which is a (pointed) decomposition of the *interior* of a polygon into pseudo-triangles.

To *flip* an interior edge in a pseudo-triangulation ([**19**]) means to remove it, thus merging two faces into one, and then to add the *unique* edge different from the removed one that would pseudo-triangulate again this merged face. See Fig. 1.

Henneberg constructions for pseudo-triangulations. As a generalization of the known Henneberg constructions for Laman graphs ([**14**], [**21**], see also [**12**]), pseudo-triangulations can be constructed inductively as follows (cf. [**20**]).

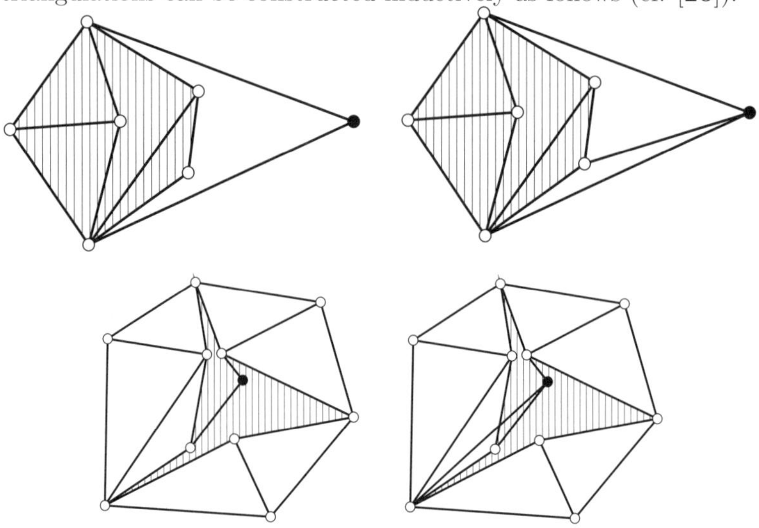

FIGURE 3. Henneberg construction for pseudo-triangulations. Left: type I, right: type II. On the right, the flipped edge is still slightly visible for reference, but it does not belong to the pseudo-triangulation. The new vertex is added on the outside face (top) or on an inner face (bottom). The two extreme edges adjacent to the new vertex are tangent to the convex hull (top, shaded) or to one or two of the inner chains of a pseudo-triangular face (bottom, shaded).

Start with the base case of a triangle. At each step, add a vertex v in one of the existing faces F. There are two types of steps (see Figure 3):

- **Henneberg I**: add two tangents from v to side chains of the face. This splits the face into two faces.
- **Henneberg II**: perform a Henneberg I step and choose the newly created face where the new vertex is a corner. Then pick up one edge on the side chain of this face which is visible from the corner (there always exists one), and flip it to obtain a third edge incident to v.

A Henneberg I step creates a new vertex of degree 2 and a Henneberg II step creates a vertex of degree 3. A pseudo-triangulation which can be obtained by applying only type I steps is called a *Henneberg I pseudo-triangulation.* [3] It must have at least one vertex of degree 2. When the edge lengths are known, they can be constructed inductively using only ruler and compass.

Henneberg steps can be generalized to the expansive 1DOF mechanisms obtained from pseudo-triangulations using the same types of steps. A 1DOF mechanism which can be obtained by applying only type I steps is called a *Henneberg I mechanism*, and can be constructed inductively with ruler and compass only.

[3]In the Rigidity Theory literature, Henneberg I graphs (general, not pseudo-triangulation) are called 2-simple.

Combinatorial Pseudo-Triangulations. Pseudo-triangulations are plane graphs (i.e. embedded planar graphs) with additional (partial) oriented matroid information for pointedness. Any plane embedding of a planar graph induces a *topological embedding* of the underlying graph G. This is independent of point coordinates, and captures only the combinatorial information about the faces of the embedding and their adjacencies. Equivalently, this information is captured by the *rotations* at each vertex: the ccw circular order of the adjacent edges at each vertex in the embedding. A *topological embedding* of a planar graph is such a system of rotations (or equivalently, the faces and their adjacencies) which can be realized in the plane with non-crossing curves (pseudo-segments). A *topological embedding with a marked face* contains, in addition, information about which face of the embedding is unbounded (the *outer* or *exterior* face).

An assignment of a label $\{C, R\}$ (standing for *convex* and *reflex*) to each angle of a face F of a topologically embedded planar graph is called a *combinatorial (pointed) pseudo-triangulation* if:

(1) For each vertex, exactly one of its adjacent angles is labelled R.
(2) There exists one face (the *outer face*) whose angles are all labelled R.
(3) All other faces have exactly three angles labelled C.

Combinatorial pseudo-triangulations are relevant to our work because they provide a combinatorial way of distinguishing among solutions to algebraic equations describing embeddings, with given edge lengths, of Laman graphs. In particular, the desired solutions for both the Local and the Global Problems described in Section 3 have a known combinatorial type.

The Pseudo-Triangulation Roadmap Algorithm. The properties of pseudo-triangulations induce the following simple approach for constructing roadmaps in configuration spaces of simple planar polygons.

ALGORITHM 2.3. (The pseudo-triangulation roadmap algorithm)

(1) *Construct a pseudo-triangulation of the polygon.*
(2) *Choose a convex hull edge which is not one of the original edges of the polygon and remove it to obtain an expansive mechanism.*
(3) *Move the mechanism along its unique trajectory which increases the distance between the endpoints of the removed edge.*
(4) *Stop when the framework is no longer pointed: at some vertex, the two extreme adjacent edges align. Call this an* alignment event, *or simply an* event.
(5) *Perform a local flip to restore the pseudo-triangulation (at least one pseudo-triangle will be semi-simple at this moment) and continue the motion using a new expansive mechanism induced by it.*

How to perform the flips to restore the pseudo-triangulation and how to control it so that the total number of such readjustments is kept small is shown in [**20**]. In particular, for a scheme based on the shortest-path tree from a vertex, the algorithm will end after at most $\mathcal{O}(n^3)$ such events.

A *shortest-path pseudo-triangulation* of a polygon is obtained as follows. Take the convex hull of the polygon. This induces a number of *pockets*: the complements with respect to the convex hull of the interior of the polygon. Each pocket is bounded by a simple polygon. Inside each polygon (the original and the pockets) add the edges of the shortest path tree from an inner convex vertex to all the other inner convex vertices of the polygon. It is shown in [20] that this produces a pointed pseudo-triangulation.

In the sequel we are interested only in how to perform steps 3 (Simulation of Motion) and 4 (Detection of next Event).

3. The Algebraic Components of the Pseudo-triangulation Roadmap Algorithm

In addition to the combinatorial part described above, a complete implementation of the algorithm involves the choice of a time parametrization of the motion and algebraic computations in steps 3 and 4, as follows.

- **Parametrization.** Parametrize the motion of one pseudo-triangulation mechanism between two consecutive events, so that, if t_0 and t_f are the start and finish times, δt is the time step, then the motion of the mechanism is simulated as a sequence of frames, one for each time $t = t_0 + k\delta t$, $t_0 \leq t \leq t_f$.
- **Simulation of the Motion of a Pseudo-Triangulation Mechanism.** Given a time step t, $t_0 \leq t \leq t_f$ and the embedding $P(t)$ at time t, find the coordinates $P(t + \delta t)$ of the embedded mechanism $G(P(t + \delta t))$ at time $t + \delta t$.
- **Detection of the Next Event.** For each mechanism corresponding to one step of the Algorithm, compute the final value $t = t_f$ of the time parameter (the time of the next alignment event) and the index k of the vertex at which the next alignment event happens.

In this section we formulate these problems in algebraic terms.

Parametrization. Originally our systems of equations will be thought of as having $2n$ variables $x_1, \cdots, x_n, y_1, \cdots, y_n$, standing for the unknown positions of the points $p_i = (x_i, y_i)$ at some moment in time t. For the Local Problem, t is a constant and for the Global Problem it is a variable.

To factor out the trivial plane motions assume without loss of generality (i.e after a suitable relabelling) that the convex hull edge missing from the pseudo-triangulation is the edge between vertices 1 and 2. After a rigid transformation of the system of coordinates, assume that the origin is at vertex 1, the edge itself lies on the x-axis in the positive direction and the second endpoint is constrained to move along the axis from the initial position $t = t_0$ to increasing values. This parametrization adds to each system the extra equations $x_1 = 0, y_1 = 0, y_2 = 0$, and uses x_2 is the *time parameter*. Substituting in all equations x_1, y_1 and y_2 by 0 and using x_2 as time parameter t we are reduced to $2n - 3$ variables.

A reminder that all happens during the lifetime of only one mechanism. The parametrization may change after an event, because it is possible that another convex hull edge will be chosen to be dropped when the pseudo-triangulation is locally reconfigured. In addition, sometimes it may happen that the expanding (missing) convex hull edge is "crossed" by some other part of the pseudo-triangulation during the motion of the mechanism: while this does not invalidate the expansive property, it may be necessary to readjust the parametrization to record that a different edge is now the missing convex hull edge, so that we maintain the format of the equations and their desired invariant properties.

Simulation of the Motion of a Pseudo-Triangulation Mechanism and the **Local Problem.** The values t_0 and t_f are computed before the simulation starts: t_0 is simply the x-coordinate of the second point p_2 at the beginning of the simulation, and t_f is computed as part of the *Detection of the Next Event.* At each step, our goal is to compute the new coordinates of the vertices, from which (say) a drawing is produced by superimposing the edges. The $2n-4$ edge lengths of the pseudo-triangulation mechanism are fixed. Since we have chosen a parametrization where the length of the removed convex hull edge at time t is t, it follows that at each time step we have a full set of $2n-3$ edge lengths for a minimally rigid framework (Laman graph), and we want to compute the *realization* of this framework which is the *closest* (in a sense that must be defined satisfactorily) to the previous instance. In particular, it must be a pseudo-triangulation with the same combinatorial structure as the one at the previous step.

Writing down the edge length constraints of these edges we get a system of $2n-3$ quadratic equations of two types:

- $2n-4$ quadratic equations, stating that the $2n-4$ edges ij of the mechanism have their prescribed lengths:

$$(x_i - x_j)^2 + (y_i - y_j)^2 = l_{ij}^2 \tag{3.1}$$

- The additional equation:

$$x_2 = t \tag{3.2}$$

This last equation corresponds to the length condition for the convex hull edge. We can use this simplified, linear version, which quadratically would be $(x_1-x_2)^2+(y_1-y_2)^2 = t^2$ (but we eliminated $x_1 = y_1 = y_2 = 0$). In this case, the value t is a constant, more precisely it is equal to $t_0 + k\delta t$ at the $k+1$st step of the simulation.

We now formulate the **Local Problem:** *Find the solution of the algebraic system (3.1)+(3.2) which corresponds to the correct position of the pseudo-triangulation mechanism at time t.*

The algebraic system (3.1)+(3.2) will be called a *Laman graph embedding system*, because the underlying graph of n vertices and $2n-3$ edges is a Laman graph: equation (3.2) amounts to adding back to the pseudo-triangulation mechanism the convex hull edge that was removed.

A Laman graph system has $2n-3$ quadratic equations in $2n-3$ unknowns. It may have exponentially many real solutions (see [**5**]): we are interested in exactly one of them, corresponding to the next position of the mechanism. This is the unique point in the configuration space of the mechanism which lies on the expansive part of the trajectory at time step t. Since a good starting point for an iterative solution is always available, as we already know the solution $G(P(t))$ to an algebraic system that is *close* to the one we want to solve at time $t+\delta t$, the natural approach is to use the Homotopy method ([**10**], [**22**]). This will compute *only one solution* (the one closer to the previous instance) via numerical iterative methods (e.g. Newton's method). The idea is simple and implementable and we have experimented with it for our algorithm. We found that it works very well unless one gets close to the alignment event, when sometimes (and not rarely) it fails to converge in a reasonable, pre-defined (as a global constant) number of iterations or it gives a wrong answer (such as a solution *on the other side of the alignment event*). We have noticed that sometimes this happens when complex solutions would be obtained, should the motion extend past the alignment event. But these phenomena await a more systematic investigation.

OPEN PROBLEM 1. *Investigate the nature of those alignment events which cause Newton's method to fail as one gets close to them, and try to find good combinatorial or algebraic predictors for such a performance.*

On the theoretical side and based on our experimentation in Mathematica with this method, we expect the following to be true.

CONJECTURE 2. *For a generic instance of the Simulation of Motion problem (where* generic *will have to be defined appropriately, see section 4), there exist time moments* $t_Z \geq t_0$ *and* $t_F \leq t_f$ *sufficiently close to the event moments* t_0 *and* t_f*, such that if* $t_Z \leq t \leq t_F$*, Newton's method produces reliable solutions. I.e. in a* polynomially bounded (in n) *number of iterations, it computes a solution (point coordinates) lying on the trajectory of the pseudo-triangulation mechanism.*

The following stronger conjecture may facilitate this proof. It is also related to an alternate approach to the Local Problem based on Elimination: compute all the solutions and select *the* particular solution having the desired combinatorial pseudo-triangulation structure. For this to work, the following must be true.

CONJECTURE 3. *Among the many possible embeddings of a planar Laman graph with given edge lengths, no two are pseudo-triangulations with the same underlying combinatorial pseudo-triangulation structure.*

If this is true, then the problem could be formulated in semi-algebraic terms by adding inequality constraints to distinguish the solution that corresponds to the desired pseudo-triangulation. These can be expressed easily semi-algebraically as convex/reflex constraints on the angles between consecutive edges in the topological embedding of the planar graph. They are similar to *order type* ([**11**]) or *oriented matroid* (see [**2**]) constraints. Indeed, the combinatorial pseudo-triangulation concept is just *partial* oriented matroid information. For instance, for two edges ik and jk around a vertex k, the condition for convexity of the ccw oriented angle $\angle ikj$ is captured by a **semi-algebraic** constraint of the form:

$$\det \begin{pmatrix} x_i & y_i & 1 \\ x_k & y_k & 1 \\ x_j & y_j & 1 \end{pmatrix} < 0 \tag{3.3}$$

Assuming that Conjecture 3 holds, this formulation of the Local Problem becomes in principle solvable using techniques from Semi-Algebraic Geometry [**4**], perhaps by making use of combinatorial properties specific to our problem. We have not investigated this possibility yet. Instead we will discuss some possible heuristic approaches in section 5.

A weaker version of Conjecture 3 is:

CONJECTURE 4. *Among the many possible embeddings of a planar Laman graph with given edge lengths, no two are pseudo-triangulations with the same underlying oriented matroid.*

It is conceivable that two embeddings may have distinct oriented matroids, but that they would differ in the orientation of triplets of points that would not affect a common underlying combinatorial pseudo-triangulation. Thus while Conjecture 3 implies Conjecture 4, the opposite may not be true.

A *side remark*: to the best of our knowledge, there are very few such results of a *global rigidity* nature. One of the most conspicuous ones is the famous Cauchy Rigidity proof ([**7**], see [**1**]).

We end the paragraph by mentioning an example which shows that Conjecture 3 would fail if we dropped *combinatorial* and asked instead just for the uniqueness of a *pseudo-triangulation* embedding. Indeed, the two pseudo-triangulations in Figure 4 have identical edge lengths and differ *only* in the combinatorial part (assignment of convex/reflex angles), not in the topological embedding as planar graphs.

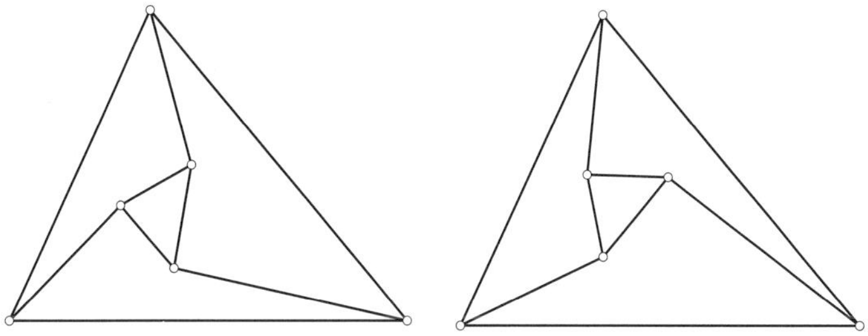

FIGURE 4. Example of a framework with two distinct pseudo-triangulation embeddings differing only in the combinatorial information of which vertices are corners.

Detection of the Next Event. We will refer to the following as the **Global Problem**: *For a pseudo-triangulation mechanism in expansive motion, compute the final value $t = t_f$ for the time parameter (the time of the next event) and the index k of the vertex at which the next alignment event happens.*

We first formulate the problem algebraically to show that it contains several instances of the Local Problem. We write down n systems of equations capturing the potential alignment of each vertex. They will be called *alignment systems* and have a common part, which expresses the fact that the $2n - 4$ edge lengths of the mechanism are fixed. They differ in one extra equation expressing the alignment of the extreme edges at the vertex.

- The $2n - 4$ quadratic *edge length* equations are the same as (3.1).
- The *alignment event equation* at vertex k expresses the condition that the (constant) sum of the lengths of the two extreme edges ik and jk of k equals the (unknown) distance between the other endpoints i and j:

$$(x_i - x_j)^2 + (y_i - y_j)^2 = (l_{ik} + l_{jk})^2 \tag{3.4}$$

Thus the algebraic part of each alignment system is similar to the Local Problem, except that we replaced an equation corresponding to a convex hull edge with an equation corresponding to an edge between the endpoint vertices i and j of the extreme edges ik and jk of vertex k. We will show in section 5 that, if i and j are not in the same rigid component, the underlying graph is still a Laman graph (but not a pseudo-triangulation, since it misses a convex hull edge and has one non-pointed (aligned) vertex). We will also show that we may not need to solve all n systems, since for some of them we can decide a priori that they have no solution.

The challenge of the *Global Problem* is to distinguish between the solutions to the alignment events systems produced by solving each system in turn: which corresponds to the next event?

The nature of the problem suggests the use of the Homotopy method. We can consider using it in two ways:

(1) As an iterative method for solving each alignment system, given as starting point of the iteration the position of the mechanism at time t_0. The drawback is that, in this case, the solution we seek may *not* be the closest to the one we start with. In practice, we have observed that an implementation of this approach found sometimes solutions that indeed align the vertices, but going in the opposite of the expansive direction! However, if we assume that somehow we get, for each system, the *closest* solution *in the direction of the motion* of the mechanism, then it is easy to detect the next event: it corresponds to the value k of the vertex index which produced a solution (alignment event) at the *smallest* value of the time parameter t. Indeed, after this time all the other alignments will contain k as a non-pointed vertex, corresponding to the fact that the mechanism moved past its aligning event.

(2) As an iterative method applied in conjunction with the Simulation of Motion. Instead of providing a final value for t_f, we compute it as we move along. At each step in the Simulation, we must check whether we have reached an Alignment event. This can be done by *verifying* whether equation 3.4 holds for the current values. This can only be done if we know the root values *exactly* (which is not a practical assumption except for some special situations presented in Section 5) and if we happen to use the right time step to fall exactly on the event. Most of the times, we will be either before or after the event.

Here are the main questions that must be researched for efficient solutions to the Homotopy method applied to the Global Problem.

Given an embedded Laman graph, verify whether it is a pseudo-triangulation. Moreover, verify that it has a given combinatorial structure. We call this the **Pseudo-Triangulation Verification Problem.**

Since this may have to be done at each step of the simulation, we want it to be very fast.

OPEN PROBLEM 5. *What is the complexity of the Pseudo-triangulation Verification Problem? Can it be done in linear time?*

We might get better solutions under the assumption that we *know* something about which edges might cross or which vertices might become non-pointed, since we are using it within the Simulation of Motion phase for a specific pseudo-triangulation.

OPEN PROBLEM 6. *Is it possible to detect in sublinear time the violation of pseudo-triangulation properties if we already know the Laman graph structure and (possibly) additional information (such as the infinitesimal velocities of the vertices at the previous time step t of the simulation)?*

Another problem with this approach is caused by the fact that the time step δt might be too big: we might accidentally pass several alignment events, and even if we detect this, we may not see which one we passed. When the final value t_f is known, the trivial choice is to simply divide the interval $t_f - t_0$ into a constant number of steps, but with this approach we must choose δt in another way: either some constant, or a value to be decided dynamically, as we move along the trajectory.

OPEN PROBLEM 7. *Develop criteria for choosing the value of the time step to avoid that we accidentally pass over events and to guarantee that we make progress towards the next event.*

Alternately, we might want to solve the following problem instead.

OPEN PROBLEM 8. *Develop criteria for recognizing that the Simulation passed over several events.*

If these problems would be solved, the Homotopy method might be the right approach in practice, as it will lead directly to the aligned vertex. We are still interested, though, in the possibility of a semi-algebraic formulation of the Global Problem, as well as in the use of non-homotopy based methods (e.g. Elimination). This leads to an extension of Conjecture 3, formulated for the *combinatorial type* of what would be the Laman graph behind each alignment event. We skip the details here, as this definition is easy to formulate and is a simple variation on the combinatorial pseudo-triangulation.

CONJECTURE 9. *Among the many possible embeddings of a planar Laman graph with given edge lengths, no two have the same underlying combinatorial structure, which is the "almost pseudo-triangulation" structure of the underlying graph of an alignment event.*

In addition, we need:

CONJECTURE 10. *Assuming that no two vertices align simultaneously, exactly one of the Alignment systems will have a solution compatible with the combinatorial structure of the pseudo-triangulation mechanism before the alignment event.*

We skip the details of how this compatibility is defined, as it is straightforward and the reader can reconstruct it easily. If this conjecture holds, the **semi-algebraic** formulation of the *Global Problem* is similar to what we had for the local problem, i.e. using orientation constraints of the type expressed by the equations (3.3).

On an even more ambitious level, one may ask whether the partial oriented matroid information of *any* Laman graph embedding suffices to guarantee the uniqueness of the embedding.

OPEN PROBLEM 11. *Investigate whether the partial oriented matroid information of a Laman graph embedding suffices to guarantee uniqueness.*

4. Problems Encountered with Standard Techniques for Solving Algebraic Equations

In this section we identify three problems that may be encountered when using one or the other of the two approaches for solving algebraic systems of equations, Elimination or Homotopy. Our examples will show why using blindly some standard functions already available in computational algebraic systems (such as Mathematica) may not work or may be very inefficient (exponential). This will motivate the need for using heuristics, such as those presented in section 6, and will provide intuitions to the related sequence of open questions and problems requiring further investigation.

4.1. Problems with Elimination. Elimination (Gröbner bases) is a notoriously time intensive algorithmic approach, but in our case it would have the advantage of better controlled precision. We would like to understand when it would or would not work, and leave as a future goal to seek techniques for fine-tuning it to the specifics of the problem to possibly get efficiency and the desirable degree of accuracy.

We first prove that a quadratic system of equations modelling the algebraic part of the *Local Problem*, i.e. the realizability problem for a Laman graph, may not have a finite set of solutions, and when it has one, it may be exponentially large.

THEOREM 4.1. *There exists a pseudo-triangulation, and even a Hamiltonian pseudo-triangulation, whose configuration space has a component of arbitrarily large dimension.*

PROOF. Based on the examples in Fig. 5. The underlying graph of the image on the left is Laman (a Henneberg I graph, in fact). The given embedding is generic (non-infinitesimally rigid Henneberg I graphs occur only when a vertex of degree 2 is aligned), but the edge lengths are very special. All the vertices (except the two at the bottom) lie on the perpendicular bisector of the segment joining the two bottom ones. Taking the mirror image of the right half of this embedding in the axis of symmetry containing the vertical vertices produces another realization, where the points and edges on the right overlap over those on the left hand side. It is now obvious that all the vertical points can be moved continuously and (except for one of the two top connected ones) independently, while keeping the edge lengths fixed, to obtain other embeddings. The dimension of the component of the realization space containing this embedding is very large ($n - 3$).

An example of a Hamiltonian pseudo-triangulation with a similar symmetry inducing a one-dimensional component of the configuration space is given in the rightmost picture of Fig. 5. □

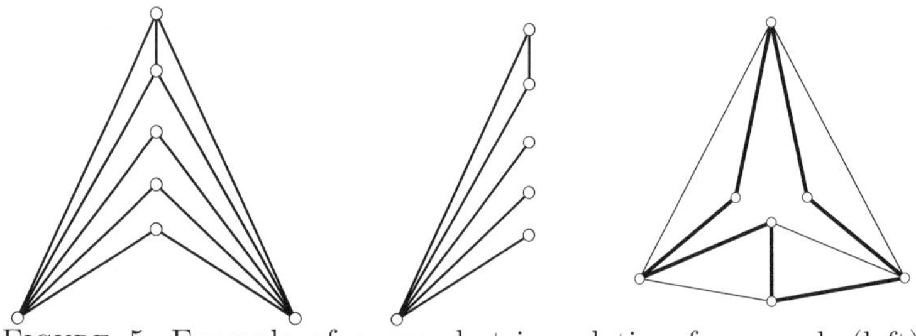

FIGURE 5. Example of a pseudo-triangulation framework (left) which has another embedding (middle) which is non-rigid, with a large number of degrees of freedom. Right: a Hamiltonian pseudo-triangulation with a similar vertical axis of symmetry inducing an embedding which is 1DOF.

A consequence of this example is that the Gröbner bases elimination algorithm may not eliminate all the variables. If such an example could be constructed for a Hamiltonian pseudo-triangulation, then elimination may fail to give a univariate equation in, say, the time t for one alignment system, when searching for the time of the next event t_f in the global problem.

The "bad" examples from Fig. 5 arise from graphs with an axis of symmetry (which in general may affect only some part of the graph), and which have a pseudo-triangular embedding conforming with this symmetry. Are all bad examples of this nature?

OPEN PROBLEM 12. *Characterize the Laman graphs with "bad" (non zero-dimensional) components in their configuration spaces. More precisely, characterize those which have a (Hamiltonian) pseudo-triangulation embedding* and *a bad component.*

In particular, this question contains as a sub-problem the characterization of those planar Laman graphs which have a pseudo-triangular embedding. This has been answered recently in [**13**]: *all* planar Laman graphs have pseudo-triangular embeddings. Hence for a more extensive collection of "bad" examples we may start with planar Laman graphs having a "combinatorial axis of symmetry" on some subgraph (*how to characterize these?*) and see if we can always realize it as a pseudo-triangulation conforming to the symmetry. We do not know whether these are the only situations to consider.

THEOREM 4.2. *There exists a pseudo-triangulation whose configuration space has an exponential number of zero-dimensional components. The statement holds also for Hamiltonian pseudo-triangulations.*

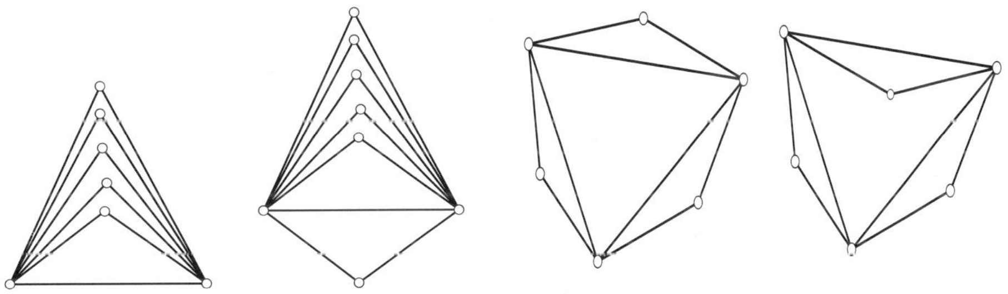

FIGURE 6. Two examples of frameworks with exponentially many embeddings (real solutions). *Flips over* are illustrated. Left: 2^{n-2} embeddings. Right: a Hamiltonian pseudo-triangulation with $n/2$ ears inducing $2^{n/2}$ embeddings, all pseudo-triangulations.

PROOF. The proof relies on the examples in Figures 6. For the first example (left two pictures), the bottom two vertices are fixed, the others may be independently *flipped over* the bottom edge to obtain another pseudo-triangular embedding. This yields 2^{n-2} possible embeddings. Note that although each of them is a pseudo-triangulation, they have distinct combinatorial types. For the second example (right two pictures) we have a convex polygon with an even number n of vertices and $n/2$ ears which can be flipped over, yielding $2^{n/2}$ embeddings. □

Another annoying problem may occur when several vertices align simultaneously. In this case, the embedding of the Laman graph for the global problem may have a space of infinitesimal motions of higher dimension. Let's see first how bad this can be.

THEOREM 4.3. *There exists a 1DOF expansive mechanism obtained from a pseudo-triangulation for which a linear number of vertices will align simultaneously.*

PROOF. Based on the obvious generalization to $n = 3k$ vertices and $n/3$ alignments of the example in Fig. 7 (done for $k = 5$). The picture shows the mechanism at the moment when the bottom two edges align. Simultaneously, all the $k - 1 = n/3 - 1$ vertices above it also align. Of course, for this to happen the edge lengths must be very special. As an aside, notice that the underlying graph is a pseudo-triangulation with a Henneberg I construction (for which the heuristics of section 6 can be applied). But it is easy to turn this into a general Henneberg example by flipping edges in the triangles adjacent to the degree 2 vertices. □

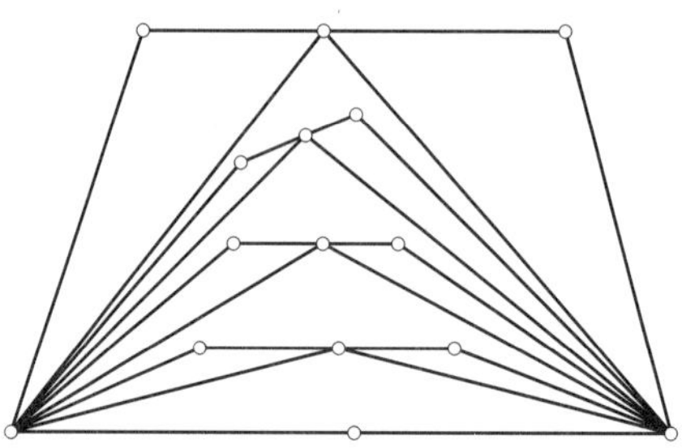

FIGURE 7. Example of a pseudo-triangulation mechanism with a linear number of vertices aligning simultaneously (shown at the moment when the alignments occur.)

A consequence of this example is that solving the alignment systems approximately may lead to numerical inaccuracies which in turn will lead to combinatorially unreliable decisions. If the times of alignment of several vertices coincide, but the numerical approximations return slightly different values, then we won't be able to detect that and might make incorrect decisions regarding which edges to flip to get the next pseudo-triangulation mechanism.

We now define our concept of **generic pseudo-triangulation mechanism**. It must satisfy two conditions:

(1) The underlying Laman graph, with edge constraints given by the $2n - 4$ mechanism bar lengths plus the convex hull edge length at time t, has a *zero-dimensional configuration space*, for any time t, $t_0 \leq t \leq t_f$.
(2) The mechanism has no multiple alignments.

In other words, a pseudo-triangulation mechanism is *generic* if the extreme examples of this section do not apply to it. The definition is not very satisfactory because it refers to a continuum of positions that have to be tested for the verification of the property. At each position the property can be in principle verified algorithmically (although not very efficiently) by computing the dimension of the configuration space.

OPEN PROBLEM 13. *Characterize in finite terms the* genericity *of pseudo-triangulation mechanisms and devise good algorithms for testing the property.*

Non-generic pseudo-triangulations have very special edge lengths and very special symmetries. Could it be the case that we can avoid them altogether for the original motivating problem (Carpenter's Rule unfolding)?

OPEN PROBLEM 14. *Given a simple planar polygon, does it always support a generic pseudo-triangulation mechanism? I.e., among the many ways of placing additional bars to obtain a pseudo-triangulation, and of removing convex hull edges, is there one that is guaranteed to yield a generic mechanism?*

More generally, the problem may be formulated for arbitrary mechanisms.

OPEN PROBLEM 15. *Define and characterize genericity of arbitrary 1DOF mechanisms.*

Notice that Open Problem 12 appears as a subproblem.

4.2. Problems with Homotopy. In this approach we attempt to find only one root of the algebraic system, starting from a solution at the previous step that is, hopefully, close enough to guarantee that Newton's method will converge to the right solution in a reasonable amount of time.

We have experimented with this technique in Mathematica and found that there are severe drawbacks even on relatively small examples. For the Local Problem, Newton's method works well until we get close to the alignment event. In that case, either the method fails to converge within a number of iterations or (even worse) it ends up giving a solution *going beyond the alignment event.* Both behaviors are impediments to a full automation of the algorithm. In connection with Open Problem 1, we venture a candidate for a good predictor. The infinitesimal velocities of the vertices of the mechanism may be computed via linear programming and used to assist with this decision.

CONJECTURE 16. *The relative speed of motion of the vertices is a possible predictor for bad behavior of the Homotopy method on the Local Problem around the alignment events.*

For the Global problem Homotopy causes more severe problems, because the starting point for the iteration (given by the position of the mechanism at time t_0) may be quite far from the desired solution t_f. Our experiments have shown that:

- We do not have to compute roots of systems associated to all the vertices. In fact, for those vertices which yield no real roots because they belong to some rigid component, the Homotopy method gives very strange "solutions". In Section 5 we show how to systematically avoid looking for solutions when they do not exists, via pre-computations of rigid components.
- It is hard to distinguish between alignment events that are very close to each other. When several are close, Newton's method may not converge or may give wrong solutions.

OPEN PROBLEM 17. *Look for predictors of Newton's method failure for the Global Problem.*

We conclude with the comment that it might be possible that, instead of looking for the final value t_f *at the beginning of the Simulation of Motion phase* (which may cause the described bad behavior with the Homotopy method applied to the Global Problem), we may try to apply it later on in the simulation. This would work if we guaranteed that we do not accidentally go beyond the event, and if we could predict that we get closer (e.g., if Newton's method would fail in a limited number of iterations on the Local problem). We have not yet experimented with this approach.

5. Elimination-based Approaches to the Local and Global Problems

The common thread in this section is the rigidity of some graphs and subgraphs occurring in the lifetime of our algorithm. We start this section by proving some properties related to the correct formulation of the Local and Global Problems, for generic pseudo-triangulations. Then we discuss a potential approach based on Elimination (as opposed to Homotopy) methods and present heuristics for simplifying the search for the alignment event.

THEOREM 5.1. *Let G be a pseudo-triangulation with a convex hull edge e removed. Let k be an arbitrary vertex and let ik and jk be the extreme edges adjacent with this vertex. Then the graph G' obtained by adding the edge ij is a Laman graph if and only if i and j do not belong to the same r-component of the mechanism $G - \{e\}$.*

The theorem is a consequence of the following more general one.

THEOREM 5.2. *Let G be a Laman graph with some edge e removed. Then the graph G' obtained by adding another edge ij is a Laman graph if and only if i and j do not belong to the same r-component of the mechanism $G - \{e\}$.*

PROOF. r-components satisfy the Laman count, so adding an edge will violate the Laman property on the set of vertices of the component. Conversely, adding an edge whose endpoints lie on two different r-components does not increase the Laman count for any subset of edges which was already saturated to $2k - 3$. But it makes the global count $2n - 3$, hence we get a Laman graph. □

The resulting graph G', although Laman, may not be infinitesimally rigid, because of the alignment of adjacent edges and the addition of an overlapping edge. See Figure 8.

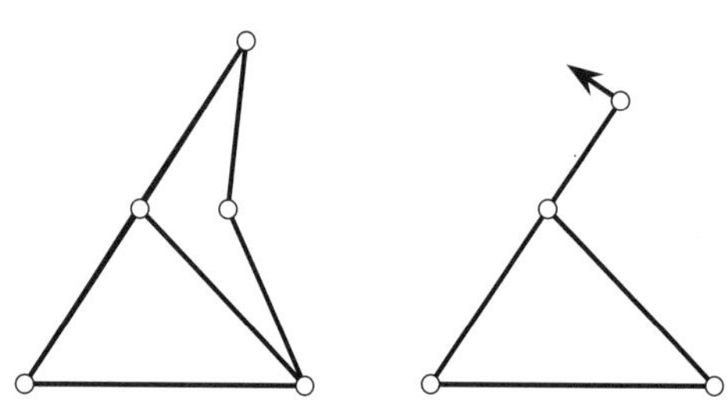

FIGURE 8. Examples of pseudo-triangulation mechanisms with one aligned vertex. The overlapping added bar between the endpoints of the two aligned vertices can't be seen. Left: infinitesimally rigid. Right: with a non-trivial infinitesimal velocity vector assignment.

However, the following may not be hard to prove:

CONJECTURE 18. *Let G be a pseudo-triangulation with a convex hull edge e removed. Let k be an arbitrary vertex and let ik and jk be the extreme edges adjacent with this vertex. Let G' be the graph embedding obtained from $G - \{e\}$ by adding the edge ij of length equal to the sum of the lengths of the edges ik and jk and whose combinatorial embedding is the same as it was for G except for the aligned edges ik*

and jk. If k is the only vertex that aligns with this modification, then G' is rigid (but may not be in general infinitesimally rigid).

A stronger conjecture is that the same may be true even when dropping the condition that no other vertices align.

We now discuss Elimination-based approaches to the Local and Global problems. Their correctness relies heavily on the fact that we deal with Laman graphs and on expected positive answers to Conjectures 3 and 9. Their practical applicability may be restricted if elimination would behave exponentially, and one would have to use the special structure of pseudo-triangulations to gain efficiency. We may think of pruning, in a semi-algebraic fashion, solutions to the Laman embedding problem that do not satisfy the semi-algebraic constraints of the desired pseudo-triangulation embedding. On the other hand, we expect better accuracy through such a method. The accuracy is not so much needed for the Simulation of Motion (which is very stable and recovers easily from errors at some previous step). It is essential, however, when deciding the next event.

For reducing the number of alignment systems to be solved when seeking the solution for the Global Problem via Elimination, we can use the following heuristic. We first detect the r-components of the 1DOF mechanism. An alignment will never occur at a vertex whose extreme edges lie in the same r-component (as they never move).

Although not explicitly designed for this purpose, algorithms for detecting r-components, based on matroidal properties of Laman graphs, can be traced in the literature. See [**12**] and the references given there. They are polynomial, but we would like better than this, at least in the case of pseudo-triangulations, where r-components have a nice structure, as they are themselves pseudo-triangulations. In particular, they contain their convex hull edges. We expect that efficient detection of rigid pseudo-triangular components is possible.

OPEN PROBLEM 19. *Find an* efficient *algorithm to detect r-components of a pseudo triangulation with an edge removed. Can this be done in linear time (like finding connected and biconnected components of an arbitrary graph)?*

6. Heuristics based on Henneberg Constructions

A consequence of Lemma 4.2 is that it may be inefficient to rely on root finding functions (such as `Solve` in Mathematica) which attempt to find all the solutions of the algebraic systems (3.1)+(3.2) or (3.1)+(3.4). To simplify, in some cases dramatically, the complexity of the system and the accuracy of finding its roots, we will use in this section the particular structure given by the *Henneberg constructions* for pseudo-triangulations. Instead of solving the associated quadratic system of equations and obtain all the solutions simultaneously, we can (sometimes) compute them sequentially. As we move on, we prune those which do not satisfy the order-type semi-algebraic constraints. This is possible when the pseudo-triangulation has a Henneberg I construction.

In this case, though, we will change the parametrization slightly: the variables to be fixed will correspond to the first edge of the Henneberg construction, which is not necessarily the missing edge of the pseudo-triangulation mechanism. But is this really necessary? Lemma 6.1 shows that it is not, for general Henneberg constructions. However, Henneberg I constructions may even be unique, see Figure

9. This example (which is also Hamiltonian) has a unique Henneberg I construction, because at each step there exists a unique vertex of degree 2. (It has many Henneberg II constructions, though.) Hence the starting edge cannot be fixed a priori in such a case.

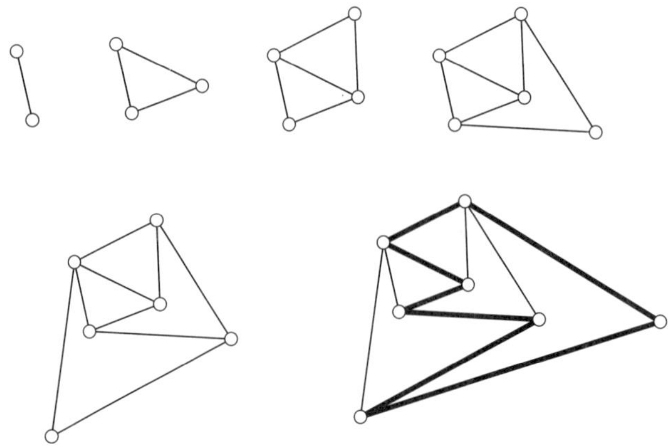

FIGURE 9. Example of a Hamiltonian pseudo-triangulation with a unique Henneberg I construction.

LEMMA 6.1. *For an arbitrary Laman graph, one can always choose any edge to be the first one in a Henneberg construction that never removes it afterwards.*

PROOF. This simple argument is due to Walter Whiteley [**25**]. The Laman count guarantees that there are at least 3 vertices of degree at most 3. Therefore, we can fix an edge and, no matter what the degree of its two endpoints, never use it in a Henneberg step, because there will be other eligible vertices. □

To simplify the root finding, follow the Henneberg I construction. Start by fixing the coordinates of the first triangle in the construction and assume (for the sake of the analysis) that we have relabelled the vertices as $1, 2$ and 3. Then at each step i, the coordinates of the previously added vertices (labelled up to $i+2$) would have been computed. Adding the next vertex using the edge lengths of its adjacent edges amounts to intersecting two circles of known centers and given radii. We can solve this system exactly, using only square roots, and retain the solution which verifies the combinatorial pseudo-triangulation constraints. Indeed, it is trivial to show that for Henneberg I graphs, Conjecture 3 holds.

Notice that there is nothing special about pseudo-triangulations in this step. Laman graphs constructible with standard Henneberg I steps (see e.g. [**12**] or [**23**] for Henneberg constructions on Laman graphs) can be solved equally efficiently. We also notice that for given edge lengths, all their embeddings (which may be up to 2^{n-2}, when all the roots are real) have distinct oriented matroids (even partial oriented matroids induced by the rotations of lines through edges around each vertex).

The case when the pseudo-triangulation does not have a Henneberg I construction is more complex. When a type II step is used, an old edge is erased. But this means that in the construction so far we have used an edge length that did not exist. We could do this parametrically. Preprocess the Henneberg construction and mark all the edges that will be eventually removed by adding new variables (parameters) for their unknown lengths. Proceed with the Henneberg construction as long as only type I steps are used and compute parametrically the coordinates of the added points (so far, only square roots should be necessary).

Then when a type II step occurs, eliminate the parameter corresponding to the removed edge by solving a system of three quadratic equations (for the three circles that must cross at the added vertex).

The problem with this approach is that we may have to carry with us, all the way to the end, *all* the parametric solutions, as there is no way of knowing which one will match the combinatorial information of the desired embedding before all the parameters have been eliminated. This part needs further investigation.

OPEN PROBLEM 20. *Investigate the feasibility of the previous approach based on parametric computation of roots for Henneberg II graphs.*

Finally, we state several open problems motivated by this heuristic. Depending on the answers, one might be able to improve and analyze formally its efficiency.

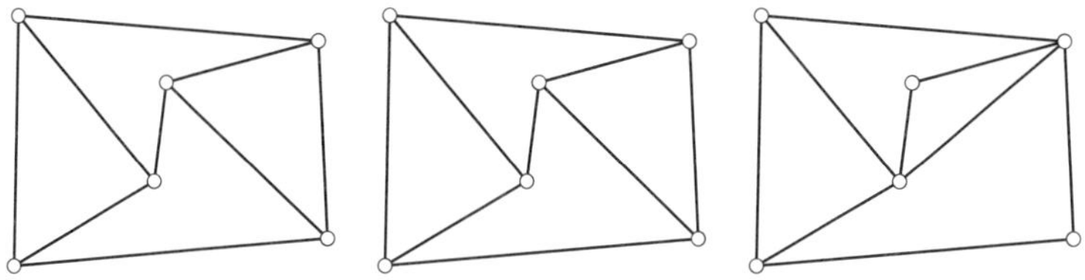

FIGURE 10. The smallest pseudo-triangulation which is not Henneberg I. The highlighted underlying polygon in the middle has a different, Henneberg I, pseudo-triangulation.

The smallest example of a pseudo-triangulation (of a point set) which is not Henneberg I is shown in Fig. 10 (left). Since it is also a Hamiltonian pseudo-triangulation (middle), this example gives the smallest pseudo-triangulation of a polygon with no Henneberg I construction. But this polygon has another pseudo-triangulation (right) which is Henneberg I. A natural question to ask is: *Does every polygon have a Henneberg I pseudo-triangulation?* That this is not true is shown by the example in Fig. 11 (left) and the following argument. [4] The proof relies on the fact that all the possible pseudo-triangulations have no vertex of degree 2. Indeed, every pseudo-triangulation of the polygon must include the three convex hull edge (middle): this already makes the vertices $1, 2, 4, 5, 7, 8$ be of degree 3. Moreover, not two of the three remaining edges adjacent to an interior vertex may not be simultaneously in the pseudo-triangulation, because of pointedness. Hence there must be an edge adjacent to each of them, making them all of degree at least 3.

This ruins the hope that there might be some simple general way (i.e. via Henneberg I heuristics) around the algebraic problems induced by the Local and the Global Problems for the Combinatorial Roadmap Algorithm.

A related problem is that a pseudo-triangulation may have several Henneberg constructions. Here's a general procedure to produce one. A simple degree count shows that every Laman graph must have either a vertex of degree 2 or of degree 3. Start working backwards, plucking off degree 2 and 3 vertices. The problem is

[4] We are grateful to the anonimous referee for this example.

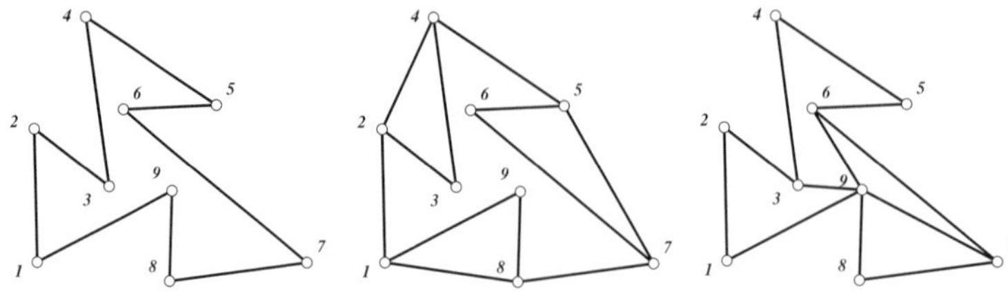

FIGURE 11. A polygon with no Henneberg I pseudo-triangulation.

that there might be several vertices which can be chosen at one step. Are there pseudo-triangulations with exponentially many Henneberg I constructions? Indeed, the example in Fig. 6 has this property: any of the degree 2 vertices can be chosen, yielding $(n-2)!$ orders in which they can be plucked off. But this is a happy situation, since any order yields a Henneberg I construction. Call it a *winning degree 2 construction.*

Are there pseudo-triangulatios with a winning degree 2 constuction, but having an option along the way so that when another degree 2 vertex is chosen, one can't continue without using a degree 3 vertex? The answer is NO, because vertices of degree 2 can be removed independently. The two edges adjacent to a degree 2 vertex are not shared with another degree 2 vertex, or else the graph won't be rigid if it has at least 4 vertices. Therefore, if there exists a Henneberg I construction, then any sequence of choices of degree 2 vertices produces one.

Another natural question, *Is there a Hamiltonian pseudo-triangulation with exponentially many Henneberg I sequences?* is answered via the same example from Figure 6 (right). Indeed, the $n/2$ ears can be plucked off independently, leading to $(n/2)!$ Henneberg I sequences.[5]

OPEN PROBLEM 21. *Characterize pseudo-triangulations which have Henneberg I constructions (other than "those for which the plucking-off-degree-2-vertices algorthms works").*

The recent result that planar Laman graphs have pseudo-triangular embeddings [**13**] reduces the problem to: *Characterize planar Laman graphs which have Henneberg I constructions.*

Assuming that we have a polygon and that we can find a Henneberg I pseudo-triangulation for it, we may want to use this one, not the shortest-path-tree approach from [**20**]. Indeed, a shortest-path pseudo-triangulation may not be Henneberg I. The simple example in Fig. 12 shows a hexagon, which has only two possible pseudo-triangulations, both of which are shortest-path pseudo-triangulations, but one is Henneberg I while the other is not.

OPEN PROBLEM 22. *Analyze the number of steps (events) of the pseudo-triangulation roadmap algorithm in other cases than the shortest path tree. In particular, one may want to see how the global number of steps (alignment events) is affected if (say) only Henneberg I pseudo-triangulations are used all the way (e.g. when this is possible) : could we guarantee that the algorithm ends after a finite number of events?*

[5]This example was found by the anonimous referee.

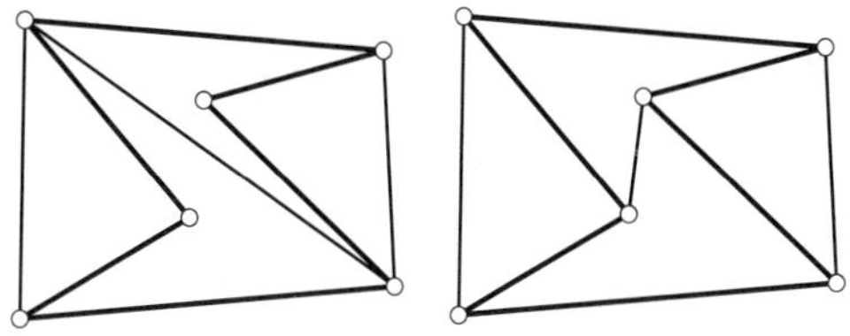

FIGURE 12. Polygon and its two possible pseudo-triangulations, both of which are shortest-path. One is Henneberg I (left), the other is not (right).

7. Further Issues and Open Problems

Our work was motivated by the practical goal of finding an efficient fully-automated implementation of the Pseudo-triangulation Roadmap Algorithm. We have experimented with existing algebraic systems such as Mathematica and found that relying on the built-in functions alone was not practical for the two problems discussed in this paper. But applying the Henneberg I heuristic was very efficient, and we could easily solve large systems. Some further issues related to this approach are included in this final section.

The parameter δt used in the simulation of *one* mechanism depends on the choice of a *constant* number of time steps. Ideally, all these choices should be synchronized and their calculation automated.

OPEN PROBLEM 23. *Develop criteria for computing the time steps δt of all the mechanisms occurring during the lifetime of an unfolding process, so that the motion looks locally smooth and at the same time maintains global esthetic qualities achieved, for instance, through a uniform global average speed.*

This calculation may happen, for instance, after all the alignment events have been computed and the *combinatorial* structure of the whole unfolding process has been determined. To achieve a global uniform speed of *one vertex*, one has to compute the lengths of all the trajectory pieces for that vertex. For a global effect of smoothness, one may have to average over all vertices.

One of the most basic questions induced by the Local Problem to which we briefly referred earlier is:

Characterize and recognize efficiently those graphs which have an embedding as a pseudo-triangulation.

An understanding of the specifics of these graphs is hoped to give insights into possible ways of solving the algebraic systems for the Local and Global Problems more efficiently. Since pseudo-triangulations are planar Laman graphs, a natural question is whether the reverse is also true. This has been answered recently, see [**13**]: *pseudo-triangulation graphs coincide with planar Laman graphs.* This still leaves open the question, naturally related to Open Problem 19.

OPEN PROBLEM 24. *Develop efficient algorithms for recognizing planar Laman graphs. Can this be done in linear time?*

A number of remaining open questions concern Hamiltonian pseudo-triangulations:

OPEN PROBLEM 25. *Characterize and recognize efficiently the Hamiltonian pseudo-triangulations.*

OPEN PROBLEM 26. *Is there a Henneberg construction for Hamiltonian pseudo-triangulations (i.e. one that would maintain Hamiltonicity throughout the construction)?*

Further examples and results of our current and future experiments may be accessed through the author's web page, http://cs.smith.edu/ streinu/research.html.

Acknowledgements. The presentation benefited from the comments of a careful anonimous referee, who also answered some of the questions posed in the first version of the paper and suggested the terminology *Hamiltonian pseudo-triangulations.* I acknowledge useful discussions on computational algebraic geometry with Günter Rote, Bernd Sturmfels, Laureano Gonzalez-Vega and Frank Sottile. Eleanor Farrington implemented a number of Mathematica experiments during a Summer 2001 REU at Smith College, sponsored by the author's NSF RUI-grant CCR-0105507. Some ideas reported in this paper emerged during the Barbados Workshops on *Pseudo-triangulations*, January 2001 and on *Rigidity Theory and Scene Analysis* (organized by the author with partial NSF support under grants CCR-0104370 and CCR-0203224), the DIMACS Workshop on Algorithmic and Quantitative Aspects of Real Algebraic Geometry in Mathematics and Computer Science (organized by Richard Pollack, Marie-Françoise Roy and Micha Sharir) at DIMACS/Rutgers University in March 2001, and several Combinatorists of New England (CONE) one-day meetings held at Smith College (organized by Michael Albertson, Karen Collins and Ruth Haas). All of these enjoyed partial support from NSF.

References

[1] M. Aigner and G. Ziegler. *Proofs from THE BOOK*, Springer, 2nd edition, 2002.

[2] A. Björner, M. Las Vergnas, B. Sturmfels, N. White and G. Ziegler, *Oriented Matroids*, Cambridge University Press, 1993.

[3] S. Basu, R. Pollack and M.-F. Roy. *Computing Roadmaps of Semi-algebraic Sets on a Variety*, JAMS, 13, pp. 55-82, 2000. prelim. version in Proc. 28th ACM Symp. Theory of Computing, pp. 168-173, 1996.

[4] S. Basu, R. Pollack and M.-F. Roy. *Algorithms in Real Algebraic Geometry*, book, in preparation.

[5] C. Borcea and I. Streinu. *On the Number of Embeddings of Minimally Rigid Graphs*, Proc. ACM Symp. Comput. Geometry (SoCG), Barcelona, pp. 25-32, 2002.

[6] J. Canny. *Complexity of Robot Motion Planning*, MIT Press, Cambridge, MA, 1988.

[7] A. Cauchy. *Sur les polygones et les polyédres, seconde mémoire*, J. École Polytechnique XVIe Cahier, Tome IX (1813), 87; Oevres Complétes, IIe Série, Vol. I, paris, 1905, 26-38.

[8] R. Connelly, E. Demaine and G. Rote. *Straightening Polygonal Arcs and Convexifying Polygonal Cycles*, Proc. 41st Symp. on Foundations of Computer Science (FOCS), 2000, pp. 432-442.

[9] D. Cox, J. Little and D. O'Shea. *Ideals, Varieties and Algorithms. An Introduction to Computational Algebraic Geometry and Commutative Algebra.* Springer Verlag, 1992.

[10] D. Cox, J. Little and D. O'Shea. *Using Algebraic Geometry.* Springer Verlag, 1998.

[11] J.E. Goodman and R. Pollack, *Semispaces of configurations, cell complexes of arrangements*, J. of Combinatorial Theory, Series A, 37 (1984) 257–293.

[12] J. Graver, B. Servatius and H. Servatius. *Combinatorial Rigidity*, Amer. Math. Soc., Graduate Studies in Mathematics vol.2, 1993.

[13] R. Haas, F. Santos, B. Servatius, D. Souvaine, I. Streinu and W. Whiteley. *Planar Minimally Rigid Graphs have Pseudo-Triangular Embeddings*, manuscript, 2002.

[14] L. Henneberg, *Die graphische Statik der starren Systeme*, Leipzig,1911.

[15] M. Kapovich and J. Millson. *On the moduli spaces of polygons in the Euclidian plane*, Journ. of Diff. Geometry, vol. 42, N 1, 1995, pp. 133-164.

[16] G. Laman, *On graphs and rigidity of plane skeletal structures.* Eng. Math., 4, 331-340, 1970.

[17] M. Pocchiola and G. Vegter *Topologically Sweeping Visibility Complexes via Pseudo-Triangulations*, Discrete Comput. Geometry 16(4), 419-453, 1996.

[18] A. Recski. *Matroid Theory and Its Applications*, Springer Verlag, 1989.
[19] G. Rote, F. Santos and I. Streinu, *Expansive Motions and the Polytope of Pointed Pseudo-Triangulations*, in *Discrete and Computational Geometry – The Goodman-Pollack Festschrift*, (B. Aronov, S. Basu, J. Pach and M. Sharir, eds.), Algorithms and Combinatorics Series, Springer Verlag, Berlin, to appear 2002.
[20] I. Streinu. *A Combinatorial Approach to Planar Non-colliding Robot Arm Motion Planning*, Proc. 41st Symp. on Foundations of Computer Science (FOCS), 2000, pp. 443-453.
[21] T.S. Tay and W. Whiteley. *Generating Isostatic Frameworks*, Structural Topology 11, pp. 20-69, 1985.
[22] J. Verschelde, P. Verlinden and R. Cools. *Homotopies Exploiting Newton Polytopes for Solving Sparse Polynomial Systems*, SIAM J. Numer. Anal. 31, pp. 915-930, 1994.
[23] W. Whiteley. *Some Matroids from Discrete Applied Geometry*, Contemporary Mathematics, vol. 197, pp. 171-311, Amer. Math. Soc. 1996.
[24] W. Whiteley. *Rigidity and Scene Analysis*, in *Handbook of Discrete and Computational Geometry*, (J.E. Goodman, J. O'Rourke, eds.), 1997, pp. 893-916.
[25] W. Whiteley. *Personal communication, 2002.*

Department of Computer Science, Smith College, Northampton, MA 01063, USA.
Current address: Department of Computer Science, Smith College, Northampton, MA 01063, USA.
E-mail address: streinu@cs.smith.edu

DIMACS Series in Discrete Mathematics
and Theoretical Computer Science
Volume **60**, 2003

Visibility Computations: From Discrete Algorithms to Real Algebraic Geometry

Thorsten Theobald

ABSTRACT. We investigate visibility computations with moving viewpoints. The initial problems are of discrete and algorithmic nature, but even for simple classes of objects (such as balls and polytopes), they lead to interesting and difficult problems from real algebraic geometry. Namely, it is necessary to characterize and compute the common tangent lines to a given set of convex bodies.

In particular, we present a new sweep algorithm in dimension 2, as well as survey and extend recent algebraic-geometric results on the tangent problems in dimension 3.

1. Introduction

Visibility computations belong to the central tasks in computer graphics and computational geometry [**19**]. Depending on the particular framework and on the particular question to be answered, there is a large variety of mathematical and algorithmic core problems behind these questions.

In the simplest case of a visibility problem, we are given a fixed viewpoint $v \in \mathbb{R}^n$ (say, $n \in \{2,3\}$), and the scene consists of a set $\mathcal{B}$ of bodies. The task is to compute a suitable two-dimensional projection of the scene ("to render the scene") that reflects which part of the scene is *visible* from the viewpoint v. In a more dynamic and interactive setting, the viewpoint can be moved interactively (see, e.g., [**4, 15**]). However, in general, after each movement of the viewpoint a new rendering process is necessary. In order to speed up this process, commercial renderers apply caching techniques [**34**].

From the algorithmic and geometric point of view it is desirable to establish a more global view of the scene in advance and answer questions like: Which of these bodies can be partially seen from *some* viewpoint within a given viewpoint area? Those bodies which are not visible from any of these viewpoints can be removed from the whole visualization process in advance. In case of dense crystals whose atoms are visualized as sufficiently large balls, this can reduce the time consumption of the rendering process significantly (see Fig. 1).

Key words and phrases. visibility, tangent lines, spheres, transversals, enumerative geometry, real algebraic geometry, computational geometry.

2000 Mathematics Subject Classification. Primary 26C10, 68U05; Secondary 14N10, 51M30.

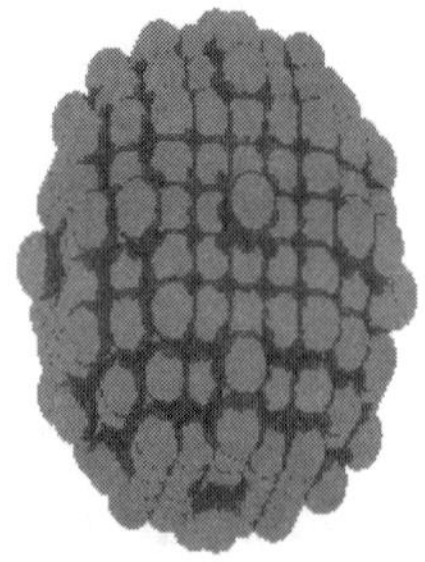

FIGURE 1. Dense arrangement of atoms

So far, there exists only few literature on visibility questions with moving viewpoints in dimension at least three (see [**4, 15, 33**]). A major reason for this can be seen by various intrinsic difficulties in the underlying geometric questions.

From the *algebraic-geometric point of view*, visibility computations with moving viewpoints require to study the interaction of the geometric bodies with lines. In particular, it is essential to characterize certain extreme situations which correspond to common tangent lines to a given number of bodies. In dimension 2, the resulting geometric questions typically remain rather elementary, and the main focus is on the algorithmic side. In contrast to this, in dimension 3, even for simple types of bodies, such as spheres, the underlying geometric problems have a high algebraic degree and hence give rise to interesting and difficult questions of real algebraic geometry. Several recent papers deal with these algebraic-geometric core problems [**17, 18, 31, 29, 32**]. In particular, the relevant real enumerative aspects constitute a rich and natural class of geometric examples within the general framework of real enumerative geometry (see the survey [**28**]).

In the present paper, we want to explore this range between algorithmic and real algebraic-geometric issues by exemplifying several corresponding aspects. We proceed along the following outline. In Section 2, the necessary notation is introduced. Then, in Section 3, we discuss algorithmic aspects of the visibility problems and present a new sweep algorithm in $\mathbb{R}^2$. Section 4 surveys recent results on the algebraic-geometric tangent problems in $\mathbb{R}^3$. By discussing an optimization aspect of the tangent problems in Section 5, we illustrate some of the proof techniques for the tangent results, and also establish a connection to the optimization of polynomial functions over the real numbers.

2. Notation

As an example for the applications mentioned in the introduction, consider the following problem from ray-tracing with moving viewpoints. Here, we want to compute information on the viewpoint positions where the visibility topology of the scene changes. As a special case, this includes tackling the following core problem of partial visibility.

A set $B \subset \mathbb{R}^n$ (say, $n \in \{2, 3\}$) is called a *(convex) body* if it is bounded, closed, convex, and contains an interior point. Now we consider a scene consisting of a set $\mathcal{B}$ of (not necessarily disjoint) bodies from a specific class $\mathcal{X}$ in $\mathbb{R}^n$ ($\mathcal{X}$ might be the set of all balls or the set of all polytopes). A body $B \in \mathcal{B}$ is called *partially visible from a viewpoint* v if there exists a line segment connecting v and B not intersecting with the interior of any other body in $\mathcal{B}$. A body $B \in \mathcal{B}$ is called *partially visible* if B can be seen from some viewpoint "outside" of the scene, i.e., if there exists a ray starting at B not intersecting with the interior of any other body in $\mathcal{B}$. We call such a ray a *visibility ray* for B. Bodies which are not partially visible can be immediately removed from the scene, which reduces the complexity of the visualization process.

3. A Sweep Algorithm for the Two-Dimensional Case

We present efficient algorithms for solving the partial visibility problem in $\mathbb{R}^2$. Here, we are not only interested in checking partial visibility of *one* of the bodies but also in computing *all* bodies which are not partially visible. In order to avoid several special cases we assume that the bodies are pairwise disjoint.

Let $\mathcal{B} := \{B_1, \ldots, B_m\}$ be a set of disjoint bodies in the plane. In the two-dimensional case, checking partial visibility of a body $B \in \mathcal{B}$ can be reduced to a finite number of geometric problems as follows (cf. the treatment of stabbing lines in [**11**]). Without loss of generality let $|\mathcal{B}| \geq 2$ and assume $B = B_1$. If there exists a visibility ray for B then we can continuously transform (i.e., rotate and translate) the visibility ray until we reach a situation where the underlying line is tangent to at least two of the bodies (one of them might be B_1 itself). Hence, it suffices to compute the set of all common tangent lines to a pair of bodies in $\mathcal{B}$ and check whether one of these lines contains a visibility ray. For any pair of disjoint bodies, the number of common tangent lines is exactly 4 (which can be seen as a very special case of the results in [**6, 16**] on the number of common supporting hyperplanes in general dimension).

In order to handle any class of bodies in the plane algorithmically, we have to assume that we can perform the following operations on this class.

(1) Compute the 4 common tangent lines to 2 bodies B_i, B_j.
(2) Compute the at most 2 intersections of a ray or a line with a body B_i.

In the following, we assume that we have access to two oracles performing these operations. Obviously, for the class of discs, the class of polygons, and the class combining discs and polygons these oracles can be realized quite easily. In particular, if the maximum number of vertices of any polygon is bounded by a constant then both oracles can be implemented in constant time.

DEFINITION 1. *A straight line is called* critical *if it is tangent to at least two bodies* B_i, B_j *with* $1 \leq i \neq j \leq m$. *A ray is called* critical *if it is contained in a critical straight line.*

Hence, the body B_1 is partially visible if and only if there exists a critical visibility ray for B_1. Consequently, it suffices to compute the set of critical lines and to check whether a critical line contains a visibility ray for B_1. Obviously, checking whether a given straight line contains a visibility ray for B_1 can be achieved with $O(m)$ calls to oracle 2.

THEOREM 2. *In dimension 2, the set of* all *partially visible bodies can be computed with* $O(m^3)$ *arithmetic steps,* $O(m^2)$ *calls to the first oracle and* $O(m^3)$ *calls to the second oracle.*

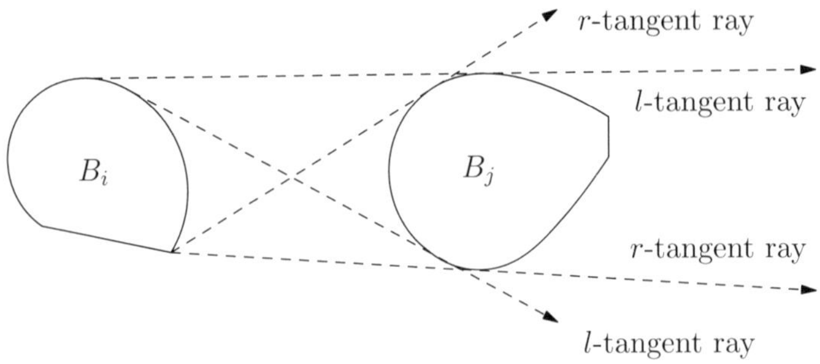

FIGURE 2. r- and l-tangent rays from B_i to B_j

PROOF. There are $4 \cdot \binom{m}{2}$ (not necessarily different) critical lines. For each critical line l it can be computed with $O(m)$ arithmetic steps and $O(m)$ calls to the second oracle which bodies intersect with l and which bodies are visible with regard to the line l. $\square$

The algorithm of Theorem 2 computes the set of *all* partially visible bodies in cubic time. However, the straighforward idea to modify it to a quadratic time algorithm for checking partial visibility of one specific body does *not* work. The reason is that it is a priori not clear which of the $O(m^2)$ critical lines can be omitted. If we are only interested in partial visibility of one specific body, say B_1, we can do better by using the following plane sweep algorithm requiring $O(m^2 \log m)$ time and $O(m)$ space. (For extensive material on sweep techniques we refer to [**9**].)

We interpret the 4 common tangent lines of two bodies B_i and B_j as rays starting in some boundary point of B_i. As a consequence of the results in [**6, 16**], there are two tangent rays such that B_i is on the left side of these tangent rays ("r-tangent rays of B_i"); and there are two tangent rays such that B_i is on the right side of the tangent rays ("l-tangent rays of B_i"), see Figure 2.

For checking visibility of B_1 we first investigate the $2(m-1)$ (not necessarily different) r-tangent rays touching B_1 and some other body B_i, $2 \leq i \leq m$. For each r-tangent ray we consider the outer normal $u \in \mathbb{S}^1$ where $\mathbb{S}^1$ denotes the unit sphere in $\mathbb{R}^2$; with each of these normals $u \in \mathbb{S}^1$ we associate the corresponding angle $0 \leq \alpha < 2\pi$ measured from the positive x-axis. As described in the following algorithm, we sweep the r-tangents according to increasing angles.

Subalgorithm for sweeping the r-tangents of B_1 and B_i, $2 \leq i \leq m$:

(1) Compute the set of r-tangents of B_1 and B_i, $2 \leq i \leq m$, and sort them by increasing angles.
(2) Compute the number of intersections of the first r-tangent with bodies B_i, $i > 1$.
(3) Consider the r-tangents successively in the order of increasing angles. In each of these steps do:
    (a) Update the number of intersections with bodies B_i, $i > 1$.
    (b) If the number of intersections is 0, then B_1 is partially visible; STOP.

For the update step we use the following lemma.

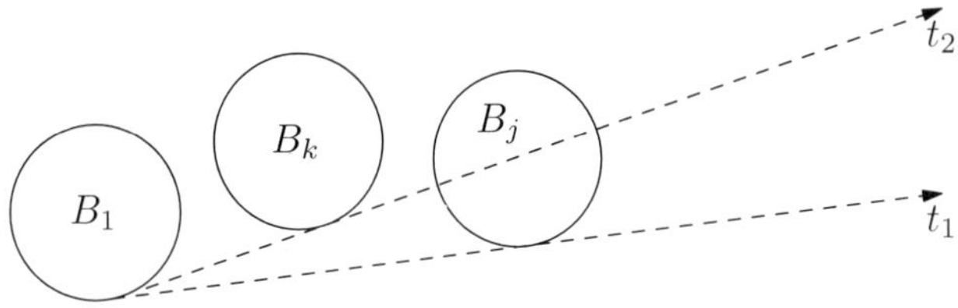

FIGURE 3. Update step during the sweep

LEMMA 3. *Let t_1 and t_2 be r-tangent rays of B_1 with angles $0 \leq \alpha_1 < \alpha_2 < 2\pi$, and let C be some body with $t_1 \cap C \neq \emptyset$, $t_2 \cap C = \emptyset$. Then there exists an r-tangent ray to B_1 with angle $\alpha_0 \in [\alpha_1, \alpha_2)$ which is tangent to C.*

PROOF. For any $\alpha \in [0, 2\pi)$ there exists some oriented tangent to B_1 with angle α (see, e.g., [**5**]) and therefore some r-tangent ray to B_1 with angle α. Let α_0 be the supremum of $\alpha \in [\alpha_1, \alpha_2)$ such that the tangent with angle α intersects with C. Since C is compact the tangent with angle α_0 is tangent to B_1 and C, i.e., the supremum is a maximum. □

In each step of the sweep we update the number of intersections of the sweep ray with bodies B_i, $i \geq 2$, in the following way. Let us first consider the case where the new angle α_2 is strictly larger than the current angle α_1 and where the r-tangent rays with angle α_1 and α_2 are each tangent to exactly two bodies. Let the r-tangent ray with angle α_1 be tangent to B_1 and B_j, and let the r-rangent ray with angle α_2 be tangent to B_1 and B_k, $2 \leq j \neq k \leq m$. Then we only have to check whether the ray with angle α_2 intersects with B_j (i.e., if the sweep ray is just "entering" B_j) and if the ray with angle α_1 intersects with B_k (i.e., if the sweep ray is just "leaving" B_k); see Figure 3. Due to Lemma 3 any additional change would imply the existence of some r-tangent with angle $\alpha_0 \in (\alpha_1, \alpha_2)$. Consequently, the update step can be done in constant time. If there are several r-tangent rays with the same angle we can combine these update steps. The amortized costs for the update step are not larger than in the case of different angles. If during the sweep we reach a situation where the number of intersections is 0 then B_1 is partially visible and we can stop immediately. After the inspection of the r-tangent rays of B_1 the l-tangent rays of B_1 are swept in the same way.

So far, we have inspected the r- and l-tangent rays of B_1. However, a visibility ray of B_1 is not necessarily tangent to B_1, and we also have to investigate the common tangents of bodies B_i, B_j with $2 \leq i \neq j \leq m$. More precisely, for every fixed $i \in \{2, \ldots, n\}$, we consider the critical rays which are tangent to B_i. Here, we start the sweep with that r-tangent ray from B_i to B_1 that has B_1 on the left side (see Figure 4). For this ray we count the number of intersections between B_i and B_1, and separately we count the number of intersections of the backward ray with other bodies. Now we sweep the r-tangent rays of B_i according to increasing angles and update the number of intersections between B_i and B_1 as well as the number of intersections of the backward ray. If we reach a situation where both numbers of intersections are simultaneously zero then B_1 is partially visible and we can stop immediately. In any case, the algorithm can stop if the r-tangent ray to B_i has B_1

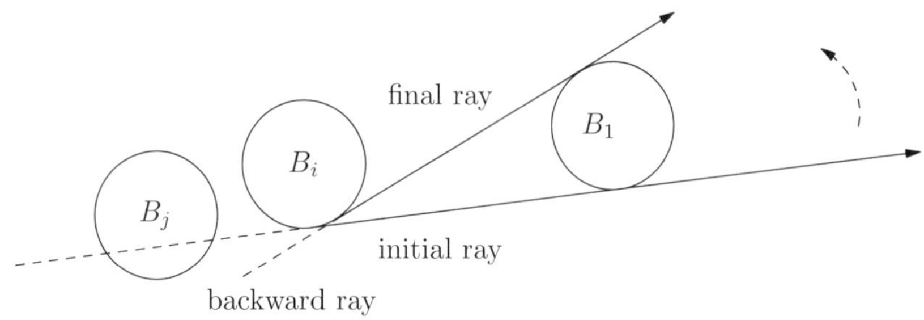

FIGURE 4. Initial and final ray for sweeping the r-tangent rays of B_i, $i \geq 2$

on its right side; see the illustration in Figure 4. After sweeping the l-tangent rays of B_i the r-tangent rays of B_i are investigated in the same way.

The correctness of the whole algorithm follows from the fact that the sweep inspects all $O(m^2)$ critical visibility lines and that the update step is correct due to Lemma 3.

For sweeping the tangent rays of some given body B_i, $1 \leq i \leq m$, the time requirements are dominated by the time to sort the tangent rays according to increasing angles. We can conclude:

THEOREM 4. *Let the dimension be $d = 2$. Then checking partial visibility of a body B_1 can be done with $O(m^2 \log m)$ arithmetic steps, $O(m^2)$ calls to the first oracle, and $O(m^2)$ calls to the second oracle, as well as $O(m)$ space.*

For other recent results on visibility computations in $\mathbb{R}^2$ see [**3**] and the references therein.

4. Real enumerative geometry for the three-dimensional case

In the three-dimensional case the basic algorithmic paradigm from Section 3 still holds. Since a line in $\mathbb{R}^3$ has four degrees of freedom, the core problem is now to compute the common tangent lines to four bodies in $\mathbb{R}^3$ (cf. [**22, 1**]). However, in the three-dimensional case, there are also some special cases where we can transform a visibility ray only to a situation with 2 or 3 bodies, or where a configuration with four bodies has an infinite number of common tangents.

In particular, an algorithmic treatment of the latter situations (depending on the class $\mathcal{X}$ of bodies) does not only require to characterize the maximum possible number of tangents, but also asks for an *a priori* characterization of the configurations with infinitely many common tangents. In contrast to some other problems in computational geometry, characterizing these situations cannot be neglected (say, by applying perturbation techniques [**10**]), since the large algebraic degree involved makes it usually highly nontrivial to guarantee a correct perturbation.

If the bodies are polytopes, the common tangents are common transversals of edges [**22**]; so, in fact, the main geometric task is to compute the common transversals to four given lines in $\mathbb{R}^3$. This geometric problem has been well-known for many years (see, e.g., [**13**]). In particular, if a configuration has only finitely many common transversals, then this number is bounded by 2; and it is well-known how to characterize the configurations with infinitely many common transversals.

However, the subsequent sections show that this situation completely changes if the set of bodies contains some non-linearly bounded bodies, such as balls.

4.1. Common tangent lines to four unit spheres. The following theorem in [**17**] characterizes the real algebraic-geometric situation when all bodies under investigation are unit balls.

PROPOSITION 5. *Four unit spheres in $\mathbb{R}^3$ have at most 12 common tangent lines unless their centers are located on the same line. Furthermore, there exists a configuration with 12 different real tangents, i.e., the upper bound is tight.*

The proof of the upper bound provides a complex bound in the sense that it bounds the complex solutions of a system of polynomial equation. So, essentially, the theorem states that algebraically this tangent problem is of degree 12. Note that due to this high degree, proving the characterization of the configurations with infinitely many common tangents is a highly nontrivial task.

From the viewpoint of projective geometry, this result is particularly interesting for the following reason. A line ℓ in three-dimensional projective space $\mathbb{P}^3$ can be represented in terms of its Plücker coordinates $p_\ell \in \mathbb{P}^5$ (see, e.g., [**13, 8**]). In particular, it is well-known that the set of vectors in $\mathbb{P}^5$ satisfying the Plücker relation

$$p_{01}p_{23} - p_{02}p_{13} + p_{03}p_{12} = 0 \tag{1}$$

is in 1-1-correspondence with the set of lines in $\mathbb{P}^3$.

As described in the following, the condition that a line ℓ is tangent to a sphere gives a quadratic equation. Namely, for a given matrix $A \in \mathbb{R}^{m,n}$ we can use the wedge operator

$$\wedge^2 : \mathbb{R}^{m,n} \quad \to \quad \mathbb{R}^{\binom{m}{2},\binom{n}{2}}$$

(cf. [**29**]). The row and column indices of the resulting matrix are subsets of cardinality 2 of $\{1,\ldots,m\}$ and $\{1,\ldots,n\}$, respectively. For $I \subset \{1,\ldots,m\}$ and $J \subset \{1,\ldots,n\}$ with $|I| = |J| = 2$,

$$(\wedge^2 A)_{I,J} := \det A_{[I,J]},$$

where $A_{[I,J]}$ denotes the 2×2 submatrix of the matrix A with row indices I and column indices J. Describing a sphere with center $c \in \mathbb{R}^3$ and radius r by the symmetric matrix

$$Q = \begin{pmatrix} c_1^2 + c_2^2 + c_3^2 - r^2 & -c_1 & -c_2 & -c_3 \\ -c_1 & 1 & 0 & 0 \\ -c_2 & 0 & 1 & 0 \\ -c_3 & 0 & 0 & 1 \end{pmatrix}$$

of the corresponding quadratic form $x^T Q x$, the following characterization of the tangent condition is applicable (see, e.g., [**29**]).

LEMMA 6. *A line $\ell \subset \mathbb{P}^3$ is tangent to a quadric Q if and only if the Plücker vector p_ℓ of ℓ satisfies*

$$p_\ell^T (\wedge^2 Q) p_\ell = 0.$$

Hence, the set of lines tangent to four given unit spheres can be formulated by a system of five quadratic equations in $\mathbb{P}^5$. The reason why there are at most 12 solution in affine space instead of the (by Bézout's Theorem) expected $2^5 = 32$ solutions is that there is a one-dimensional component of solutions at infinity. It can be shown that this one-dimensional component of solutions accounts for the "missing" $2^5 - 12 = 20$ solutions. Quite remarkably, as observed in this intersection-theoretical analysis [**2**], the excess component cannot be resolved by a single blow-up.

Now let us have a look at the situation when the centers of the four spheres are affinely dependent. In [**18**] it is shown that under this condition the maximum number of real tangent lines goes down.

PROPOSITION 7. *If the four centers of four unit spheres in* $\mathbb{R}^3$ *are coplanar but not collinear, then there are at most 8 common tangent lines, and this bound is tight.*

The proof of this statement uses some purely real algebraic arguments in the following sense. A line tangent to the four unit spheres can be seen as the axis of a unit cylinder whose surface passes through the four centers. Unless the axis is parallel to the plane of the four centers, its surface intersects this plane in an ellipse with semi-minor axis of length 1. Each such ellipse corresponds to two cylinders. Using an appropriate algebraic description, it is shown in [**17**] that there can be at most 6 candidate conic sections satisfying the relevant algebraic properties, hence giving at most 12 common tangents. In addition to this (complex) algebraic argument, it is shown in [**18**] that some of the candidate conic sections are not ellipses but hyperbolas.

4.2. Realization issues. In order to find the common tangent lines we can either start from a system of polynomial equations, or we can construct a univariate polynomial equation whose solutions encode the tangents. For both approaches the numerical computation of the tangents may become instable, especially for configurations of centers which are close to singular configurations (e.g., configurations corresponding to *reducible* polynomials in an algebraic description). Hence, it is essential to treat realization questions, such as: For which numbers $k \in \{0, \ldots, 12\}$ do there exist configurations with exactly k different common tangents in *real* space?

If not all numbers $k \in \{0, \ldots, 12\}$ can be established in real space this offers the possibility of strong and valuable consistency checks within a program. If, however, all numbers can be realized then this would prove the non-existence of such a control mechanism. Indeed, the latter result is true, as stated in the following theorem in [**31**].

THEOREM 8. *For any number* $k \in \{0, \ldots, 12\}$ *there exists a configuration of 4 unit spheres in* $\mathbb{R}^3$ *which have exactly* k *different common tangents in* $\mathbb{R}^3$.

This result contrasts the 4-sphere-problem to some well-studied problems in classical and enumerative geometry. For example, concerning one of the most famous problems from enumerative geometry, the number of 27 lines on a smooth cubic surface, the question of real solutions has already been studied long time ago ([**25, 27**], see also [**24**], p. 188). In particular, for a cubic surface in $\mathbb{R}^3$ only the numbers 3, 7, 15, and 27 can be established with *real* lines. Another famous example in geometry is Apollonius' problem which asks for the circles tangent to three given circles. For this problem, there exist configurations with $k \in \{0, 1, \ldots, 6, 8\}$ real tangent circles but provably no configuration with 7 real tangent circles [**21**].

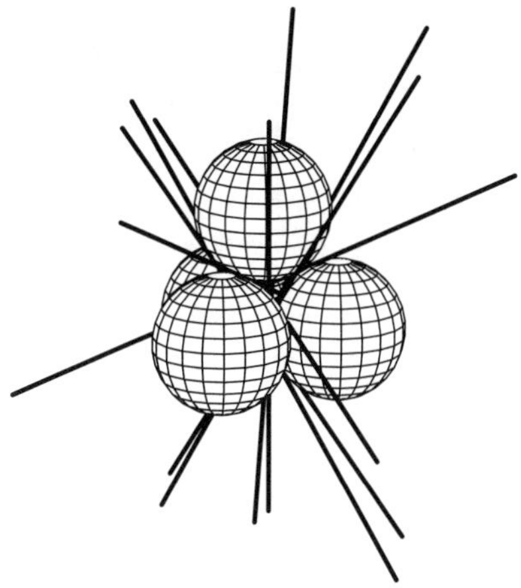

FIGURE 5. Construction of four unit spheres with nine common tangent lines

As an example of the geometry of the constructions, Figure 5 shows a configuration with four unit spheres and nine common tangent lines. For configurations with twelve common tangent lines see [**17, 31**].

4.3. Combinations of spheres and polytopes. The next question is, what happens when the class of admissible bodies in the scene consists of both balls and polytopes. From the algebraic-geometric viewpoint, this immediately leads to the problem of finding the common tangents/transversals to k spheres and $n-k$ lines, $k \in \{0, \ldots, 4\}$. For convenience of notation, we consider a transversal of a line as a tangent to the line.

Tight upper bounds for the number of common tangents to k spheres and $n-k$ lines are known in the finite case, $k \in \{0, \ldots, 4\}$. Here, as in Section 4.1, *tightness* refers to the following (quite strong) sense of real algebraic geometry (cf. [**30**]): On the one hand, for each k we bound the number of solutions by algebraic methods, say, by some number m. Then, on the other hand, constructions are known which indeed leads to m solutions in *real* space $\mathbb{R}^3$ (which would not be possible if, say, any polynomial formulation contained some complex solutions).

The general difficulty of proving tight bounds of this kind may be seen by the following two aspects. For the classical enumerative geometry problem of conics tangent to five given conics (dating back to Steiner in 1848) the existence problem of 3264 *real* solutions had not been solved until few years ago ([**23**] and [**12**], §7.2). Furthermore, as pointed out in [**30**] (cf. also [**28**]), there are nearly no criteria or general techniques for tackling questions of this type. For these reasons, it is even more remarkable that in *all* of the situations there exists a construction matching the upper bound.

Table 4.3 summarizes the results from [**32**]. It shows the upper bounds for the number of solutions and the matching numbers of real solutions in the best constructions. The last column shows that in only a few cases, explicit characterizations of the configurations with an infinite number of common tangents are known so far. Namely, besides the already mentioned results for 4 lines and for

	upper bound # solutions	# real solutions of a construction	characterization of degenerate instances
4 lines	2	2	yes
3 lines, 1 sphere	4	4	yes
2 lines, 2 spheres	8	8	–
1 line, 3 spheres	12	12	–
4 unit spheres	12	12	yes
4 spheres	12	12	–

TABLE 1. Summary of results

4 unit spheres, this characterization is only known for the case of three lines and one sphere. In entries with a "–" such a characterization is not known yet.

Let N_k denote the maximum number of common tangents/transversals to k spheres and $4-k$ lines in the finite case, $0 \le k \le 4$. The proofs for the tight upper bounds of $N_0, \dots, N_4$ in [**32**] are of quite different flavors. For $k \in \{1,2\}$, the upper bounds immediately follow from Bézout's Theorem. Whereas for $k=1$ it is easy to give a construction matching this bound, the correctness proof of the construction for $k=2$ is based on the computer-algebraic technique of using standard bases. For $N_3 = 12$, the Bézout bound of a projective formulation is 16 instead of 12. In order to prove the bound of 12, by analyzing the underlying algebraic geometry of the problem, two double solutions at infinity can be spotted. Finally, for the proof for N_4 (for general radii) there are two possibilities. An appropriate elementary geometric description can be used to deduce a polynomial formulation with Bézout bound 12, or one can use that the intersection theory technique mentioned in Section 4.1 also holds for general radii.

5. On the smallest radius giving common tangent lines

In the previous section we dealt with the enumerative geometry problem to count, characterize, and find the common transversals/tangent lines to given bodies. In this section we would like to illustrate another relationship between these problems and real algebraic geometry: that of optimizing polynomial functions. This aspect is not only relevant for optimization aspects with regard to the visibility/tangent problems, but in particular is also related to the problem of finding smallest enclosing/circumscribing cylinders of arbitrary point sets in $\mathbb{R}^3$ (cf. [**1, 7, 26**]).

Let $c_1, \dots, c_4$ be affinely independent. Then, of course, there does not exist a line passing through all of these points. However, for every edge of the tetrahedron $c_1c_2c_3c_4$, there exists a line which is parallel to that edge and which is tangent to the four spheres $S(c_i, r)$ for *some* radius $r > 0$ (see [**17**]). Hence, there exists a smallest radius $r > 0$ such that the four spheres $S(c_i, r)$ have a common tangent line. A tangent line for this radius r is called a *tangent line with minimal radius*. In general, this problem is a polynomial optimization problem over the real numbers, and – by the results in Section 4 on the algebraic degree involved – it is hard to find the exact global optima (cf. the discussion in [**20**]). In particular, the latter paper deals with new ideas to find global optima of certain unconstrained real polynomial optimization problems. Generally, we think that for geometric problems like finding the tangent line of minimal radius it is an important and challenging question to combine general methods for polynomial optimization (like the one mentioned) with specific geometric insights.

To provide an example for combinations of this kind, let us again look at the 4-sphere-problem. Of course, it is hard to find good geometric characterizations in the general case. However, for an interesting subclass of 4-sphere-configurations, we can prove the following geometric characterization.

THEOREM 9. *Let $c_1c_2c_3c_4$ define an equifacial tetrahedron, and let ℓ be a tangent line to $S(c_1, r), \ldots, S(c_4, r)$ with minimal radius r. Then the direction vector of ℓ is perpendicular to the direction vectors of two opposite edges, and each of these two edges has length $2r$.*

In order to prove this theorem, the following notation of [**17**] is used. Let $c_1, \ldots, c_4$ be affinely independent. Then the four centers define a tetrahedron T in $\mathbb{R}^3$. Further, let $\ell = \{p + \mu s : \mu \in \mathbb{R}\}$ with $p, s \in \mathbb{R}^3$, $s \neq 0$, $p \perp s$, be a line. ℓ is tangent to the four spheres $S(c_i, r)$ with radius $r > 0$ if the line $\ell - p$ (which passes through the origin) has distance r from the four points $c_i - p$, i.e, if and only if

$$((c_i - p) \times s)^2 = r^2 s^2, \qquad 1 \leq i \leq 4. \tag{2}$$

PROOF. It is well-known that the vertices of an equifacial tetrahedron T can be regarded as four pairwise non-adjacent vertices of a rectangular box (see, e.g., [**14**]). Hence, there exists a representation $c_1 = (\lambda_1, \lambda_2, \lambda_3)^T$, $c_2 = (\lambda_1, -\lambda_2, -\lambda_3)^T$, $c_3 = (-\lambda_1, \lambda_2, -\lambda_3)^T$, $c_4 = (-\lambda_1, -\lambda_2, \lambda_3)^T$ with $\lambda_1, \lambda_2, \lambda_3 > 0$.

Assuming without loss of generality $s^2 - 1$, application of Lagrange's identity on (2) gives

$$((c_i - p) \times s)^2 = c_i^2 - \langle c_i, s\rangle^2 - 2\langle c_i, p\rangle + p^2.$$

Hence, Equation (2) takes the form

$$\langle c_i, s\rangle^2 + 2\langle c_i, p\rangle = \sum_{j=1}^{3} \lambda_j^2 + p^2 - r^2, \qquad 1 \leq i \leq 4. \tag{3}$$

Subtracting these equations pairwise gives

$$4(\lambda_2 p_2 + \lambda_3 p_3) = -4(\lambda_1\lambda_3 s_1 s_3 + \lambda_1\lambda_2 s_1 s_2)$$

(for indices 1, 2) and analogous equations, so that

$$\lambda_1 p_1 = -\lambda_2\lambda_3 s_2 s_3, \quad \lambda_2 p_2 = -\lambda_1\lambda_3 s_1 s_3, \quad \lambda_3 p_3 = -\lambda_1\lambda_2 s_1 s_2.$$

Since $\langle p, s\rangle = 0$, this yields $s_1 s_2 s_3 = 0$. Without loss of generality we can assume $s_1 = 0$. In this case,

$$p = \left(-\frac{\lambda_2\lambda_3}{\lambda_1} s_2 s_3, 0, 0\right)^T.$$

So we can express (3) in terms of the direction vector s,

$$\lambda_2^2 s_2^2 + \lambda_3^2 s_3^2 = \sum_{j=1}^{3} \lambda_j^2 + \left(-\frac{\lambda_2\lambda_3}{\lambda_1} s_2 s_3\right)^2 - r^2,$$

which, by using $s_2^2 + s_3^2 = 1$, gives

$$\lambda_2^2\lambda_3^2 s_2^4 + (\lambda_1^2\lambda_2^2 - \lambda_1^2\lambda_3^2 - \lambda_2^2\lambda_3^2) s_2^2 + \lambda_1^2(r^2 - \lambda_1^2 - \lambda_2^2) = 0. \tag{4}$$

In particular, $s = (0,1,0)^T$ gives $r^2 = \lambda_1^2 + \lambda_3^2$, and $s = (0,0,1)^T$ gives $r^2 = \lambda_1^2 + \lambda_2^2$. By setting $z_2 = s_2^2$, Equation (4) defines a quadratic function $r^2(z_2)$,

$$r^2(z_2) = \frac{1}{\lambda_1^2}\left(-\lambda_2^2\lambda_3^2 z_2^2 - (\lambda_1^2\lambda_2^2 - \lambda_1^2\lambda_3^2 - \lambda_2^2\lambda_3^2)z_2\right) + \lambda_1^2 + \lambda_2^2 .$$

Since this function is concave in $z_2 \in [0,1]$, one of the given boundary values of r^2 is the minimum value. Moreover, since both $2\sqrt{\lambda_1^2 + \lambda_2^2}$ and $2\sqrt{\lambda_1^2 + \lambda_3^2}$ are the lengths of a pair of opposite edges, the statement follows. □

ACKNOWLEDGMENTS. Thanks to Abhi Dattasharma for his helpful comments.

References

[1] P.K. Agarwal, B. Aronov, and M Sharir. Line transversals of balls and smallest enclosing cylinders in three dimensions. *Discrete Comput. Geom.*, 21:373–388, 1999.
[2] P. Aluffi and W. Fulton. Personal communication.
[3] P. Angelier and M. Pocchiola. A sum of squares theorem for visibility complexes. In *Proc. Symposium on Computational Geometry (Medford, MA)*, 2001.
[4] M. Bern, D. Dobkin, D. Eppstein, and R. Grossman. Visibility with a moving point of view. *Algorithmica*, 11:360–378, 1994.
[5] T. Bonnesen and W. Fenchel. *Theorie der konvexen Körper.* Springer-Verlag, 1934. English translation: BCS Associates, Moscow, Idaho, 1987.
[6] S.E. Cappell, J.E. Goodman, J. Pach, R. Pollack, M. Sharir, and R. Wenger. Common tangents and common transversals. *Adv. Math.*, 106:198–215, 1994.
[7] T.M. Chan. Approximating the diameter, smallest enclosing cylinder, and minimum-width annulus. In *Proc. Symposium on Computational Geometry (Clear Water Bay, Hong Kong)*, pages 300–309, 2000.
[8] D. Cox, J. Little, and D. O'Shea. *Ideals, Varieties, and Algorithms.* Springer-Verlag, New York, 2nd edition, 1996.
[9] M. de Berg, M. van Kreveld, M. Overmars, and O. Schwarzkopf. *Computational Geometry.* Springer-Verlag, Berlin, 2nd edition, 2000.
[10] H. Edelsbrunner and E.P. Mücke. Simulation of simplicity: A technique to cope with degeneracy. *ACM Trans. Graph.*, 9(1):43–72, 1990.
[11] H. Edelsbrunner and M. Sharir. The maximum number of ways to stab n convex nonintersecting sets in the plane is $2n - 2$. *Discrete Comput. Geom.*, 5:35–42, 1990.
[12] W. Fulton. *Introduction to Intersection Theory in Algebraic Geometry.* CBMS 54. AMS, 2nd edition, 1996.
[13] W. Hodge and D. Pedoe. *Methods of Algebraic Geometry*, volume 2. Cambridge University Press, 1952.
[14] Y.S. Kupitz and H. Martini. Equifacial tetrahedra and a famous location problem. *Math. Gazette*, 83:464–467, 1999.
[15] H.-P. Lenhof and M. Smid. Maintaining the visibility map of spheres while moving the viewpoint on a circle at infinity. *Algorithmica*, 13:301–312, 1995.
[16] T. Lewis, B. von Hohenbalken, and V. Klee. Common supports as fixed points. *Geom. Dedicata*, 60:277–281, 1996.
[17] I.G. Macdonald, J. Pach, and T. Theobald. Common tangents to four unit balls in $\mathbb{R}^3$. *Discrete Comput. Geom.*, 26:1–17, 2001.
[18] G. Megyesi. Lines tangent to 4 spheres with affinely dependent centres. *Discrete Comput. Geom.*, 26:493–497, 2001.
[19] J. O'Rourke. Visibility. In J.E. Goodman and J. O'Rourke, editors, *Handbook of Discrete and Computational Geometry*, pages 467–479. CRC Press, Boca Raton/NY (USA), 1997.
[20] P.A. Parrilo and B. Sturmfels. Minimizing polynomial functions. This volume.
[21] D. Pedoe. The missing seventh circle. *Elem. Math.*, 25:14–15, 1970.
[22] M. Pellegrini and P.W. Shor. Finding stabbing lines in 3-space. *Discrete Comput. Geom.*, pages 191–208, 1992.

[23] F. Ronga, A. Tognoli, and T. Vust. The number of conics tangent to 5 given conics: the real case. *Rev. Mat. Univ. Complut. Madrid*, 10:391–421, 1997.

[24] G. Salmon. *A Treatise on the Analytic Geometry of Three Dimensions*, volume 1. Longmans and Green, London, 7th edition, 1928. Reprinted by Chelsea Publishing Company, New York, 1958.

[25] L. Schläfli. On the distributions of surfaces of the third order into species, in reference to the absence or presence of singular points, and the reality of their lines. *Phil. Trans. Royal Society of London*, 153:193–241, 1863.

[26] E. Schömer, J. Sellen, M. Teichmann, and C. Yap. Smallest enclosing cylinders. *Algorithmica*, 27:170–186, 2000.

[27] B. Segre. *The Non-Singular Cubic Surfaces*. Oxford University Press, 1942.

[28] F. Sottile. Enumerative real algebraic geometry. This volume.

[29] F. Sottile. From enumerative geometry to solving systems of polynomial equations with Macaulay 2. In *Computations in Algebraic Geometry with Macaulay 2*, ed. by D. Eisenbud, D. Grayson, M. Stillman, and B. Sturmfels. Springer-Verlag, Berlin, 2001.

[30] F. Sottile. Enumerative geometry for the real Grassmannian of lines in projective space. *Duke Math. J.*, 87:59–85, 1997.

[31] T. Theobald. How to realize a given number of tangents to four unit balls in $\mathbb{R}^3$. To appear in *Mathematika*.

[32] T. Theobald. An enumerative geometry framework for algorithmic line problems in $\mathbb{R}^3$. To appear in SIAM J. Computing.

[33] C.A. Wang and B. Zhu. Three-dimensional weak visibility: Complexity and applications. *Theor. Comp. Sci.*, 234:219–232, 2000.

[34] J. Wernecke. *The Inventor Mentor: Programming Object-Oriented 3D Graphics with Open Inventor*. Addison-Wesley, Reading, MA, 1994.

Thorsten Theobald, Zentrum Mathematik, Technische Universität München, D-80290 München, Germany

E-mail address: `theobald@ma.tum.de`

Titles in This Series

60 **Saugata Basu and Laureano Gonzalez-Vega, Editors,** Algorithmic and Quantitative Real Algebraic Geometry

59 **Michael H. Goldwasser, David S. Johnson, and Catherine C. McGeoch, Editors,** Data Structures, Near Neighbor Searches, and Methodology: Fifth and Sixth DIMACS Implementation Challenges

58 **Simon Thomas, Editor,** Set Theory: The Hajnal Conference

57 **Eugene C. Freuder and Richard J. Wallace, Editors,** Constraint Programming and Large Scale Discrete Optimization

56 **Alexander Barg and Simon Litsyn, Editors,** Codes and Association Schemes

55 **Ding-Zhu Du, Panos M. Pardalos, and Jie Wang, Editors,** Discrete Mathematical Problems with Medical Applications

54 **Erik Winfree and David K. Gifford, Editors,** DNA Based Computers V

53 **Nathaniel Dean, D. Frank Hsu, and R. Ravi, Editors,** Robust Communication Networks: Interconnection and Survivability

52 **Sanguthevar Rajasekaran, Panos Pardalos, and D. Frank Hsu, Editors,** Mobile Networks and Computing

51 **Pierre Hansen, Patrick Fowler, and Maolin Zheng, Editors,** Discrete Mathematical Chemistry

50 **James M. Abello and Jeffrey Scott Vitter, Editors,** External Memory Algorithms

49 **Ronald L. Graham, Jan Kratochvíl, Jaroslav Nešetřil, and Fred S. Roberts, Editors,** Contemporary Trends in Discrete Mathematics

48 **Harvey Rubin and David Harlan Wood, Editors,** DNA Based Computers III

47 **Martin Farach-Colton, Fred S. Roberts, Martin Vingron, and Michael Waterman, Editors,** Mathematical Support for Molecular Biology

46 **Peng-Jun Wan, Ding-Zhu Du, and Panos M. Pardalos, Editors,** Multichannel Optical Networks: Theory and Practice

45 **Marios Mavronicolas, Michael Merritt, and Nir Shavit, Editors,** Networks in Distributed Computing

44 **Laura F. Landweber and Eric B. Baum, Editors,** DNA Based Computers II

43 **Panos Pardalos, Sanguthevar Rajasekaran, and José Rolim, Editors,** Randomization Methods in Algorithm Design

42 **Ding-Zhu Du and Frank K. Hwang, Editors,** Advances in Switching Networks

41 **David Aldous and James Propp, Editors,** Microsurveys in Discrete Probability

40 **Panos M. Pardalos and Dingzhu Du, Editors,** Network Design: Connectivity and Facilities Location

39 **Paul W. Beame and Samuel R Buss, Editors,** Proof Complexity and Feasible Arithmetics

38 **Rebecca N. Wright and Peter G. Neumann, Editors,** Network Threats

37 **Boris Mirkin, F. R. McMorris, Fred S. Roberts, and Andrey Rzhetsky, Editors,** Mathematical Hierarchies and Biology

36 **Joseph G. Rosenstein, Deborah S. Franzblau, and Fred S. Roberts, Editors,** Discrete Mathematics in the Schools

35 **Dingzhu Du, Jun Gu, and Panos M. Pardalos, Editors,** Satisfiability Problem: Theory and Applications

34 **Nathaniel Dean, Editor,** African Americans in Mathematics

33 **Ravi B. Boppana and James F. Lynch, Editors,** Logic and Random Structures

32 **Jean-Charles Grégoire, Gerard J. Holzmann, and Doron A. Peled, Editors,** The SPIN Verification System

TITLES IN THIS SERIES

31 **Neil Immerman and Phokion G. Kolaitis, Editors,** Descriptive Complexity and Finite Models

30 **Sandeep N. Bhatt, Editor,** Parallel Algorithms: Third DIMACS Implementation Challenge

29 **Doron A. Peled, Vaughan R. Pratt, and Gerard J. Holzmann, Editors,** Partial Order Methods in Verification

28 **Larry Finkelstein and William M. Kantor, Editors,** Groups and Computation II

27 **Richard J. Lipton and Eric B. Baum, Editors,** DNA Based Computers

26 **David S. Johnson and Michael A. Trick, Editors,** Cliques, Coloring, and Satisfiability: Second DIMACS Implementation Challenge

25 **Gilbert Baumslag, David Epstein, Robert Gilman, Hamish Short, and Charles Sims, Editors,** Geometric and Computational Perspectives on Infinite Groups

24 **Louis J. Billera, Curtis Greene, Rodica Simion, and Richard P. Stanley, Editors,** Formal Power Series and Algebraic Combinatorics/Séries Formelles et Combinatoire Algébrique, 1994

23 **Panos M. Pardalos, David I. Shalloway, and Guoliang Xue, Editors,** Global Minimization of Nonconvex Energy Functions: Molecular Conformation and Protein Folding

22 **Panos M. Pardalos, Mauricio G. C. Resende, and K. G. Ramakrishnan, Editors,** Parallel Processing of Discrete Optimization Problems

21 **D. Frank Hsu, Arnold L. Rosenberg, and Dominique Sotteau, Editors,** Interconnection Networks and Mapping and Scheduling Parallel Computations

20 **William Cook, László Lovász, and Paul Seymour, Editors,** Combinatorial Optimization

19 **Ingemar J. Cox, Pierre Hansen, and Bela Julesz, Editors,** Partitioning Data Sets

18 **Guy E. Blelloch, K. Mani Chandy, and Suresh Jagannathan, Editors,** Specification of Parallel Algorithms

17 **Eric Sven Ristad, Editor,** Language Computations

16 **Panos M. Pardalos and Henry Wolkowicz, Editors,** Quadratic Assignment and Related Problems

15 **Nathaniel Dean and Gregory E. Shannon, Editors,** Computational Support for Discrete Mathematics

14 **Robert Calderbank, G. David Forney, Jr., and Nader Moayeri, Editors,** Coding and Quantization: DIMACS/IEEE Workshop

13 **Jin-Yi Cai, Editor,** Advances in Computational Complexity Theory

12 **David S. Johnson and Catherine C. McGeoch, Editors,** Network Flows and Matching: First DIMACS Implementation Challenge

11 **Larry Finkelstein and William M. Kantor, Editors,** Groups and Computation

10 **Joel Friedman, Editor,** Expanding Graphs

9 **William T. Trotter, Editor,** Planar Graphs

8 **Simon Gindikin, Editor,** Mathematical Methods of Analysis of Biopolymer Sequences

7 **Lyle A. McGeoch and Daniel D. Sleator, Editors,** On-Line Algorithms

6 **Jacob E. Goodman, Richard Pollack, and William Steiger, Editors,** Discrete and Computational Geometry: Papers from the DIMACS Special Year

5 **Frank Hwang, Fred Roberts, and Clyde Monma, Editors,** Reliability of Computer and Communication Networks

For a complete list of titles in this series, visit the AMS Bookstore at **www.ams.org/bookstore/**.